Intelligent Techniques for Cyber-Physical Systems

Intelligent Techniques for Cyber-Physical Systems covers challenges, opportunities, and open research directions for cyber-physical systems (CPS). It focuses on the design and development of machine learning and metaheuristics-enabled methods as well as blockchain for various challenges like security, resource management, computation offloading, trust management, and others in edge, fog, and cloud computing, Internet of Things (IoT), Internet of Everything (IoE), and smart cities. It also includes the design and analysis of deep learning-based models, sensing technologies, metaheuristics, and blockchain for complex real-life systems for CPS.

- Offers perspectives on the research directions in CPS;
- Provides state-of-the-art reviews on intelligent techniques, machine learning, deep learning, and reinforcement learning-based models for cloud-enabled IoT environment;
- Discusses intelligent techniques for complex real-life problems in different CPS scenarios;
- Reviews advancements in blockchain technology and smart cities;
- Explores machine learning-based intelligent models for combinatorial optimization problems.

This book is aimed at researchers and graduate students in computer science, engineering, and electrical and electronics engineering.

Computational Intelligence Techniques

Series Editor: Vishal Jain

The objective of this series is to provide researchers a platform to present state-of-the-art innovations, research, and design and implement methodological and algorithmic solutions to data processing problems, designing and analyzing evolving trends in health informatics and computer-aided diagnosis. This series provides support and aid to researchers involved in designing decision support systems that will permit societal acceptance of ambient intelligence. The overall goal of this series is to present the latest snapshot of ongoing research as well as to shed further light on future directions in this space. The series presents novel technical studies as well as position and vision papers comprising hypothetical/speculative scenarios. The book series seeks to compile all aspects of computational intelligence techniques from fundamental principles to current advanced concepts. For this series, we invite researchers, academicians, and professionals to contribute, expressing their ideas and research in the application of intelligent techniques to the field of engineering in handbook, reference, or monograph volumes.

Smart Computing and Self-Adaptive Systems
Simar Preet Singh, Arun Solanki, Anju Sharma, Zdzislaw Polkowski and Rajesh Kumar

Advancing Computational Intelligence Techniques for Security Systems Design
Uzzal Sharma, Parmanand Astya, Anupam Baliyan, Salah-ddine Krit, Vishal Jain and Mohammad Zubair Kha

Graph Learning and Network Science for Natural Language Processing
Edited by Muskan Garg, Amit Kumar Gupta and Rajesh Prasad

Computational Intelligence in Medical Decision Making and Diagnosis Techniques and Applications
Edited by Sitendra Tamrakar, Shruti Bhargava Choubey and Abhishek Choubey

Applications of 5G and Beyond in Smart Cities
Edited by Ambar Bajpai and Arun Balodi

Healthcare Industry 4.0: Computer Vision-Aided Data Analytics
Edited by P. Karthikeyan, Polinpapilinho F. Katina, and R. Rajaagopal

Blockchain Technology for IoE: Security and Privacy Perspectives
Edited by Arun Solanki, Anuj Kumar Singh, and Sudeep Tanwar

Intelligent Techniques for Cyber-Physical Systems
Edited by Mohammad Sajid, Anil Kumar Sagar, Jagendra Singh, Osamah Ibrahim Khalaf, and Mukesh Prasad

For more information about this series, please visit: www.routledge.com/Computational-Intelligence-Techniques/book-series/CIT

Intelligent Techniques for Cyber-Physical Systems

Edited by
Mohammad Sajid, Anil Kumar Sagar,
Jagendra Singh, Osamah Ibrahim Khalaf,
and Mukesh Prasad

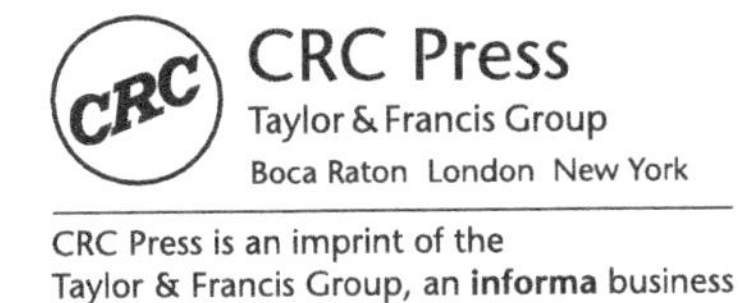

CRC Press is an imprint of the
Taylor & Francis Group, an **informa** business

Designed cover image: © Mohammad Sajid, Anil Kumar Sagar, Jagendra Singh, Osamah Ibrahim Khalaf and Mukesh Prasad

First edition published 2024
by CRC Press
2385 NW Executive Center Drive, Suite 320, Boca Raton FL 33431

and by CRC Press
4 Park Square, Milton Park, Abingdon, Oxon, OX14 4RN

CRC Press is an imprint of Taylor & Francis Group, LLC

ISBN: 978-1-032-45286-9 (hbk)
ISBN: 978-1-032-57258-1 (pbk)
ISBN: 978-1-003-43858-8 (ebk)

DOI: 10.1201/9781003438588

Typeset in Times
by codeMantra

Contents

Preface

Cyber-physical systems (CPS) integrate computation, networking, and physical processes with minimal human interventions for automated systems. CPS and related technologies include Internet of Things (IoT), industrial IoT, smart cities, smart grid, and "smart" anything (e.g., homes, buildings, appliances, cars, transportation systems, hospitals, manufacturing, distribution systems). The advancement in CPS will lead to progress in emergency response, healthcare, transportation and traffic flow systems, home care, education, electric power generation, delivery systems, and other areas. In the modern era of CPS, 25 billion devices, including machines, sensors, and cameras, are connected and continuously growing at high speed. 41.6 billion IoT devices are expected to be connected, generating around 79.4 zettabytes of data in 2025 due to the CPS advancements. CPS and related technologies intersect in various developments and generate massive data.

CPS is required to handle the challenges of many cutting-edge technologies, including IoT communication and computation protocols, interoperability, development, and management of distributed, context-aware, and self-adaptive IoT applications, containerization, automation and dynamic orchestration mechanisms for cloud, fog, and edge computing paradigms, 5G, massive IoT deployments, human-centric solutions, efficient IoT data storage, knowledge graphs, federated learning, data privacy and security, blockchain, and many others. CPS challenges occur due to the combinations of hybrid systems, heterogeneous and distributed components, massive components, dynamic environments, and human-in-the-Loop. The most challenging issue is developing and designing concepts, methods, and tools to handle CPS interoperability with the above characteristics. The developed concepts, techniques, and tools must be cross-domain, application- and learning-centric, adaptive, secure, efficient, and human-centric.

Intelligent techniques consist of designing and applying biologically and linguistically motivated computational paradigms. It comprises three main constituents, i.e., neural networks, fuzzy systems, evolutionary computation, and other evolving systems such as ambient intelligence, artificial life, cultural learning, artificial endocrine networks, social reasoning, and artificial hormone networks. Intelligent techniques combine different techniques and paradigms to address the current and upcoming challenges of complex real-world problems of CPS. The objective of the intelligent methods is to develop intelligent systems, including healthcare, games, optimization, and cognitive systems. Intelligent methods can solve problems that are not efficiently solvable by traditional methods or whose models are too complex for mathematical reasoning. Intelligent techniques employ inexact and incomplete information to resolve the problem in acceptable computational time. Hence, intelligent methods can analyze CPS data and act accordingly to create better values in our daily life.

About the Editors

Mohammad Sajid is Assistant Professor in the Department of Computer Science at Aligarh Muslim University, India. He has completed his Ph.D., M.Tech., and MCA degrees at the School of Computer and Systems Sciences, Jawaharlal Nehru University (JNU), New Delhi. His research interests include parallel and distributed computing, cloud computing, bio-inspired computation, and combinatorial optimization problems. He has published one patent and was awarded a research start-up grant in 2017 from University Grants Commission (UGC), India.

Anil Kumar Sagar is Professor in the Department of Computer Science and Engineering at Sharda University, Greater Noida, India. He completed his B.E., M.Tech., and Ph.D. in Computer Science. His research interests include mobile ad hoc networks and vehicular ad hoc networks, IoT, and artificial intelligence. He has received a Young Scientist Award for the year 2018–2019 from the Computer Society of India and the Best Faculty Award for the years 2006 and 2007 from SGI, Agra.

Jagendra Singh is Associate Professor in the School of Computer Science, Engineering and Technology, Bennett University, Greater Noida. He received his Ph.D. in Computer Science from Jawaharlal Nehru University, New Delhi. His areas of interest are natural language processing (information retrieval system, recommendation system, and sentiment analysis) and machine learning (deep learning, neural network, and data analytics).

Osamah Ibrahim Khalaf is Senior Assistant Professor in Engineering and Telecommunications at Al-Nahrain University/College of Information Engineering. He holds 10 years of university-level teaching experience in computer science and network technology, holds patents, and has received several medals and awards due to his innovative work and research activities. He earned his Ph.D. in Computer Networks from the Faculty of Computer Systems and Software Engineering, University of Malaysia Pahang. He has overseas work experience at Binary University in Malaysia and University of Malaysia Pahang.

Mukesh Prasad is Senior Lecturer at the School of Computer Science in the Faculty of Engineering and IT at the University of Technology Sydney. His research interests include big data, computer vision, brain–computer interface, and evolutionary computation. He is also working in the field of image processing, data analytics, and edge computing. His research is backed by industry experience, specifically in Taiwan, where he was the principal engineer (2016–2017) at the Taiwan Semiconductor Manufacturing Company (TSMC). He received a Ph.D. from the Department of Computer Science at the National Chiao Tung University in Taiwan (2015).

Contributors

Shafiqul Abidin
Department of Computer Science
Aligarh Muslim University
Aligarh, India

Shakeel Ahamad
Software Quality Assurance Lab
School of Computer and Systems Sciences
Jawaharlal Nehru University
New Delhi, India

Md Ahateshaam
Department of Computer Science and Engineering
Sharda University
Greater Noida, India

Kazi Nahian Haider Amlan
Department of Business and Technology Management
Islamic University of Technology
Gazipur, Bangladesh

Elena Basan
Institute of Computer Technologies and Information Security
Southern Federal University
Taganrog, Russia

Anmol Chaddha
Department of Electrical Engineering
Indian Institute of Technology Patna
Patna, India

Pushpa Chaoudhary
Department of Master of Computer Applications
JSS Academy of Technical Education
Noida, India

Shailesh Kumar Gupta
Department of Electrical and Electronics Engineering
IIMT Engineering College
Meerut, India

Ratneshwer Gupta
Software Quality Assurance Lab
School of Computer and Systems Sciences
Jawaharlal Nehru University
New Delhi, India

B. Iswarya
Department of Computer Applications
SNMV College of Arts and Science
Coimbatore, India

Md Saquib Jawed
Department of Computer Science
Aligarh Muslim University
Aligarh, India

S. Julie Violet Joyslin
Department of Computer Technology
SNMV College of Arts and Science
Coimbatore, India

Bùi Huy Khôi
Department of Business Administration
Industrial University of Ho Chi Minh City
Ho Chi Minh City, Vietnam

Dinesh Kumar
Computer Science & Engineering Department
National Institute of Technology Jamshedpur
Jamshedpur, India

J.P. Josh Kumar
Department of Electronics and Communication Engineering
Agni College of Technology
Chennai, India

Satyam Kumar
Department of Computer Science and Engineering
Sharda University
Greater Noida, India

Sudhir Kumar
Department of Electrical Engineering
Indian Institute of Technology Patna
Patna, India

Abhisweta Lahiri
Department of Computer Science and Engineering
Sharda University
Greater Noida, India

Maria Lapina
Information Security of Automated Systems
Institute of Digital Development
North Caucasus Federal University
Stavropol, Russia

Tazwar Mahmud
Department of Business and Technology Management
Islamic University of Technology
Gazipur, Bangladesh

V. Manjuladevi
Department of Computer Technology
SNMV College of Arts and Science
Coimbatore, India

Dayima Musharaf
Department of Computer Science and Engineering
Sharda University
Greater Noida, India

Seerat Musharaf
Department of Computer Science and Engineering
Sharda University
Greater Noida, India

C. Nandhini
Department of Information Technology
Sri Ramakrishna College of Arts & Science
Coimbatore, India

Ankur Pandey
Electrical and Electronics Engineering
Rajiv Gandhi Institute of Petroleum Technology
Amethi, India

Olga Peskova
Institute of Computer Technologies and Information Security
Southern Federal University
Taganrog, Russia

B. Radha
Department of Information Technology
Sri Krishna Arts and Science College
Coimbatore, India

J. Rajeswari
Department of Electronics and Communication Engineering
Agni College of Technology
Chennai, India

Nahiyan Bin Riyan
Department of Business and Technology Management
Islamic University of Technology
Gazipur, Bangladesh

Mohit Kumar Saxena
Department of Electrical Engineering
Indian Institute of Technology Patna
Patna, India

S. Selvaraj
Department of Electronics and Communication Engineering
Agni College of Technology
Chennai, India

Mohammad Shahid
Department of Commerce
Aligarh Muslim University
Aligarh, India

Shivam
Department of Computer Science & Engineering
Motilal Nehru National Institute of Technology Allahabad
Prayagraj, India

Arun Kumar Singh
Department of Computer Science and Engineering
Greater Noida Institute of Technology
Greater Noida, India

Balvinder Singh
Department of Computer Science and Engineering
Sharda University
Greater Noida, India

Mahesh Kumar Singh
Department of Computer Science
ABES Engineering College
Ghaziabad, India

Pushpendra Singh
Department of Information Technology
Raj Kumar Goel Institute of Technology
Ghaziabad, India

M. Sreedhar
Department of Electronics and Communication Engineering
Agni College of Technology
Chennai, India

Nikita Sushkin
Institute of Computer Technologies and Information Security
Southern Federal University
Taganrog, Russia

S. Thavamani
Department of Computer Applications
Sri Ramakrishna College of Arts & Science
Coimbatore, India

Mohammad Shamsu Uddin
Department of Business and Technology Management
Islamic University of Technology
Gazipur, Bangladesh

Mohammad Zeeshan
Department of Electrical Engineering
Indian Institute of Technology Patna
Patna, India

1 Delay-Aware Partial Computational Offloading and Resource Allocation in Fog-Enabled Cyber-Physical Systems

Mohit Kumar Saxena, Anmol Chaddha, and Sudhir Kumar

CONTENTS

DOI: 10.1201/9781003438588-1

1.1 INTRODUCTION AND BACKGROUND

The Internet of Things (IoT) is a promising computing paradigm that allows "things" to be connected and can act without the intervention of humans. Smart devices can sense their surroundings and may generate a massive amount of data, which need to be transferred to the cloud [1]. IoT applications are the core of modern industrial and urban development. Similarly, robotic automation and artificial intelligence along with IoT can enable a prosperous platform for the industrial community. However, the rural setup has a few challenges, such as intermittent Internet connectivity. Thus, the present model of cloud-dependent IoT motivates bringing new solutions. The primary function of IoT-enabled systems is to collect and exchange information [2]. The deployment of devices that need Internet connectivity is expanding from smart cities, industrial applications, smart homes, smart agriculture, e-healthcare, e-education, and many others. Handling such massive amounts of smart devices, which generate data up to zettabytes, is one of the primary challenges with cloud-based IoT applications. Further, device heterogeneity, various wireless protocols, and lack of Internet-based communication infrastructure restrict the dependency on cloud servers. The high uplink latency to the cloud serves in IoT applications is critical for many real-time IoT applications. In spite of massive resources in cloud computing, there are still various challenges that limit the quality of service (QoS) experience of IoT applications. Integrating computing, networking, and storage capabilities makes the system more intelligent. These physical systems capable of interacting with cyber are termed cyber-physical systems (CPSs). In everyday life, CPS is increasingly found in many industries, such as health care, consumer electronics, manufacturing, smart automotive, and robotics. All CPSs require computational resources in addition to the physical processes. The next-generation CPS demands high computing resources that can interact in dynamic environments. Achieving these requirements is challenging in cloud-based environments. These technologies may provide new business, industry, and economic paradigms. Industrial robots can also leverage the IoT for the required computational resources and a network of robots talking to each other. Thus, the convergence of robotics and IoT emerges as the Internet of Robotic Things (IoRT).

To address the challenges of the cloud-based IoT, Computer Information System Company (CISCO) introduced a new distributed computing paradigm, which enables computational, storage, and networking resources near the origin of the tasks [3]. Fog computing cannot replace cloud computing. However, it can reduce the dependency of IoT applications on cloud resources. The fog resources can be considered a mini-data center, distributed along a small geographical area. Under this context, future CPS and IoT applications with many interconnected smart devices can leverage distributed and scalable access to the heterogeneous computational resources near the physical system.

The rest of the chapter is arranged as follows: The second section focuses on the three-layer architecture of fog-enabled IoT networks, fog computing deployment models, use cases of fog networks, and the research direction in fog computing. The third section presents an overview of task offloading and resource allocation. After this, we present the computation model, which includes total latency and energy requirements, in the next section. We obtain the cost function of task processing in fog networks, which mainly includes the total latency and energy consumption. The cost function may have different QoS requirements based on IoT applications. In the fifth section, we introduce the knapsack optimization-based resource allocation strategy, which ensures the efficient distribution of the task among the fog resources.

1.2 FOG COMPUTING IN IoT NETWORKS

Fog networks are highly virtualized networks that enable computing, storage, and networking services for IoT applications. Thus, IoT can leverage the resources of fog networks to achieve the QoS criteria of various IoT applications. Hence, integrating IoT and fog computing opens new avenues benefiting the various latency-sensitive IoT applications. Efficient resource allocation of fog networks is a significant challenge in fog computing-enabled IoT networks. In general, fog computing is an intermediate computing and storage layer between things and cloud servers. Fog enables low latency, context-aware decentralized cloud services to a better extent. The proximity of fog resources to end users ensures real-time services for IoT applications. The Long Term Evolution (LTE)/5G mobile network can be the backbone for the deployment of fog computing. Considering the massive IoT applications in the 6G era, managing the network, storage, and computing resources while ensuring QoS is challenging.

1.2.1 Fog Computing Network Scenario

Some IoT applications require powerful computational resources to deploy services. Fog computing can enable IoT applications, along with cloud servers. However, fog computing cannot replace the requirements of the cloud computing paradigm. Fog computing brings down the cloud computing resources to the network's edge. The computation offloading allows the IoT devices to lay off their computation to the nearby fog resources. This way, the IoT device can achieve ultra-low latency. The computation offloading can be a binary decision where the smart device offloads the task to the fog layer. Similarly, after offloading, the task may be scheduled among the multiple fog resources. Task partitioning allows efficient allocation of fog resources to achieve the QoS criteria, load balancing, and cooperation among the fog nodes to share their resources. Thus, IoT applications can fulfill their requirements of real-time response in health care, industrial control, smart cities, and many more areas using fog computing. The primary goal of fog computing networks is to minimize the end-to-end latency in IoT applications. The fog-enabled IoT is implemented in three-layer structures: the smart device layer, the fog layer, and the cloud layer. Figure 1.1 shows the proposed framework for fog-enabled CPS and IoT applications. Here, we consider that the smart device can offload the task to fog networks and then fog networks may offload the task to the central cloud. The primary functional layer contains IoT devices. With the help of IoT devices, the smart device layer collects the raw data from the surroundings through various intelligent sensors and devices with sensing abilities.

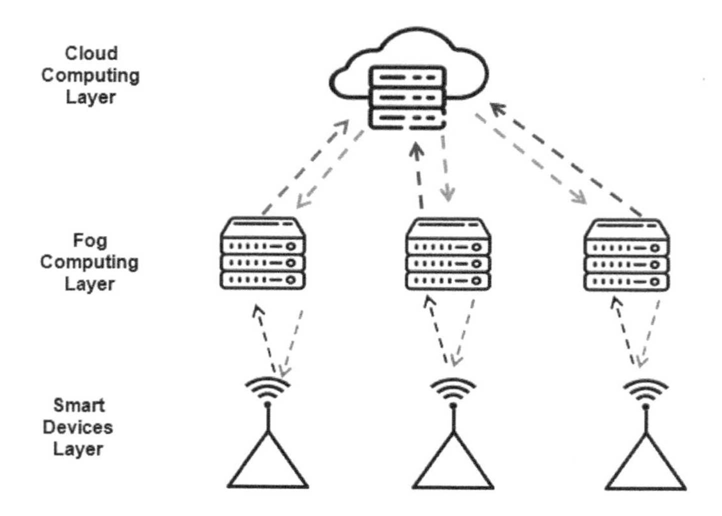

FIGURE 1.1 Three-layer framework for fog-enabled CPS and IoT applications.

The collected data are needed to be analyzed and computed for further action in real time. The fog computing layer enables the services and tools for data processing that can benefit IoT applications such as home automation, e-healthcare, and smart cities. Most IoT applications demand the real-time response of the sensed data.

The smart devices aggregate these data to the nearest fog resources for processing. Fog nodes provide computing and storage services for smart devices. The data processing results are then transferred to the designated IoT device through wireless communication protocols. In the three-layer framework, cloud servers have the highest computing, storage, and networking resources. In general, resource-constrained smart devices leverage these resources. However, smart devices need Internet connectivity to communicate with remotely located cloud servers. The limitation of Internet connectivity for cloud servers makes fog computing a prominent solution for future CPS and IoT applications. However, in case of higher queue delay at the fog resources, the computational assignment can be further forwarded to the centralized cloud layer.

1.2.2 Fog Network Characteristics

The distinguishing characteristic of fog networks includes the following:

- **Heterogeneity**: Fog networks consist of heterogeneous devices, such as base stations (BS), routers, network adapters, and LTE network component. These devices have distinct computing, storage, and networking capabilities, making the fog layer heterogeneous. This makes fog networks a highly virtualized paradigm.
- **Geographical Distribution**: The geographical distribution of fog resources between smart devices and cloud data centers makes location awareness, which is essential for many applications such as augmented reality, gaming,

and high-quality video streaming. This makes fog networks more suitable than cloud computing.
- **Mobility Support**: The direct link between fog resources and smart devices supports mobility in many IoT applications.

The above characteristics are essential for deploying many IoT applications where emergency response is critical.

1.3 LITERATURE OVERVIEW OF OFFLOADING AND TASK SCHEDULING

Future industrial development cannot be untouched by the IoT, cyber-physical systems (CPSs), and big data analytics. CPSs are the new ecosystem for IoT applications. CPSs require computing, communication, storage, and control ability [4]. CPSs use the new generation of smart sensors that can sense and transmit information to the physical world. The applications that need to incorporate the CPSs, including health care, energy management, and smart city, are facing the communication network problem. The deployment of CPS applications involves a communication network that acts as an intermediate connection between the physical world and the cyberworld. A CPS-enabled medical system distributes the raw data collected by smart healthcare devices to the fog layer; here, the fog resources are distributed in a virtual machine placement environment, and this work minimizes the overall cost of task processing and achieves the QoS criteria [5]. The overall cost of task execution at the fog resources includes end-to-end latency, energy consumption, and cost of network usage. In Ref. [6], a resource distribution scheme is presented, which achieves the energy-delay trade-off in fog networks. Big data analytics generate a large amount of data that are needed to be processed in the cloud. However, it is inefficient to communicate all the data to the cloud due to the high redundancy of data, which may increase the cost of network usage. Pre-filtering the data can be done in the fog layer before sending all the data to the cloud. The primary purpose of fog computing is to process task requests with ultra-low latency and send the filtered data to the cloud, which is needed to store for a long duration [7]. In modern industry, smart sensors are inherent in different processes, and the collected raw data from their surrounding are needed to process. Smart devices leverage distributed computing paradigm to drain their workload to some resource-enriched devices in a local area [8]. Several papers have emphasized computational offloading decisions, which are based on resource availability and QoS requirements of IoT applications in fog computing. Latency-sensitive computation-intensive IoT applications such as health care, virtual reality, and games need to process massive amounts of data in real time [9]. The primary function of fog-enabled IoT systems is to achieve QoS criteria. A fog node placement scheme for IoT networks is proposed, which achieves QoS satisfaction [10]. Resource monitoring in edge servers and cloud data centers is considered for offloading computational tasks [11]. A coordinated resource allocation and computation offloading to a fog node for multiple users are investigated [12]. An offloading problem in fog computing, which considers a small cell of fog node and a remote cloud server, is presented [13]. For fog nodes, the author considers the orthogonal multiple access technique. However, this increases the delay and interference. The resource allocation strategies are categorized into central and distributed allocation [14]. A service delay-minimizing approach is presented in Ref. [15], which enables fog-to-cloud cooperation. Hence, the offloaded tasks to the fog nodes may be offloaded to the cloud server. However, the author does not consider task partitioning among the fog nodes. In central allocation techniques, a fog server collects real-time information, such as computing, storage, and power consumption capacity of the associated fog nodes. However, in distributed allocation, the computational task can be allocated to any available fog node in the current time slot. Hence, the distributed allocation may fail to utilize resources efficiently, and obtaining the globally optimized allocation is also challenging. Thus, centralized resource allocation strategies are preferable for fog-enabled IoT networks. Different industrial processes have different smart sensors, which may support a variety of communication protocols. The heterogeneity of fog resources can communicate with these sensors directly as opposed to Internet-based cloud services. However, the heterogeneity in IoT devices and fog resources may raise challenges in the resource allocation of fog networks. Hence, various CPSs and IoT applications require a dynamic resource allocation strategy according to the requirements of QoS criteria.

The main contribution of this chapter is to model and evaluate the total latency of CPS and IoT applications in fog-enabled IoT networks. First, the resource-constrained smart devices take the offloading decision and offload the tasks to fog networks where the fog controller leverages the knapsack optimization to allocate the fog resources according to the latency requirements and available processing capacity.

1.4 COMPUTATION MODEL

Here, we describe the proposed three-layer architecture for fog-enabled IoT networks. Further, we formulate the end-to-end latency of computational tasks for IoT devices and fog layers. In a three-layer architecture, first, we define the objective of each layer. The smart device layer comprises heterogeneous devices such as sensors, actuators, and computation hardware. The heterogeneity of these devices poses various challenges, and the data generated by these devices need a different computation, networking, and storage facility under different constraints, which are based on required QoS criteria. Furthermore, energy consumption is an important criterion for these devices as most of them use battery for their operations, such as smart watches, wearable sensors, and multimedia devices. Thus, the tasks generated by such smart devices are needed to process somewhere else. Cloud computing is the general solution to the problem; however, all devices may not have the feature of the Internet. We consider that the fog resources support different wireless communication protocols, which are essential in the case of heterogeneous smart devices, and each fog resource can serve a limited geographical area. However, the dense deployment of fog nodes and the feature of mobility make it possible to ensure uninterrupted services. Thus, fog computing can meet the requirements of smart devices.

We consider fog nodes with varying computational capacities in terms of the maximum allowed workload in fog networks. In fog networks, we consider the deployment of fog controllers as shown in Figure 1.2. We assume that the smart devices cannot

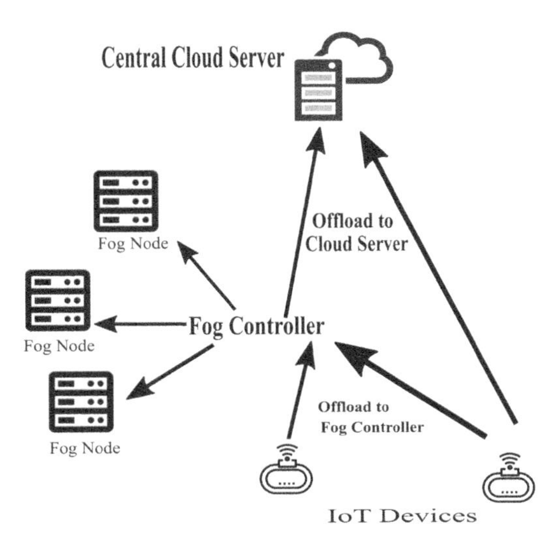

FIGURE 1.2 Network model with the fog controller in the fog computing layer.

offload their computational tasks to the fog node directly. The fog controller receives the tasks from smart devices. Each fog controller has a few associated heterogeneous fog nodes with them. We assume that the fog controller can monitor the workload of associated fog nodes in the current time slot and take the task partitioning decision. The IoT devices lay off the task to the nearby fog controller instead of fog nodes. The fog controller takes the computational task scheduling decision based on the available resources in the current time slot. Thus, we can reduce the wait time in the queue at the fog controllers and the dependency on central cloud servers. In the fog layer, the dense deployment of fog resources is considered, ensuring that each computational task request from the smart devices is served. Thus, we can leverage the existing LTE network resources as fog nodes. This way, the LTE service providers can generate revenue for their existing computational, storage, and networking resources.

We denote the P smart devices as $L=\{L_1, L_2, \ldots, P\}$, where $i=1, 2, 3, \ldots, P$, which generate the computational tasks $t=\{t_1, t_2, \ldots, P\}$ with different data sizes and maximum allowed latency. Similarly, the Q fog nodes are denoted by $F=\{F_1, F_2, \ldots, Q\}$, where $j=1, 2, 3, \ldots, Q$, which are associated with a set of fog controllers. A feasible scheduling strategy ensures the constraint as follows:

$$\sum_{j=1}^{Q} F_j \leq F_j^{\max} \qquad \forall\ j \in N \tag{1.1}$$

where $F_j^{\max}$ is the maximum number of fog nodes associated with any fog controller. Therefore, in the case of task partitioning, the required number of fog nodes to process the task must be upper bound with the total number of associated fog nodes to any fog controller. Thus, we can ensure the efficient processing of the offloaded task in the fog layer instead of being sent to the cloud servers. We also consider centralized cloud servers, which have ample storage, computing, and networking resources.

We can leverage central cloud servers in case of higher latency at the fog layer. The fog controller can allocate each associated fog nodes, which is available in the current time slot. However, if fewer fog nodes are available to process a new task, the fog controller can schedule the task to another nearby fog controller or offload the task to the cloud server. We characterize each computational task generated from the smart devices with the data size of the task D_t, required central processing unit (CPU) resource, and maximum allowed latency (T^{allowed}).

1.4.1 Latency Model in Smart Device

The computational tasks generated by smart devices can be processed either on the smart device or remotely through computation offloading. Here, the task processing latency (T_i^L) is proportional to the data size of the task and processing density (PR_i^{SD}); however, it is inversely proportional to the CPU frequency (f_i^{SD}) of the smart device.

1.4.2 Latency Model in Fog Networks

Under resource-constrained IoT environments, the IoT devices offload their computational tasks to the fog controller through wireless links. Then, the fog nodes with the required computation capabilities are allocated to process the task and send back the processed results to the same smart device. The offloaded tasks are subjected to additional delays such as task uplink delay, queuing delay at the fog controller, and latency in downloading the computation result. Now, we obtain the latency in task processing at the fog layer.

1.4.2.1 Task Up-Link Delay

First, the offloading decision depends on the smart device's latency and the IoT application's maximum allowed latency. Second, the uplink delay affects the offloading decision. Once the smart device takes the computational offloading decision, the task is transmitted to the nearby fog controller using the wireless network. Each generated task from the smart devices experiences a different uplink delay (T_{ij}^{up}), which largely depends on the size of the computational task and the data uplink rate from the smart devices to the fog controller. The task uplink rate depends on the network bandwidth between the IoT device and the fog controller. Now, the smart devices can offload the computational task to a single fog controller, which is in the coverage area of the smart device. The task offloading decision $O = \{\gamma_{ij} \in L\}$primarily depends on the difference between the latency on the smart device and the allowed latency for a given IoT application as follows:

- When $T_i^L \leq T^{\text{allowed}}$ or $T_{ij}^{\text{up}} \geq T^{\text{allowed}}$, the computational task cannot be offloaded to the fog layer and is processed locally at the smart device due to a higher uplink delay.
- When $T_i^L \geq T^{\text{allowed}}$ or $T_{ij}^{\text{up}} \leq T^{\text{allowed}}$, the computational task is offloaded to the fog controller.

Hence, the offloading decision is defined as $\gamma_{ij} = \{0, 1\}$. Here, $\gamma_{ij} = 1$ indicates that the task is offloaded to the fog controller, and $\gamma_{ij} = 0$ indicates that the computational task is processed on a smart device. However, an extended uplink delay may affect the QoS requirements.

1.4.2.2 Queuing Delay at Fog Controller

This delay refers to the time spent by tasks in the queue at the fog controller before scheduling to a fog node. The computational task received at the fog controller can only be scheduled on the appropriate fog node available in the current time slot. However, an appropriate fog node may not be available in the current time slot. Hence, we propose task partitioning at the fog controller. This allows scheduling an enormous data size task into numerous fog nodes instead of waiting for an alone fog node to become available. As a result, we can achieve a significantly reduced queue delay at the fog controller. In the fog controller, we assume $M / M / c$ queue model, where c is the number of associated fog nodes. At the fog controller, the task arrival rate follows the Poisson process with λ_j^{total}. We consider the average delay (T_Q^{FGC}) in the queue at the fog controller.

1.4.2.3 Task Computation Latency

This delay defines the time to compute the task after scheduling from the fog controller. The CPU frequency (f_j^F) and processing rate (PR_j^F) of the computing device decide the computation delay. Thus, the computation latency at fog nodes is given as follows:

$$\sum_{i=1}^{P}\sum_{j=1}^{Q} T_{ij}^F = \frac{\gamma_{ij} D_t \text{PR}_j^F}{f_j^F} \quad \forall\ i \in P, j \in Q \tag{1.2}$$

The CPU frequency of the fog node is upper bound (equation 1.3) by the maximum frequency ($f_j^{F\max}$) of the associated fog node with the fog controller. However, it may be possible that the fog node with the highest CPU frequency is occupied by another task in the current time slot.

$$f_j^F \leq \sum_{j=1}^{Q} f_j^{\max} \quad \forall\ j \in Q \tag{1.3}$$

Considering the maximum acceptable queue delay in conjunction with the task deadline, the fog controller offloads the task to another fog controller in the vicinity or to the central cloud. Hence, we define the acceptable queue delay time, which must incorporate the processing delay. However, before the actual scheduling decision, it is not easy to get the exact task processing delay, so we consider the upper bound on the processing delay given by the fog node with the least computation capacity. According to the acceptable queue delay, the fog controller takes the scheduling decision as follows:

- When $T_{ij}^{\text{up}} + T_Q^{\text{FGC}} + T_{ij}^F < T^{\text{allowed}}$, the computational task is scheduled to the suitable fog node.
- When $T_{ij}^{\text{up}} + T_Q^{\text{FGC}} + T_{ij}^F > T^{\text{allowed}}$, the task cannot be processed at any single fog node. Thus, the computational task is partitioned and scheduled

to multiple fog nodes. However, in case of the required fog resources are busy in the current time slot, and then, the task cannot be scheduled to the fog layer.

The result of the processed task at fog networks is transmitted to the same smart devices from which the fog controller received this. However, the computational result has less data size as compared to the tasks. Hence, we can neglect the transmission latency between fog resources and IoT devices.

1.4.2.4 Total Task Computation Latency

The task uplink delay, queuing delay at fog controller, and task computation latency are the total latency of the offloaded tasks. This defines the total delay in computing the task. The computational task is divided into subtasks to minimize the queue delay at the fog controller and improve resource allocation efficiency. These subtasks may differ in data size and offer different computation latency. Thus, we consider the highest computation latency among all fog nodes.

$$\sum_{i=1}^{P}\sum_{j=1}^{Q} T_{ij}^{\text{total}} = \sum_{i=1}^{P}\sum_{j=1}^{Q} T_{ij}^{\text{up}} + T_{Q}^{\text{FGC}} + \max \sum_{i=1}^{P}\sum_{j=1}^{Q} T_{ij}^{F} \quad \forall\ i \in P, j \in Q \tag{1.4}$$

1.5 KNAPSACK OPTIMIZATION-BASED RESOURCE ALLOCATION

Here, the task offloading decision is presented and then the knapsack-based resource allocation strategy. In task offloading decisions, the smart device decides where to offload the computation task. The task offloading decision may be based on total latency and energy consumption; however, the other factors that may be considered are networking cost and deployment cost. This work considers latency as the primary constraint in offloading decisions. Here, the computational task offloading is a binary decision problem based on the delays at the IoT device, uplink, and queue at the fog controller. In other words, the proposed offloading decision is preferred for latency-sensitive IoT applications that ensure the computational task must be finished before the deadline. Algorithm 1.1 states the offloading decision at the smart device and fog controller.

Now, we define the resource allocation optimization problem. As mentioned before, fog nodes are resource-constrained compared with cloud servers. To achieve low latency, we now define the objective function (1.5a), which gives the minimum number of fog resources in which the task can be divided in case of task partitioning.

$$\min \left[\sum_{i=1}^{P}\sum_{j=1}^{Q} T_{ij}^{\text{total}} \right] \quad \forall\ i \in P, j \in Q \tag{1.5a}$$

$$\text{s.t.} \quad \sum_{i=1}^{P}\sum_{j=1}^{Q} t_i \ \mu_j^{Fr} \leq \sum_{j=1}^{Q} \mu_j^{F\max} \quad \forall\ i \in P, j \in Q \tag{1.5b}$$

Algorithm 1.1 Task Offloading Decision

Input: P, Q, O, D_t, T_i^L, T_{ij}^{up}

Output: γ_{ij}

1 for i: 1 to P do
2 For every task on the smart device, Compute T_i^L
3 if $T_i^L \le T^{\text{allowed}}$ or $T_{ij}^{\text{up}} \ge T^{\text{allowed}}$ then
4 return $\gamma_{ij} = 0$
5 The task is executed on the device layer
6 else if $T_i^L \ge T^{\text{allowed}}$ and $T_{ij}^{\text{up}} \le T^{\text{allowed}}$ then
7 return $\gamma_{ij} = 1$
8 The tasks are offloaded to the fog controller
9 end if
10 end for

$$\sum_{i=1}^{P} t_i \le \frac{\sum_{j=1}^{Q} \mu_j^{F\max}}{\sum_{j=1}^{Q} \mu_j^{Fr}} \quad \forall\, i \in P, j \in Q \tag{1.5c}$$

$$\sum_{j=1}^{Q} x_j \le F_j^{\max} \quad \forall\, j \in Q \tag{1.5d}$$

We consider that $\sum_{j}^{Q} \mu_j^{Fr}$ and $\sum_{j}^{Q} \mu_j^{F\max}$ denote the required computing resources and the maximum available computing resources of the associated fog nodes with the fog controller at the current time slot, respectively. Similarly, tasks available in the queue are denoted as $\sum_{i=1}^{P} t_i$, which need to be scheduled on the fog nodes and must satisfy the constraint (1.5b). Constraint (1.5c) gives the upper bound on the tasks that can be scheduled in any given time slot. This shows that all the tasks in the queue cannot be scheduled simultaneously. Thus, the scheduling strategy can minimize the queue delay by forwarding the remaining tasks from the queue to the cloud. However, the tasks that are further forwarded to the cloud server may have a higher uplink delay. Thus, the dependency on the cloud needs to be minimized, and an appropriate task partitioning decision can achieve this. Therefore, the proposed strategy minimizes the end-to-end latency and determines an appropriate task scheduling decision for each upcoming task from smart devices. The task partitioning decision depends on the ratio of available and required computation capacity of associated fog resources for the offloaded task. The task partitioning decision leverages the knapsack optimization, which can schedule the task on multiple fog nodes to process it efficiently. In constraint (1.5d), the task can be partitioned among the maximum number of fog nodes associated with any fog controller.

Now, the fog controller takes resource allocation decisions and schedules the tasks for each time slot. The proposed knapsack-based resource allocation strategy ensures that a maximal workload of IoT devices can be scheduled on the fog nodes in all time slots. Based on the task partitioning, we consider the fractional knapsack problem, assuming the knapsack capacity is F_c^K. Fractional knapsack optimization aims to find the optimal number of fog nodes and to increase the number of tasks being able to schedule within the knapsack capacity. If any single appropriate fog node is busy in the current time slot, the computational tasks can be partitioned into smaller pieces so that each can be scheduled. Thus, the fog controller does not require waiting for a fog node with a massive computing capacity. This way, the proposed strategy can ensure the minimum queue delay and also reduce the dependency on the centralized cloud. The t_i task is partitioned into Q parts, and each part of the task has the weight w_j, now constraint (1.5d) is modified, and the total weight of the knapsack is given as follows:

$$\sum_{i=1}^{P}\sum_{j=1}^{Q} x_j \; w_j \leq F_c^K \qquad \forall \; i \in P, j \in Q \tag{1.6}$$

Equation (1.6) ensures that an optimal resource allocation must fill the knapsack; otherwise, the remaining part of the task may lead to a higher queue delay, which increases the total latency of the computational task. First, we need to sort the fog nodes according to the ratio of the size of partitioned task to the processing latency, which is given as follows:

$$\frac{\sum_{j=1}^{Q} D_{tj}}{\sum_{i=1}^{P}\sum_{j=1}^{Q} T_{ij}^{F}} \qquad \forall \; i \in P, j \in Q \tag{1.7}$$

The numerator of equation (1.7) ensures that a task can be partitioned among the associated fog resources to the fog controller. The detailed steps of the knapsack-based resource allocation strategy are presented in Algorithm 1.2.

Now, we analyze the time complexity of the proposed method. The complexity primarily depends on equation (1.7) and the number of fog nodes. In the worst case, the time complexity of the knapsack-based resource allocation strategy is $O(n^2)$.

This can be $O(n \log n)$ in case all fog nodes are sorted according to equation (1.7) for the given data size of the task. To improve the performance of the proposed strategy, we sort the available fog nodes based on their CPU frequency and the delay offered. Hence, the proposed knapsack-based resource allocation strategy achieves the optimal task partitioning and allocates the fog resources so that the maximum fog resources can be utilized in each time slot. Thus, a higher throughput of the fog controller can achieve the QoS criteria of IoT applications and minimize the dependency on centralized cloud servers.

Algorithm 1.2 Knapsack-Based Resource Allocation

Input: $P, Q, O, D_t, \mu_j^{Fr}, F_c^K$

Output: $(\sum_j^Q w_j)$

1 for i: 1 to P do
2 for j: 1 to Q do
3 For every offloaded task at fog controller
4 if $T_{ij}^{\text{up}} + T_Q^{\text{FGC}} + T_{ij}^F < T^{\text{allowed}}$ then
5 Task is scheduled on the suitable fog node
6 else if $T_{ij}^{\text{up}} + T_Q^{\text{FGC}} + T_{ij}^F > T^{\text{allowed}}$ then
7 Task is partitioned and scheduled to multiple fog nodes
8 Sort $\frac{\sum_{j=1}^{Q} D_{tj}}{\sum_{i=1}^{P}\sum_{j=1}^{Q} T_{ij}^F}$ in descending order
9 Allocate the fog resources $\sum_{i=1}^{P}\sum_{j=1}^{Q} x_j \; w_j \leq F_c^K$, for the first task in the queue
10 Recalculate the ratio (equation 1.7) based on remaining resources and allocate the next task from the queue.
11 end if
12 end for
13 end for

1.6 RESULTS AND DISCUSSION

In this section, first, the parameters for task offloading are presented. Subsequently, numerical results and comparisons on the number of fog nodes, varying task sizes, and average latency are described, respectively.

1.6.1 Numerical Parameters

We consider a fog network where each fog controller has ten heterogeneous fog nodes. The IoT device either processes its task or offloads it to the nearest fog controller. In case of offloading the task to fog networks, the fog controller may forward the tasks to the cloud. The fog nodes are scattered randomly within the coverage area of the fog controller. IoT devices do not have a direct link to fog resources. The tasks are offloaded to the fog network, where the uplink rate may vary from 250 Kb/s up to 100 Mb/s. To present the actual scenario, we consider the fog resource similar to the Raspberry Pi and an Intel Core i5 CPU. We assume that the maximum allowed latency that an IoT application requires to ensure the QoS criteria ranges from 50 to 500 ms. Similarly, we consider the 1–10 Mb task size range. To measure the efficacy of the knapsack-based resource allocation strategy, we analyze the average end-to-end latency on multiple fog nodes with task partitioning. Further, we compare the knapsack-based resource allocation with the other baseline strategies, such as random task offloading, maximum processing capacity offloading [16], and minimum queue delay offloading [17].

1.6.2 Total Latency for Varying Number of Fog Nodes

Figure 1.3a shows the total latency obtained from the number of available fog nodes in which a task can be partitioned. First, the total latency decreases to a minimum level. When the task is partitioned among all the fog nodes, the latency increases due to the higher queue and task processing delay. This illustrates that the computational task must not be partitioned among all the fog resources because all the fog

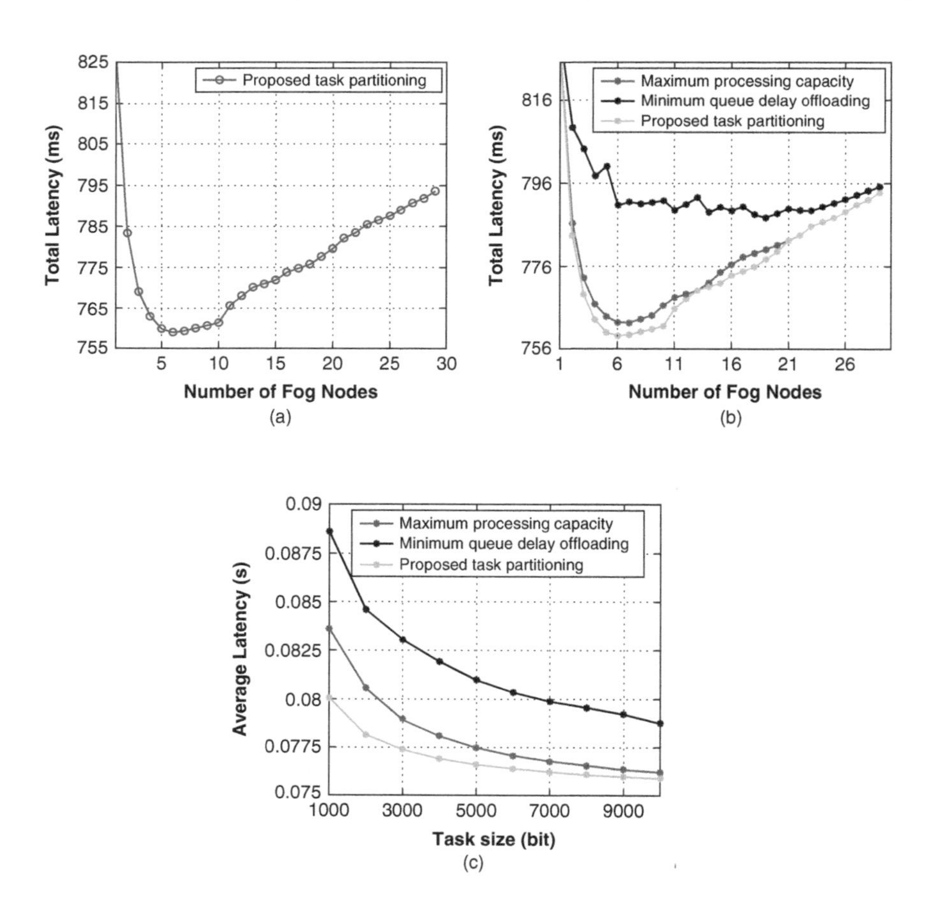

FIGURE 1.3 (a) Total latency in the proposed strategy, (b) average latency with varying sizes of tasks, and (c) total latency with the varying number of fog nodes.

nodes have different computational resources. The fog node with fewer computation resources may increase the total latency. Hence, it is optimal for the given computational task to partition it among six fog nodes to achieve minimum latency of 0.759 seconds.

1.6.3 Effect of Varying Task Sizes on End-to-End Latency

Figure 1.3b illustrates the average latency under varying task sizes. This shows that the proposed strategy offers minimum average latency as the task size increases. The proposed algorithm achieves 6% and 8% lower average latency for a task size of 2,000 bits compared with maximum processing capacity offloading and minimum offloading delay, respectively. Similarly, the proposed algorithm achieves a 5% lower average latency after 4,000 bits and a 4% lower average latency after 6,000 bits. This shows that the task partitioning-based knapsack algorithm achieves a decreased average delay for the larger sizes of tasks. Hence, a data-intensive task may simultaneously be processed on multiple fog resources, which may reduce the queue delay. The tasks are offloaded from the IoT device to the fog controller without partitioning. Then, the fog controller may partition the task by observing the queue delay and available fog resources. Thus, in the case of task partitioning, the task processing delay dominates the uplink delay and queue delay in total latency. The proposed task partitioning strategy offers average latency of 0.08 and 0759 seconds for task sizes of 1,000 and 10,000 bits, respectively, which is a reduction of nearly 7%. The comparisons with existing algorithms are stated in Table 1.1.

1.6.4 Average Latency Performance

Here, we compute the average latency of the task execution for various strategies. The average latency is essential in meeting the QoS criteria requirement and indicates efficient resource allocation. The offloaded task at the fog controller may be scheduled to a fog node with sufficient resources, or the task can be partitioned among multiple fog nodes in case of higher queue delay. Since the uplink delay depends

TABLE 1.1
Algorithm Comparison

	Without Task Partitioning	With Task Partitioning	
Algorithm	Latency at Primary Fog Node (ms)	Optimal Number of Fog Nodes	Latency at Optimal Number of Fog Nodes (ms)
Maximum processing capacity [16]	825.424	7	762.185
Minimum queue delay offloading [17]	830.442	19	787.631
Proposed knapsack-based resource allocation	825.442	6	759.015

on the task size and available network bandwidth, the queue delay is the time a task waits before it is assigned to a fog node. The queue delay primarily depends on the current workload of the associated fog nodes. In a random task offloading scheme, the tasks are scheduled to any available fog node without considering the required computing capacity for task processing. Similarly, the maximum processing capacity offloading strategy schedules the task to the fog resource with the highest CPU frequency in the current time slot. However, this may increase the task computation latency of tasks awaited in the queue because this strategy does not consider the maximum allowed latency of the tasks. Figure 1.3c compares various baseline algorithms for task offloading. The proposed algorithm can dynamically compute the optimal task partitioning among the fog nodes. Thus, we can balance the workload of fog nodes and achieve the required ultra-low latency by adjusting the task partitioning.

1.7 CONCLUSION

The increasing demand for CPS and IoT applications may not be fulfilled through cloud computing. The latency-sensitive applications restrict the deployment of cloud-based IoT applications. Fog networks have the potential to serve many industrial and domestic latency-sensitive applications, which require computation resources. The efficacy of fog networks in future CPS and IoT applications is examined in this chapter. The proposed strategy is straightforward to implement and can ensure the QoS criteria. The strategy is divided into two decisions where the first decision of task offloading is taken by the smart devices, which generate the task request. Then, the fog controller makes the second decision on resource allocation and task scheduling. The numerical results indicate the effect of task partitioning on total latency that can be considered for many real-time IoT applications. The future work will consider QoS criteria other than total latency, along with the placement of fog resources to minimize the queue delay.

REFERENCES

1. C. Mouradian, D. Naboulsi, S. Yangui, R. H. Glitho, M. J. Morrow, and P. A. Polakos, "A comprehensive survey on fog computing: State-of-the-art and research challenges," *IEEE Communications Surveys Tutorials*, vol. 20, no. 1, pp. 416–464, 2018.
2. M. Mukherjee, L. Shu, and D. Wang, "Survey of fog computing: Fundamental, network applications, and research challenges," *IEEE Communications Surveys Tutorials*, vol. 20, no. 3, pp. 1826–1857, 2018.
3. F. G Bonomi, R. A. Milito, J. Zhu, and S. Addepalli, "Fog computing and its role in the internet of things," *Proceedings of the MCC Workshop on Mobile Cloud Computing*, 08 2012, Helsinki Finland.
4. A. K. Tyagi and N. Sreenath, "Cyber physical systems: Analyses, challenges and possible solutions," *Internet of Things and Cyber-Physical Systems*, vol. 1, pp. 22–33, 2021.
5. L. Gu, D. Zeng, S. Guo, A. Barnawi, and Y. Xiang, "Cost efficient resource management in fog computing supported medical cyber-physical system," *IEEE Transactions on Emerging Topics in Computing*, vol. 5, no. 1, pp. 108–119, 2017.
6. M. K. Saxena and S. Kumar, "Differential scale based multi-objective task scheduling and computational offloading in fog networks," in *2021 National Conference on Communications (NCC)*, Kanpur, India, 2021, pp. 1–6.

7. Y.-J. Ku, D.-Y. Lin, C.-F. Lee, P.-J. Hsieh, H.-Y. Wei, C.-T. Chou, and A.- C. Pang, "5g radio access network design with the fog paradigm: Confluence of communications and computing," *IEEE Communications Magazine*, vol. 55, no. 4, pp. 46–52, 2017.
8. R. Mahmud, R. Kotagiri, and R. Buyya, *Fog Computing: A Taxonomy, Survey and Future Directions*. Singapore: Springer, 2018, pp. 103–130. [Online] Available: https://doi.org/10.1007/978-981-10-5861-5.
9. X. Peng, J. Ren, L. She, D. Zhang, J. Li, and Y. Zhang, "Boat: A block-streaming app execution scheme for lightweight IoT devices," *IEEE Internet of Things Journal*, vol. 5, no. 3, pp. 1816–1829, 2018.
10. I. Lera, C. Guerrero, and C. Juiz, "Availability-aware service placement policy in fog computing based on graph partitions," *IEEE Internet of Things Journal*, vol. 6, no. 2, pp. 3641–3651, 2019.
11. J. Shuja, A. Gani, M. Habibur Rehman, E. Ahmed, S. Madani, K. Khan, and K. Ko, "Towards native code offloading based mcc frameworks for multimedia applications: A survey," *Journal of Network and Computer Applications*, vol. 75, pp. 335–354, 2016.
12. K. Liang, L. Zhao, X. Zhao, Y. Wang, and S. Ou, "Joint resource allocation and coordinated computation offloading for fog radio access networks," *China Communications*, vol. 13, supplement 2, pp. 131–139, 2016.
13. J. Du, L. Zhao, X. Chu, F. R. Yu, J. Feng, and C.-L. I, "Enabling low-latency applications in LTE-A based mixed fog/cloud computing systems," *IEEE Transactions on Vehicular Technology*, vol. 68, no. 2, pp. 1757–1771, 2019.
14. W. Na, S. Jang, Y. Lee, L. Park, N.-N. Dao, and S. Cho, "Frequency resource allocation and interference management in mobile edge computing for an internet of things system," *IEEE Internet of Things Journal*, vol. 6, no. 3, pp. 4910–4920, 2018.
15. M. K. Saxena and S. Kumar, "Latency-aware task partitioning and resource allocation in fog networks," in *2022 IEEE 19th India Council International Conference (INDICON)*, Kochi, India, 2022, pp. 1–5.
16. M. Mukherjee, V. Kumar, J. Lloret, and Q. Zhang, "Revenue maximization in delay-aware computation offloading among service providers with fog federation," *IEEE Communications Letters*, vol. 24, no. 8, pp. 1799–1803, 2020.
17. R. Jindal, N. Kumar, and H. Nirwan, "MTFCT: A task offloading approach for fog computing and cloud computing," in *2020 10th International Conference on Cloud Computing, Data Science Engineering (Confluence)*, Noida, India, 2020, pp. 145–149.

2 Enhancing the Security of Cryptographic Algorithms in Cloud-Based Cyber-Physical Systems (CCPSs)

Md Saquib Jawed, Mohammad Sajid,
Maria Lapina, and Mukesh Prasad

CONTENTS

DOI: 10.1201/9781003438588-2

2.1 INTRODUCTION

The fourth wave of industrialization, also recognized as Industry 4.0 [1], is completely transforming how corporations manufacture, improve, and deliver their products. Industry 4.0 seeks to create smart and connected firms that are highly automated and optimized for efficiency, flexibility, and productivity [2]. The term "Industry 4.0" refers to the integration of cutting-edge technologies such as big data analytics [3], artificial intelligence and machine learning [4], cloud computing [5], and Internet of things (IoT) [6] into manufacturing operations and other aspects of business practices. Industry 4.0 relies heavily on cloud-based cyber-physical systems (CCPSs) [7] to achieve its goals. A CCPS merges physical components, communication networks, and cloud computing to create a highly connected and intelligent system that allows real-time monitoring and control of the physical process. The cloud computing aspect of CCPS provides a centralized platform for collecting, processing, and distributing data [8]. The cloud also provides scalable computing resources and high accessibility, allowing users to access the system from anywhere with an Internet connection [9]. The physical components consist of sensors, actuators, and other physical devices that collect and provide information about the real-world environment. While the communication networks are used to connect the physical components with the cloud platform, allowing for rapid response to changing conditions, this real-time control enables the system to respond quickly to potential problems and make necessary adjustments, improving the overall reliability and efficiency of the system. Together, they represent a significant transformation in the manufacturing industry, enabling companies to improve their operations, reduce costs, and deliver high-quality products more efficiently. The growing trend of technologies such as 5G, IoT, fog cloud [10] paradigm, and edge cloud [11] paradigm has boosted the demand and growth of CCPS. Today, there are a wide range of promising applications of CCPS in various domains including construction [12], robotics [13], smart grids [14], intelligent transportation [15], health care, and smart manufacturing [16]. The explosive growth of CCPS applications has also resulted in an enormous amount of data being generated and transferred to the cloud every second, which raises many concerns regarding data security and privacy [17]. Even a single data breach or tampering can result in significant financial loss for an organization or nation. Therefore, cloud data security has become one of the emerging topics for academicians and researchers. Every year, many symmetric and asymmetric cryptographic algorithms are designed and tested. All cryptographic methods revolve around their keys and their lengths. If an adversary can correctly guess the key, they will have access to all of the data; however, it will not be difficult for them to do so if the key is of a shorter length [18]. Therefore, unpredictability and complexity should constitute the key. Due to the non-deterministic polynomial-time (NP) hard nature of such keys, numerous deterministic, heuristic, and metaheuristic algorithm-based cryptographic algorithms are suggested for both symmetric and asymmetric key cryptography. Metaheuristic algorithms are a type of evolutionary computing used in the process of finding solutions to pragmatic optimization challenges [19–21]. The majority of metaheuristics attempt to discover solutions by imitating the behavior of biological systems. In cryptology, metaheuristic algorithms are utilized to find solutions to a variety of issues, including addition chains, optimization of Boolean functions, optimization of keys, creation of pseudo numbers, and substitution boxes [22]. The whale optimization algorithm (WOA) is an example of a metaheuristic algorithm, which replicates humpback whale foraging behavior. This chapter designs a pseudorandom key generator framework using metaheuristic algorithms, transfer function, and Shannon entropy. Using the framework, a key of various lengths is generated for the WOA, which can be used in emerging cryptographic algorithms namely Data Encryption Standard (DES), PRESENT, Advanced Encryption Standard (AES) 128/192/256, Triple DES (3DES), and Blowfish. The randomness of the WOA key is compared to peer algorithms. Further, the National Institute of Standards and Technology (NIST) statistical tests are also utilized to assess the quality of the keys.

The chapter is divided into the following sections: Section 2.2 enlightens some of the recent works related to the chapter. Section 2.3 explains all the algorithms used in the chapter, while Section 2.4 provides extensive attention to problem formulation. Section 2.5 shows the overall flow and methodology used in the proposed work. Section 2.6 represents all the experiments and their results. The final Section 2.7 summarizes the findings and also includes recommendations for future research.

2.2 RELATED WORK

This section discusses the recent developments that have been made in relation to the cryptographic key and cryptographic algorithms.

A comprehensive literature review of cloud computing security concerns has been provided by Alouffi et al. [23]. Eighty potential chapters were analyzed between 2010 and 2020 to perform the review. The authors have compiled a list of seven main security risks, the most discussed of which are data tampering and leakage. However, there are additional risks associated with data intrusion in cloud storage. Ahsan et al. [24] have also reviewed various bioinspired algorithms considering access control systems (ACS), user identification and authentication, network security, intrusion detection system (IDS), trust management, virtualization, and forensics as cloud security challenges. According to the authors, Particle Swarm Optimization (PSO) and Neural Network (NN) tackle maximum cloud security issues, while Ant Colony Optimization (ACO), fruit fly, and grasshopper are mainly used in deciphering specific issues. However, genetic algorithm is mainly used for network security systems. Duo et al. [25] have first explained the system models for the CPSs, i.e., time-driven and event-driven systems, and then surveyed the various emerging cyberattacks on these systems. Cyberattacks and their defense strategy are analyzed based on the three main security aspects of confidentiality, integrity, and availability. In the end, the authors discuss some of the open issues and challenges for further exploration. Priyadarshini et al. [26] have proposed a Cross-Breed Blowfish and Message Digest (MD5) approach to provide security to the healthcare data stored in the CCPS. The algorithm uses a wireless sensor network (WSN) for the transmission of data using a fuzzy trust-based routing protocol. Also, the best route for data transmission is identified using butter–ant optimization. The author claims that their algorithms have 10%–15% better latency, throughput, execution time, encryption–decryption time, and security level compared with conventional methods. Tahir et al. [27] have proposed the cryptographic algorithm CryptoGA in which a genetic algorithm has been used to generate the cryptographic key. Various cryptographic parameters are evaluated by performing encryption and decryption using CryptoGA on ten different data sets. The authors claim that their algorithm achieves higher encryption and decryption throughput compared with emerging algorithms such as AES, Blowfish, 3DES, and DES. However, some flaws in the algorithm enabled the decryption without using the key. Jawed et al. [28] have identified and removed this flaw in their algorithm XECryptoGA by introducing Exclusive-OR (XOR) and elitism. The authors claim that their algorithm successfully provides confidentiality and integrity, also having a six times faster key generation mechanism than the CryptoGA. In another publication, Jawed et al. [18] have performed the cryptanalysis using metaheuristic algorithms on lightweight block ciphers such as PRESENT, SIMON, and SPECK. It has been suggested by the authors that the cuckoo search outperforms both the Salp Swarm Algorithm (SSA) and the WOA. Also, for the weaker keys WOA can breach all three algorithms. For cloud data security, Saleh Muhammad Rubai [29] has suggested a new key generation algorithm that relies on heuristics. In this algorithm, the key generation depends on data sanitization, which is performed by considering a multi-objective problem. The switched–searched butterfly–moth flame optimization (PS-BMFO) is applied to solve the multi-objective problem and generates the key. The author claims that their model is better than grey wolf optimization (GWO), Jaya Algorithm (JA), Moth Flame Optimization (MFO), and Butterfly Optimization Algorithm (BOA). Ahamad et al. [30] have proposed a multi-objective privacy preservation model based on Jaya-based shark smell optimization (J-SSO) for cloud security. Data sanitization and restoration are two primary components of the suggested model. The optimal key generation problem, which is constructed by deriving a multi-objective function, is the primary determinant of data sanitization. According to the authors, the efficiency of the model has been found to be better than state-of-the-art models.

According to recent research, very little has been done to address the complexity of keys; instead, the majority of proposed solutions concentrate on cryptographic algorithms rather than cryptographic keys. The key is such a crucial component that if an attacker can guess it, they will have access to all the data transferred to the cloud. Even though some works generate the key to encrypt the data, for decryption, there is no need for that key, which is a big drawback of the suggested algorithm. In light of these considerations, this chapter suggests a framework that generates keys of various lengths that are highly random, unpredictable, and as complex as possible. To design such a framework, the proposed work makes use of any metaheuristic algorithm, as well as the transfer function and the Shannon entropy.

2.3 ALGORITHMS

This section provides an explanation of the fundamental algorithm that is used throughout this chapter.

2.3.1 Shannon Entropy

Shannon entropy [31], a way to measure the quantity of information in a piece of data, emerged as a result of the study of information theory. High values of Shannon entropy indicate that the data are very unpredictable. However, if the Shannon entropy is low, it suggests that the data are less random and more information leaks. Equation (2.1) gives the mathematical formula of Shannon entropy:

$$H(X) = -\sum_{i=1}^{n} P(x_i)\log_2 P(x_i) \tag{2.1}$$

where $P(x_i)$ is the chance of x_i appears in X.

2.3.2 Whale Optimization Algorithm

The WOA [32] draws inspiration from the hunting techniques of humpback whales to discover the optimal solution for any given optimization problem. There are three stages to this hunting method: encircling the prey, exploiting the prey, and searching for a prey.

2.3.2.1 Encircling Prey

During this phase, the whales locate and encircle their target. Because the optimum design location in the search space is uncertain, the WOA suggests that the best possible solution at the instant corresponds to the target prey or is near to it. Once the best search agent has been found, the remaining search agents will change their locations to align with the best agent using the following equation:

$$\vec{S} = \left| \vec{P} \otimes \vec{V}^*(t) - \vec{V}(t) \right| \tag{2.2}$$

$$\vec{V}(t+1) = \vec{V}^*(t) - \vec{Q} \otimes \vec{S} \tag{2.3}$$

where V denotes the candidate solution's location and V^* denotes the best candidate solution in t iterations. Symbols $\otimes$ and || indicate element-by-element multiplication and modulus operator, while P and Q are the coefficient vector, which can be computed as follows:

$$\vec{P} = 2 \times \vec{r} \tag{2.4}$$

$$\vec{Q} = 2\vec{a} \otimes \vec{r} - \vec{a} \tag{2.5}$$

$$\vec{a} = 2 - t\left(\frac{2}{\text{max iteration}}\right) \tag{2.6}$$

where r is a random vector with a value between 0 and 1, and a declines linearly from 2 to 0.

2.3.2.2 Exploitation Phase

In this stage, the whales move toward the optimal solution by exploiting the information gathered from the encircling prey and exploration phase. Exploitation in WOA takes place in two ways: the shrinking encircling mechanism and the spiral updating position.

2.3.2.2.1 Shrinking Encircling Mechanism

The shrinking encircling mechanism is one of the exploitation techniques used in the WOA to converge toward the optimal solution. During the algorithm's exploitation phase, the current best candidate solution is considered to be the target prey or closest to it. The objective of the shrinking encircling mechanism is to reposition the remaining search agents so that they all converge toward the best search agent.

2.3.2.2.2 Spiral Updating Position

The spiral updating position approach mimics the helix-shaped movement of humpback whales when hunting their prey. The spiral updating position starts by first determining how far a whale is from its target. After that, a spiral equation is generated between the positions of the whale itself and the target, and the equation determines where the search agents are located within the search region. The spiral equation defines a spiral path around the current best solution, enabling search agents to explore the surrounding area more effectively. The spiral equation can be formulated as follows:

$$\vec{V}(t+1) = \left|\vec{V}^*(t) - \vec{V}(t)\right| \times e^{bl} \times \cos(2\pi l) + \vec{V}^*(t) \tag{2.7}$$

where b is a logarithmic spiral shape constant and l is a random number between −1 and 1.

The humpback whales approach their prey in the form of a contracting circle while simultaneously following a spiral update position. To account for both behaviors at once, there is an equal chance for both the encircling mechanism and the spiral model to update their position while exploiting. This instance of a humpback whale can be modeled as follows:

$$\vec{V}^*(t+1) = \begin{cases} \vec{V}^*(t) - \vec{Q} \otimes \vec{S} & \text{if } p < 0.5 \\ \left|\vec{V}^*(t) - \vec{V}(t)\right| \times e^{bl} \times \cos(2\pi l) + \vec{V}^*(t) & \text{if } p \geq 0.5 \end{cases} \tag{2.8}$$

where p is a number chosen at random from 0 to 1.

2.3.2.3 Search for Prey (Exploration Phase)

The search for prey phase is a critical component of the exploration phase in the WOA. The algorithm aims to search the solution space for potential regions of the optimal solution. The goal of the exploration phase is to explore different areas of the search space to avoid being stuck in local optima.

During the search for prey phase, each search agent moves in the search space in a random direction using a random search operator. This random search operator can be modeled as follows:

$$\vec{S} = \left|\vec{P}\otimes\vec{V}_{\text{rand}}-\vec{V}\right| \tag{2.9}$$

where V_{rand} is the position vector of the random search agent. The search agents then move toward the new position determined by adding a random value to their current position:

$$\vec{V}(t+1) = \vec{V}_{\text{rand}}-\vec{Q}\otimes\vec{S} \tag{2.10}$$

2.3.3 Grey Wolf Optimization

The GWO [33] algorithm is a metaheuristic optimization technique that simulates the foraging techniques of gray wolves. Each individual wolf in the population is treated as a potential solution to the issue, and the fitness of each individual wolf is determined based on how effectively it can solve the issue. For the GWO algorithm to function properly, the location of each wolf in the population is iteratively updated. The algorithm involves three hunting strategies of the gray wolf: encircling prey, attacking prey, and searching for new prey.

- **Encircling Prey:** The encircling actions of the gray wolf can be interpreted as the exploration phase, where the algorithm explores the search space to identify the best solution. This phase randomly generates solutions, evaluates their fitness, and keeps the best ones for further exploration.
- **Attacking Prey:** The attacking actions of the gray wolf can be interpreted as the exploitation phase, where the algorithm focuses on the best solutions identified in the exploration phase and tries to improve them using different search strategies. These strategies may include making small random changes to the solutions or intensifying the search in promising regions of the search space.
- **Searching for Prey:** The searching for a prey phase corresponds to the diversification phase, where the algorithm tries to escape from local optima by exploring other regions of the search space. This phase randomly generates new solutions and evaluates their fitness to identify new promising regions.

2.3.4 Bat Algorithm

The bat algorithm (BAT) [34] is a swarm intelligence optimization algorithm inspired by the echolocation behavior of microbats. The algorithm resembles the behaviors of a swarm of bats using echolocation to search for food in the dark. In the algorithm, each bat is represented as a solution to an optimization problem, and the fitness of each bat corresponds to how well its solution performs on the problem. The algorithm works by initializing a population of bats with random solutions and velocities. Each bat then updates its position and velocity based on a set of rules meant to mimic real bats' behavior. In particular, each bat moves toward a target position that is determined by its current position and the positions of the other bats in the population. The bat's velocity is also updated by adding a random component that is meant to mimic the randomness of real bat behavior. The algorithm also includes a set of tuning parameters that control the rate at which bats move toward their target positions and the amount of randomness in their velocities. These parameters are adjusted during the optimization process to balance the exploration of the solution space with the exploitation of promising regions.

2.4 PROBLEM FORMULATION

The randomness of a signal generated from source S can be measured using Shannon entropy $H(S)$, which has been used to quantify the average degree of uncertainty embedded in a random variable. In the proposed algorithm, the Shannon entropy of the key has been maximized to demonstrate a high degree of randomness.

Suppose an n bit key $K = k_0, k_1, k_2, \ldots, k_{n-1}$ is generated using the proposed algorithm; then, the cryptographic key generation problem can be constructed as follows:

$$\text{Maximinze}: \; H(K) = -\sum_{i=1}^{n} p(k_i)\log_2 p(k_i) \tag{2.11}$$

where

$$k_j \in \mathbb{Z}_2^{(n)}$$

$$0 \le p(k_j) \le 1$$

2.5 PROPOSED WORK

In this chapter, a framework of a pseudorandom key generator is designed using metaheuristic algorithms, transfer function, and Shannon entropy, as shown in Figure 2.1. Using the framework, a key of various lengths is generated for the WOA, which can be used in various state-of-the-art cryptographic algorithms: DES, PRESENT, AES 128/192/256, 3DES, and Blowfish.

2.5.1 Proposed Framework

The proposed framework of a pseudorandom key generator is designed in such a way that any metaheuristic algorithm, in addition to the transfer function and Shannon entropy, can be used to generate highly random cryptographic keys. Also, the length of the key depends on the dimension of the initial population taken in the metaheuristic algorithms.

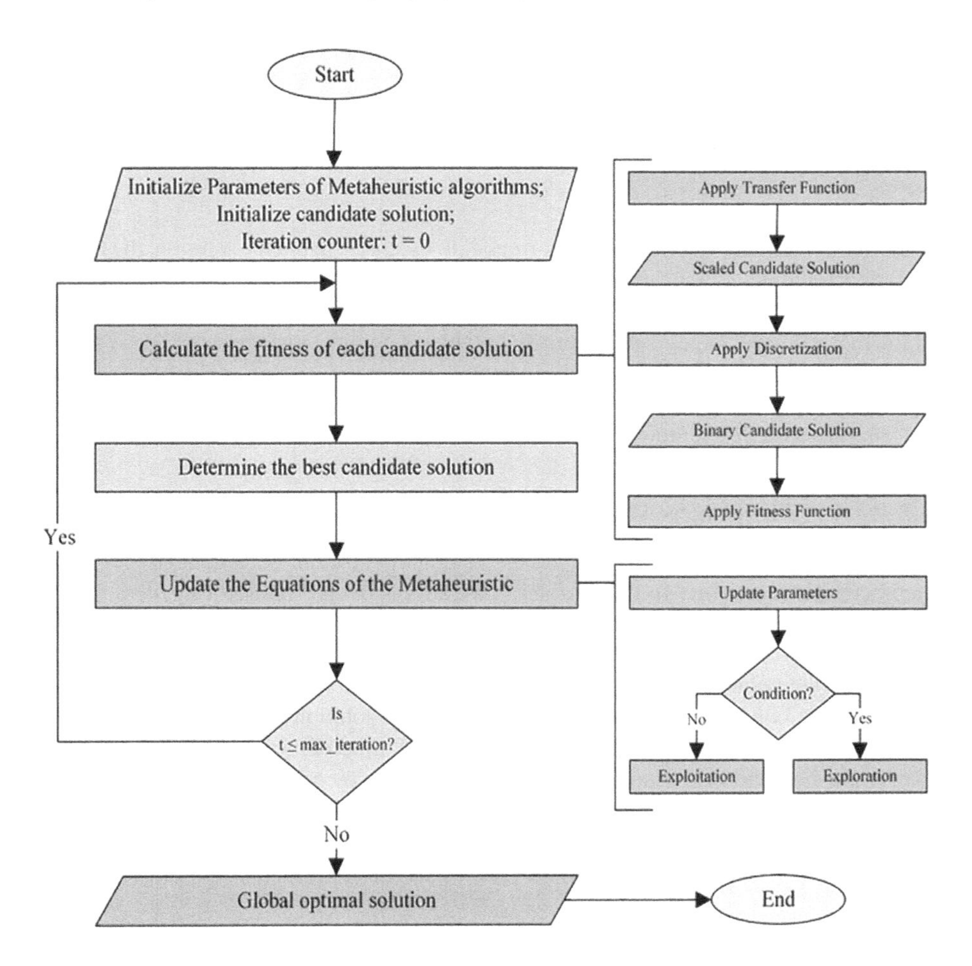

FIGURE 2.1 Proposed framework.

Similar to other real-world optimization problems, the metaheuristic algorithm for the key generation problem also starts with a random candidate solution, as shown as follows:

$$V_i = \{V_1, V_2, V_3, \ldots, V_n\} \tag{2.12}$$

Key generation is a discrete binary problem, while metaheuristic algorithms are for the real-valued continuous problem; therefore, to calculate the fitness of each candidate solution, first, search space is scaled by applying a transfer function to each candidate solution.

$$\text{TF}(V_i) = v_i \tag{2.13}$$

After scaling the search space, the discretization mechanism is used to transform the scaled candidate solution into the binary candidate solution. The discretization mechanism generates a random number r between 0 and 1 and, based on equation (2.14), applies the transformation.

$$b_i = \begin{cases} 1 & \text{if} \quad r < v_i \\ 0 & \text{if} \quad r \geq v_i \end{cases} \tag{2.14}$$

Once each candidate solution is transformed into the discrete binary solution, the fitness of each candidate solution is calculated using equation (2.1) and the fittest candidate solution is identified as the best solution for the current iteration. Now, the parameters of the metaheuristic algorithm are tuned, and based on that, a new population is generated for the next iteration by performing exploitation or exploration. The above process keeps repeating until the maximum iteration is reached. Once the iterations are over, the solution with maximum fitness is selected as a randomized cryptographic key.

2.5.2 Key Generation Using Whale Optimization Algorithm

Utilizing the proposed framework, the cryptographic key can be generated using any metaheuristic algorithm. In this chapter, a WOA has been proposed to generate such keys. WOA initializes the population using equation (2.11); then, transfer function and discretization have been used to scale and transform the real value candidate solution into a discrete binary solution. Now, the fitness of each discrete binary solution is calculated using equation (2.1) and the fittest candidate solution is identified as the best solution for the current iteration. Now, the WOA parameters are tuned, and based on that, a new population is generated for the next iteration by performing encircling prey, exploitation, or exploration using equation (2.3), (2.7), or (2.9). The same process keeps on repeating until the maximum iteration is reached. After the iterations are completed, the solution with the maximum fitness is chosen as a WOA-based cryptographic key.

2.6 SIMULATION AND RESULTS

All the simulations related to the proposed framework are performed on the device HP 15-ay007tx with a specification of 4 GB random access memory (RAM), 1 terabyte hard disk drive (TB HDD), and an Intel Core i5 processor. IDLE, a Python platform, has been used to write and execute all programs. Using the frameworks, WOA, GWO, and BAT have been chosen for the key generation. V-type transfer function V_3 has been used to transform the real-valued continuous solution into a discrete binary solution, while Shannon entropy is used to calculate the fitness of each candidate solution. Table 2.1 represents the values of various parameters of each algorithm used in the experiment.

2.6.1 Result

The lengths of 64-bit, 80-bit, 128-bit, 192-bit, 256-bit, and 448-bit keys are generated for emerging cryptographic algorithms, namely DES, PRESENT, AES 128/192/256, 3DES, and Blowfish. The experiments compare the entropy (randomness) of the keys generated using WOA, GWO, and BAT. Each experiment is carried out ten times, and their average outcome is displayed in Figures 2.2(a–f).

TABLE 2.1
Various Evaluation Parameters of the Experiments and Their Values

Algorithm	Parameter	Value
Gray wolf optimization (GWO)	Initial population size	100
	Converging constant	Decreases linearly from 2 to 0
	Coefficient vector: A	[−4, 4]
	Coefficient vector: C	[−2, 2]
	Upper bound	−100
	Lower bound	−100
Whale optimization algorithm (WOA)	Initial population size	100
	Probability factor	[0, 1]
	Coefficient vector	[−2,2]
	Converging constant	Decreases linearly from 2 to 0
	Upper bound	−100
	Lower bound	100
Bat algorithm (BA)	Initial population size	100
	Frequency	[0, 2]
	Loudness	0.5
	Pulse rate	0.5
	Upper bound	−100
	Lower bound	100
Shannon entropy	Input	$\mathbb{Z}_2$
	Entropy length	1 bit
Transfer function	X_i	$V_3(X_i) = \left\| \frac{X_i}{\sqrt{1+X_i^2}} \right\|$

The given graphs indicate that for the key generation problem, WOA works better than GWO and BAT. For 64-bit, 80-bit, 192-bit, 256-bit, and 448-bit keys, WOA is found to be 112%, 127%, 163%, 177%, 212%, and 177% more random than GWO. At the same time, WOA outperforms BAT by 140% for 64-bit keys, 194% for 80-bit keys, 244% for 128-bit keys, 245% for 256-bit keys, and 400% for 448-bit keys.

2.6.2 NIST Statistical Test

The NIST [35] has developed a statistical test suite to assess the randomness of long series of bits. The statistical test suite performs various tests on the data. Each test depends on a test statistic estimate that was determined using a data-driven function. The test statistic calculates the *P*-value and summarizes the proof against the null hypothesis. If the *P*-value is smaller than the significance threshold, then the data set is not random; otherwise, the data set is considered to be random.

To conduct statistical analysis on the WOA-based cryptographic key, a sample of one million bits is generated and tested against a significant value of 0.01. The test

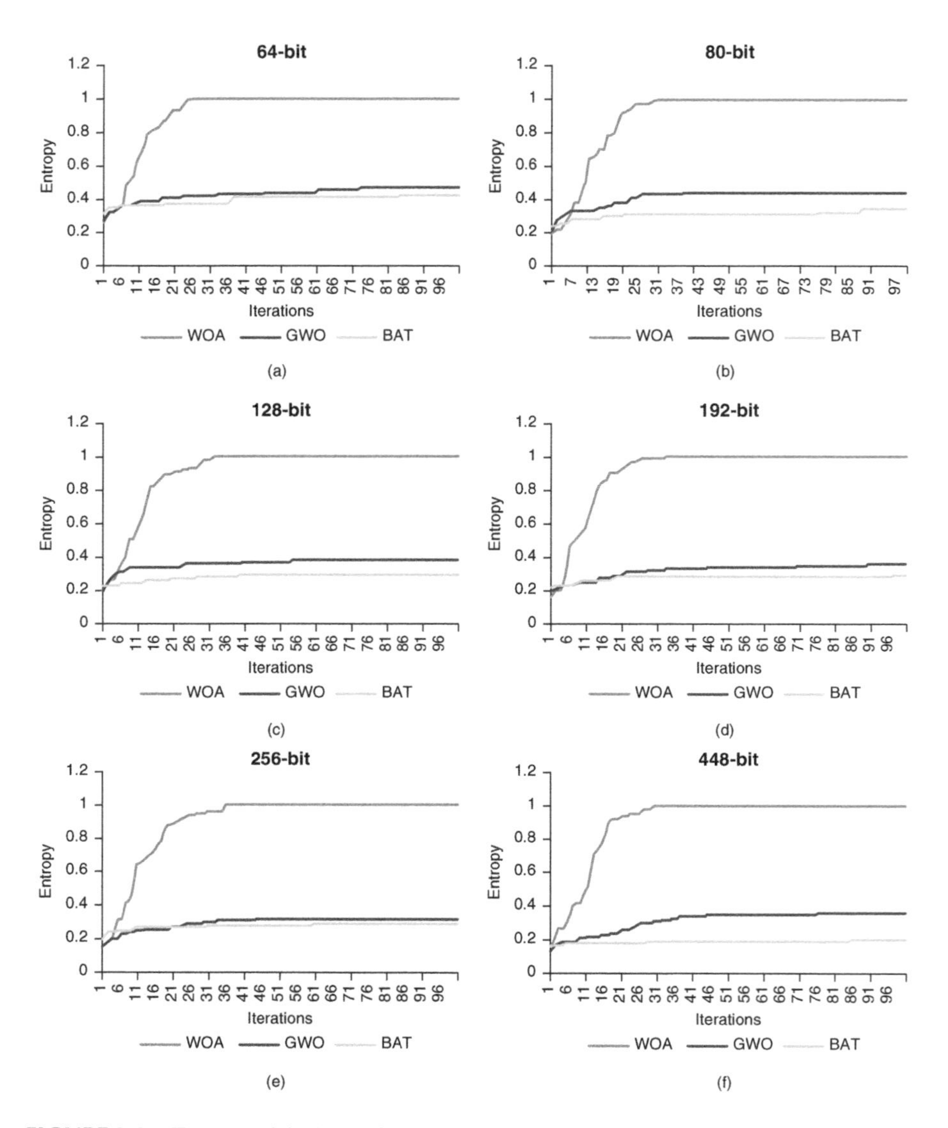

FIGURE 2.2 Entropy of the keys of various lengths generated using WOA, GWO, and BAT. (a), (b), (c), (d), (e), and (f) are graphs of 64-bit, 80-bit, 128-bit, 192-bit, 256-bit, and 448-bit, respectively.

results suggest that the WOA-based cryptographic keys pass all the tests. Table 2.2 provides a summary of both the *P*-value and the overall results of the study.

2.6.3 Observations

The findings of the experiments make it abundantly clear that WOA performs significantly better than both GWO and BAT. The WOA strikes a better balance between the exploration and exploitation phases than the GWO and BAT, which

TABLE 2.2
***P*-Value and the Final Result of Each Test**

Statistical Tests	*P*-Value	Result
Longest run test	0.872331	Successfully passes
Approximate entropy test	0.888885	Successfully passes
Block frequency test (block = 128)	0.941812	Successfully passes
Nonoverlapping template matching test	0.492067	Successfully passes
Cumulative sum test (forward)	0.653159	Successfully passes
Overlapping template matching test	0.201496	Successfully passes
Cumulative sum test (reverse)	0.649458	Successfully passes
Rank test	0.152636	Successfully passes
Fast Fourier transform (spectral) test	0.479815	Successfully passes
Run test	0.421397	Successfully passes
Frequency test	0.996808	Successfully passes
Universal test	0.939687	Successfully passes
Linear complexity test	0.062500	Successfully passes
Serial test 1	0.503557	Successfully passes
Serial test 2	0.547403	Successfully passes

focus more on the exploitation phase and result in local maximum stagnation. Additionally, WOA is simple to implement because it requires fewer parameters and is easier to fine-tune for a variety of optimization problems. At the same time, BAT has several parameters that need to be tuned, such as loudness and pulse rate. WOA also has faster convergence due to a dynamic search strategy, whereas BAT uses a fixed search strategy, which is inappropriate for the key generation problem.

2.7 CONCLUSION

Data security has always been a significant barrier to the acceptance of any new technologies. CCPS is not immune to it either. Over the past few years, numerous data breaches have caused the worst disaster for any country or organization. Therefore, new cryptographic algorithms are developed and deployed by scientists every year. The key is an essential component of any cryptographic algorithm, which should be highly random, unpredictable, and as complex as possible. This chapter has proposed a framework that utilizes any metaheuristic algorithm, Shannon entropy, and transfer function to generate such keys. Keys of various lengths are generated for the state-of-the-art cryptographic algorithms, namely DES, PRESENT, AES 128/192/256, 3DES, and Blowfish using the WOA, GWO, and BAT. The experiments suggest that the WOA-based key is highly random for every key length and outperforms GWO and BAT. Additionally, the WOA-based keys satisfy each statistical test that NIST suggests. Hence, the proposed framework is found to be promising and can be used for key generation in DES, PRESENT, AES 128/192/256, 3DES, and Blowfish. Also, in the future, the research work can be extended to lightweight cryptographic ciphers for IoT devices.

REFERENCES

1. Z. You and L. Feng, "Integration of Industry 4.0 related technologies in construction industry: A framework of cyber-physical system," *IEEE Access*, vol. 8, pp. 122908–122922, 2020. doi: 10.1109/ACCESS.2020.3007206.
2. V. Mullet, P. Sondi, and E. Ramat, "A review of cybersecurity guidelines for manufacturing factories in Industry 4.0," *IEEE Access*, vol. 9, pp. 23235–23263, 2021. doi: 10.1109/ACCESS.2021.3056650.
3. J. Lee, H. D. Ardakani, S. Yang, and B. Bagheri, "Industrial big data analytics and cyber-physical systems for future maintenance & service innovation," *Procedia CIRP*, vol. 38, pp. 3–7, 2015. doi: 10.1016/j.procir.2015.08.026.
4. F. O. Olowononi, D. B. Rawat, and C. Liu, "Resilient machine learning for networked cyber physical systems: A survey for machine learning security to securing machine learning for CPS," *IEEE Communications Surveys and Tutoria*, vol. 23, no. 1, pp. 524–552, 2021. doi: 10.1109/COMST.2020.3036778.
5. M. S. Jawed and M. Sajid, "A comprehensive survey on cloud computing: Architecture, tools, technologies, and open issues," *International Journal of Cloud Applications and Computing*, vol. 12, no. 1, pp. 1–33, 2022. doi: 10.4018/IJCAC.308277.
6. L. Da Xu, W. He, and S. Li, "Internet of Things in Industries: A survey," *IEEE Transactions on Industrial Informatics*, vol. 10, no. 4, pp. 2233–2243, 2014. doi: 10.1109/TII.2014.2300753.
7. A. Villalonga, G. Beruvides, F. Castano, and R. E. Haber, "Cloud-based industrial cyber-physical system for data-driven reasoning: A review and use case on an Industry 4.0 pilot line," *IEEE Transactions on Industrial Informatics*, vol. 16, no. 9, pp. 5975–5984, 2020. doi: 10.1109/TII.2020.2971057.
8. M. Sajid and Z. Raza, "Cloud computing: Issues & challenges," in *International Conference on Cloud, Big Data and Trust (ICCBDT) 2013*, Bhopal, India, 2013, pp. 35–41.
9. M. Qasim, M. Sajid, and M. Shahid, "Hunger games search: A scheduler for cloud computing," in *2022 International Conference on Data Analytics for Business and Industry (ICDABI)*, October 2022, pp. 170–175. doi: 10.1109/ICDABI56818.2022.10041521.
10. S. Sharma and M. Sajid, "Integrated fog and cloud computing issues and challenges," *International Journal of Cloud Applications and Computing*, vol. 11, no. 4, pp. 174–193, 2021. doi: 10.4018/IJCAC.2021100110.
11. K. Cao, S. Hu, Y. Shi, A. W. Colombo, S. Karnouskos, and X. Li, "Survey on edge and edge-cloud computing assisted cyber-physical systems," *IEEE Transactions on Industrial Informatics*, vol. 17, no. 11, pp. 7806–7819, 2021. doi: 10.1109/TII.2021.3073066.
12. C. J. Turner, J. Oyekan, L. Stergioulas, and D. Griffin, "Utilizing Industry 4.0 on the construction site: Challenges and opportunities," *IEEE Transactions on Industrial Informatics*, vol. 17, no. 2, pp. 746–756, 2021. doi: 10.1109/TII.2020.3002197.
13. N. Zhang, "A cloud-based platform for big data-driven CPS modeling of robots," *IEEE Access*, vol. 9, pp. 34667–34680, 2021. doi: 10.1109/ACCESS.2021.3061477.
14. Y. Liu, Y. Zhou, and S. Hu, "Combating coordinated pricing cyberattack and energy theft in smart home cyber-physical systems," *IEEE Transactions on Computer-Aided Design of Integrated Circuits and Systems*, vol. 37, no. 3, pp. 573–586, 2018. doi: 10.1109/TCAD.2017.2717781.
15. L. Wu and Y. Sun, "Guaranteed security and trustworthiness in transportation cyber-physical systems," in Y. Sun and H. Song (Eds.), *Secure and Trustworthy Transportation Cyber-Physical Systems*. Berlin/Heidelberg: Springer, pp. 3–22, 2017. doi: 10.1007/978-981-10-3892-1_1.
16. A. W. Colombo, M. Gepp, J. B. Oliveira, P. Leitão, J. Barbosa, and J. Wermann, "Digitalized and harmonized industrial production systems: The PERFoRM approach,"

in A. W. Colombo, M. Gepp, J. B. Oliveira, P. Leitao, J. Barbosa, and J. Wermann (Eds.), *Digitalized and Harmonized Industrial Production Systems*. Boca Raton, FL: CRC Press, pp. 1–332, 2019. doi: 10.1201/9780429263316.
17. P. Yang, N. Xiong, and J. Ren, "Data security and privacy protection for cloud storage: A survey," *IEEE Access*, vol. 8, pp. 131723–131740, 2020. doi: 10.1109/ACCESS.2020.3009876.
18. M. S. Jawed and M. Sajid, "Cryptanalysis of lightweight block ciphers using metaheuristic algorithms in Cloud of Things (CoT)," in *2022 International Conference on Data Analytics for Business and Industry (ICDABI)*, October 2022, pp. 165–169. doi: 10.1109/ICDABI56818.2022.10041583.
19. M. Sajid, H. Mittal, S. Pare, and M. Prasad, "Routing and scheduling optimization for UAV assisted delivery system: A hybrid approach," *Applied Soft Computing*, p. 109225, 2022. doi: 10.1016/j.asoc.2022.109225.
20. M. Sajid, A. Jafar, and S. Sharma, "Hybrid genetic and simulated annealing algorithm for capacitated vehicle routing problem," in *2020 Sixth International Conference on Parallel, Distributed and Grid Computing (PDGC)*, 2020, pp. 131–136. doi: 10.1109/PDGC50313.2020.9315798.
21. M. Sajid and Z. Raza, "Energy-aware stochastic scheduling model with precedence constraints on DVFS-enabled processors," *The Turkish Journal of Electrical Engineering & Computer Sciences*, vol. 24, no. 5, pp. 4117–4128, 2016. doi: 10.3906/elk-1505-112.
22. S. Picek and D. Jakobovic, "Evolutionary computation and machine learning in cryptology," in *Proceedings of the 2020 Genetic and Evolutionary Computation Conference Companion*, 2020, pp. 1147–1173. doi: 10.1145/3377929.3389886.
23. B. Alouffi, M. Hasnain, A. Alharbi, W. Alosaimi, H. Alyami, and M. Ayaz, "A systematic literature review on cloud computing security: Threats and mitigation strategies," *IEEE Access*, vol. 9, pp. 57792–57807, 2021. doi: 10.1109/ACCESS.2021.3073203.
24. M. M. Ahsan, K. D. Gupta, A. K. Nag, S. Poudyal, A. Z. Kouzani, and M. A. P. Mahmud, "Applications and evaluations of bio-inspired approaches in cloud security: A review," *IEEE Access*, vol. 8, pp. 180799–180814, 2020. doi: 10.1109/ACCESS.2020.3027841.
25. W. Duo, M. C. Zhou, and A. Abusorrah, "A survey of cyber attacks on cyber physical systems: Recent advances and challenges," *IEEE/CAA Journal of Automatica Sinica*, vol. 9, no. 5, pp. 784–800, 2022. doi: 10.1109/JAS.2022.105548.
26. R. Priyadarshini, A. Quadir Md, N. Rajendran, V. Neelanarayanan, and H. Sabireen, "An enhanced encryption-based security framework in the CPS cloud," *Journal of Cloud Computing*, vol. 11, no. 1, 2022. doi: 10.1186/s13677-022-00336-z.
27. M. Tahir, M. Sardaraz, Z. Mehmood, and S. Muhammad, "CryptoGA: A cryptosystem based on genetic algorithm for cloud data security," *Cluster Computing*, vol. 24, no. 2, pp. 739–752, 2021. doi: 10.1007/s10586-020-03157-4.
28. M. S. Jawed and M. Sajid, "XECryptoGA: A metaheuristic algorithm-based block cipher to enhance the security goals," *Evolutionary Systematics*, 2022. doi: 10.1007/s12530-022-09462-0.
29. S. M. Rubai, "Hybrid heuristic-based key generation protocol for intelligent privacy preservation in cloud sector," *Journal of Parallel and Distributed Computing*, vol. 163, pp. 166–180, 2022. doi: 10.1016/j.jpdc.2022.01.005.
30. D. Ahamad, S. Alam Hameed, and M. Akhtar, "A multi-objective privacy preservation model for cloud security using hybrid Jaya-based shark smell optimization," *Journal of King Saud University: Computer and Information Sciences*, vol. 34, no. 6, pp. 2343–2358, 2022. doi: 10.1016/j.jksuci.2020.10.015.
31. C. E. Shannon, "A mathematical theory of communication," *The Bell System Technical Journal*, vol. 27, no. 3, pp. 379–423, 1948. doi: 10.1002/j.1538-7305.1948.tb01338.x.

32. S. Mirjalili and A. Lewis, "The whale optimization algorithm," *Advances in Engineering Software*, vol. 95, pp. 51–67, 2016. doi: 10.1016/j.advengsoft.2016.01.008.
33. S. Mirjalili, S. M. Mirjalili, and A. Lewis, "Grey wolf optimizer," *Advances in Engineering Software*, vol. 69, pp. 46–61, 2014. doi: 10.1016/j.advengsoft.2013.12.007.
34. X.-S. Yang, "A new metaheuristic bat-inspired algorithm," in J. R. González, D. A. Pelta, C. Cruz, G. Terrazas, and N. Krasnogor (Eds.), *Nature Inspired Cooperative Strategies for Optimization (NICSO 2010)*. Berlin, Heidelberg: Springer, pp. 65–74, 2010. doi: 10.1007/978-3-642-12538-6_6.
35. E. Barker, "Recommendation for key management," *National Institute of Standards and Technology*, 2020. doi: 10.6028/nist.sp.800-57pt1r5.

3 Containerized Deployment of Microservices in Cloud Computing

Shivam and Dinesh Kumar

CONTENTS

DOI: 10.1201/9781003438588-3

3.1 INTRODUCTION

Before the cloud was created in March 2006, when Amazon introduced simple storage service (S3) and elastic compute cloud (EC2), organizations had to purchase a stack of servers, manage the variable traffic load, and monitor and maintain servers. This had some drawbacks, including costly server setup, conflicts with business objectives and infrastructure issues, and variable traffic that resulted in resource wastage due to inefficient resource utilization. All these drawbacks prompted the need for a sizable, centralized online technology (in this case, a collection of data centers) where one could orchestrate various operations, applications, and resource management effectively by connecting all these data centers together through a network. It also included the ability to use the resources and manage them appropriately. Later, this technology became known as the cloud computing. This enables cloud users to rent the services and pay only for the amount of resources that's needed. According to NIST, cloud computing can be defined as follows:

> Cloud computing is a model for enabling ubiquitous, convenient, on-demand network access to a shared pool of configurable computing resources (e.g., networks, servers, storage, applications, and services) that can be rapidly provisioned and released with minimal management effort or service provider interaction [1].

Although cloud computing offers a large number of computing and storage solutions for enterprises, yet there are several challenges for utilizing cloud computing for cyber physical systems that need to be addressed. Some of the challenges are real-time and scalable service provisioning, real-time response to time-critical applications, and real-time scheduling in hypervisors [2,3]. In cloud computing, physical instance of an application or resource is shared among various organizations or consumers by creating a software-based or virtual version of the instance whether that can be networks, storage, servers, or applications. This technique is referred to as virtualization. Virtualization is usually implemented using virtual machine manager (VMM) which is also referred as hypervisor. Hypervisor is a firmware/software which acts as a middleware between physical memory and virtual machines (VMs), that is, type-I hypervisor (or bare metal hypervisor) or between host OS and VMs, that is, type-II hypervisor (or hosted hypervisor), that manages to create and run VMs. These are also known as VMM.

Cloud environment supports applications to be deployed on a remote server rather than being hosted on a local server or machine. This process is called application deployment in cloud computing. Application deployment comprises the following steps [4]:

- Selection of web server, load balancer, and database services from a pool of VM images that are preconfigured.
- Configuration of the component of the VM images.
- Code to meet component's requirement for the resources.
- Handling of networking, security, and scalability issues, by the developer.

Figure 3.1 shows the difference in deployment of application via hypervisor using VM and deployment via container engine using container. In hypervisor-based VM model, isolated operating system and its supporting libraries are installed to deploy an application while Docker container model shares the kernel system as well as enables images to layer which in turn allows to share supporting libraries. Hence, Docker container model is well versed in scaling and deploying (lightweight) than hypervisor-based models.

Advancement in the area of containerization brings challenges to minimize the time of processing, computing resources, power consumption, and total cost of microservices. It also needs to scale computing resources based on the requests, as oversupply of computing resources leads to resource wastage while underprovisioning of computing resources leads to degradation of performance as well high energy consumption (a greater number of migrations of containers over servers). With more containers deployed simultaneously over servers due to containers' lighter weight than VMs, there is a need for auto-scaling and a container-conscious strategy to identify the active server's resource computing capability and trigger scaling based on predefined threshold values of server's computing resources.

At present, several studies in the area of application deployment using containers are carried out; these studies still strive to efficiently deploy containers. Efficiency of container deployment usually depends on resource wastage, power consumption, and cost of deployment. Studies in attempt to increase the efficiency of deployment are carried out, taking one parameter at a time which in turn elevates the inefficiency of other metrics, which avoids small-scale industries to adopt container-based deployment for their business model. To make containerized application affordable in terms of computing resources and to scale this technology to be adopted by every level of industries (low, medium, and big), this work aims to further increase the efficiency of performance metrics. Various scheduling algorithms are proposed to reduce the resource wastage for more efficient containerized application deployment.

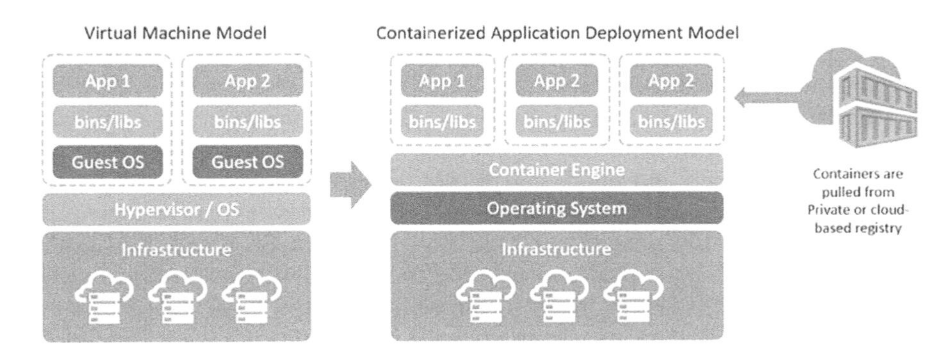

FIGURE 3.1 Hypervisor-based VM and Docker container models [5].

In this work, the research aspects are addressed by taking one requirement at a time.

- For placement of microservices on container, two different algorithms are proposed and are compared based on resource usage and cost. These two algorithms are best fit over CPU capacity of containers and minimum residual load first.
- Deployment of containers over physical machine is done using particle swarm optimization (PSO). The proposed method is compared with traditional bin-packing strategies in terms of resource wastage and power consumption.

3.2 BACKGROUND

This section provides a brief description of microservices, containers, and Dockers. A brief overview of PSO and dynamic bin packing is also discussed.

3.2.1 Microservices

Complex, large, and distributed applications were developed and deployed using various software development paradigms on to the cloud. These paradigms can commonly be categorized into two architectures including monolithic architecture and microservices-based architecture. Complex and large applications were initially deployed on a monolithic architecture by assigning physical resources to every deployed application in a tightly coupled single large code base. Due to its single code-based structure, with increase in complexity of application with time, it becomes difficult to scale resources and maintain code with change in technologies because the services were not isolated in these architectures. Studies were conducted to overcome issues caused by monolithic architecture by dismantling of large complex applications into small independent autonomous modules, where each module is based on separate functionality of the application and is independent of each other. This type of architecture was termed as microservice-based architecture, where each module of the application provides specific functionality called as microservice.

Microservices in cloud computing refer to the use of the microservices architecture in combination with cloud computing technology. In this approach, an application is divided into small, independent services, each running in its own container. These containers can be deployed and managed in a cloud computing environment, providing access to unlimited computing resources and enabling the application to scale horizontally as needed. The use of microservices and cloud computing together provides organizations with a scalable, flexible, and easy-to-manage solution for delivering modern applications, allowing them to respond quickly to changing business needs and deliver new features and functionality at a faster pace.

3.2.2 Containers

Containerization transcended the limitation of monolithic architecture. A container is defined as the standard unit of software that encapsulates code and its dependencies providing applications to run efficiently and reliably from one to another computing environment. Unlike monolithic architecture, containers share both physical memory and host OS, not requiring operating system for each application, providing server efficiencies, hence reducing server cost, and making them lightweight than standard monolithic architecture. Various available container orchestration tools such as Docker, Kubernetes, OpenShift, and Nomad create containers. This work uses Docker as orchestration tool to containerize application.

3.2.3 Docker

This section provides a basic insight of one of the container orchestration tools available for containerization of application. Primary concept and architecture of Docker is also discussed briefly.

3.2.3.1 Definition

Docker was released in the year 2013 as a container orchestration tool, and since then, it completely revolutionized the way of development, deployment, and shipment of application over cloud and made the production much easier by providing abstraction at operating system level. Docker is a management service for containers that provide an open platform for developing, shipping, and testing of applications by encapsulating software with all its dependencies into a single deliverable unit. These deliverable units are quick to deliver due to the isolation between application and infrastructure, provided by Docker containers. Since its OS as well as physical resources carry and shipment of updates in application modules between production, testing and development environment became quick, easy, and efficient.

3.2.3.2 Docker Container Architecture

Architecture of Docker comprises Docker client, Docker server, and Docker registry, where client is connected to the Docker daemon via REST API over network interface or sockets. Docker client provides a way of interaction of multiple Docker users with Docker, and client uses Docker API to send request to Docker daemon where Docker daemon listens to the requests from client and manages Docker objects comprising Docker images, networks, volumes, and containers. Docker registry holds images from where we can pull and run images as containers; Docker provides two types of registries: public and private. Docker hub is public registry provided by Docker from where we can directly pull and push images via Docker pull and Docker push command. Figure 3.2 shows the architecture of Docker container.

Docker client uses Docker API to send request to Docker daemon where Docker daemon listens to the requests from client and manages Docker objects comprising

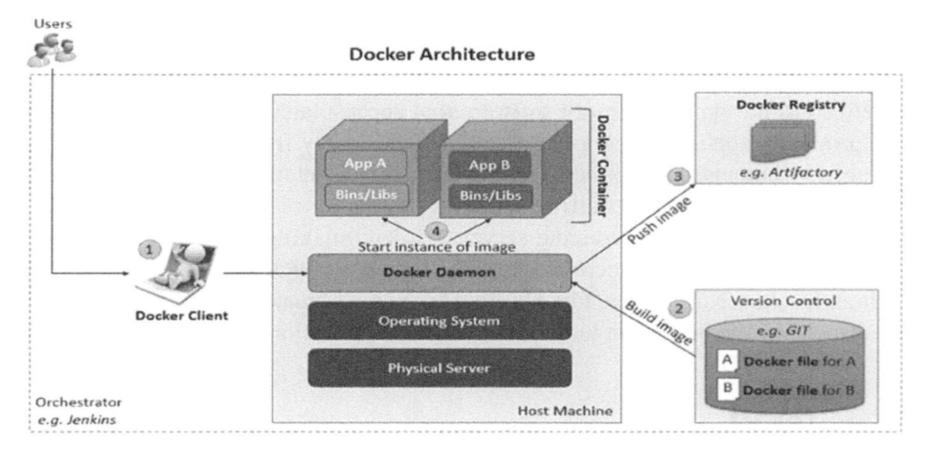

FIGURE 3.2 Docker container architecture [5].

Docker images, networks, volumes, and containers. Docker registry holds images from where we can pull and run images as containers. Docker provides two types of registries: public and private. Docker hub is public registry provided by Docker from where we can directly pull and push images via Docker pull and Docker push command.

3.2.3.3 Docker Containers and Images

A Docker container image is a standalone, lightweight, executable package of software that includes all the requirements to run an application such as system libraries, system tools, code, runtime, and settings; these Docker images become Docker containers at runtime, that is, Docker container is an instance of a Docker image running on Docker engine as shown in Figure 3.2. These containers isolate software from its environment ensuring its uniformity despite differences for instance between development and staging.

3.2.3.4 Application and Research Areas

Containers isolate software from its environment that further assists in managing workloads dynamically. It also provides ease in scaling up and tearing down of services based on the requirements. Isolation also gives an edge to the developer to manage and optimize the resources by fixating the resources for different types of containers and scheduling their placement to multiple servers.

3.2.4 Optimization Techniques

Various optimization techniques are available today, and different techniques are better for different types of problem. None of the optimization techniques can be quoted as "best for all". This work constitutes two optimization techniques, namely dynamic bin packing and PSO.

3.2.4.1 Dynamic Bin Packing

Dynamic bin packing is a classic optimization problem in computer science where the goal is to pack items of varying sizes into a finite number of containers, such that the number of containers used is minimized. It is a dynamic problem because the containers can be resized and rearranged as new items are added. This problem is used to model a variety of real-world scenarios, such as packing boxes in a warehouse, allocating tasks to servers in a cloud computing environment, or assigning network resources to users in a telecommunication network. The solution to the dynamic bin-packing problem is important in many applications, as it helps to optimize resource utilization and reduce waste, ultimately improving efficiency and reducing costs.

3.2.4.2 Particle Swarm Optimization

PSO was first introduced in year 1952 by James Kennedy and Russell Eberhart [6]. It is inspired by the social behavior of bird flocks and fish schools. It is a metaheuristic technique used to solve complex optimization problems in various fields, including engineering, finance, and computer science. PSO works by maintaining a population of candidate solutions, called particles, that move through the search space guided by their own personal best solution and the best solution found by the entire population. The movement of particles is updated at each iteration based on a set of velocity and position update rules, which incorporate the particle's current velocity, personal best position, and the global best position. The algorithm terminates when a satisfactory solution is found or a specified stopping criterion is met. PSO is a simple, easy-to-implement, and computationally efficient algorithm that has been applied to a wide range of optimization problems with good results.

There are different versions of PSO that have been introduced, in the original version of PSO. The particles are scattered in a D-dimensional solution space where each particle has position and velocity vector associated with them as shown in Figure 3.3. In each iteration, velocity and position vector of each particle is updated based on local best and global best of all the particles so that particle starts moving in a desired direction. For example, in Figure 3.3, every particle has random velocity and is desired to move to the star in the figure; after every iteration, the velocity vector gets more directed toward the star and position vector nearer to it, so after certain number of iterations, particle gets converge to a desired point providing most relevant and optimized particle position. PSO flowchart is shown in Figure 3.4.

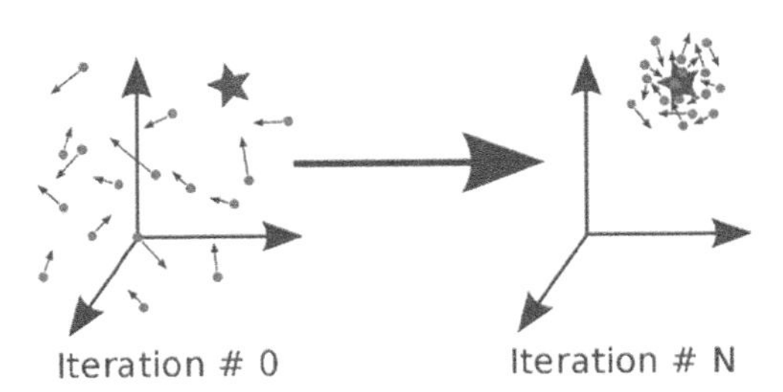

FIGURE 3.3 PSO optimization strategy.

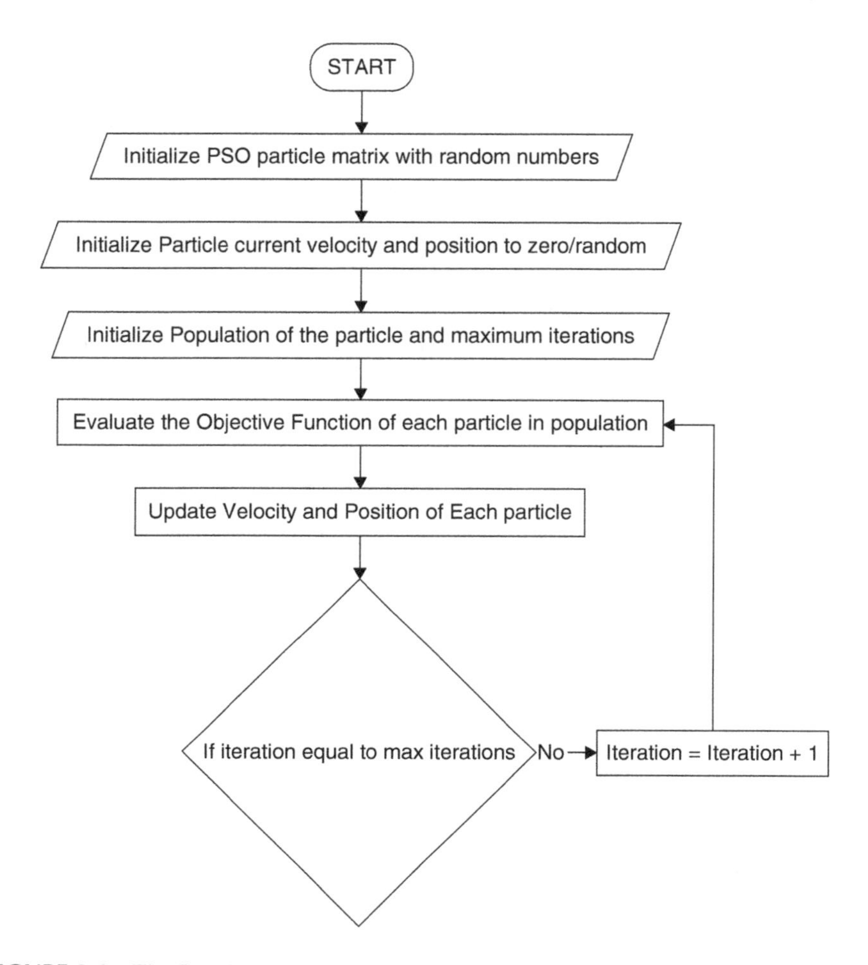

FIGURE 3.4 The flowchart to represent the PSO model.

3.3 RELATED WORK

Multiple studies have been done in the field of application deployment in cloud, since monolithic architecture that was based on hypervisors communicating with heavy-weighted VMs was not feasible for deploying complex and large applications on to the cloud due to its lack in scaling of resources and inefficient processing and deploying cost. This opened up an area to proliferate distributed systems performing different functionalities, to make it efficiently deploy while taking account of processing cost and time.

3.3.1 Application Deployment

Initially, MapReduce, a programming model, was introduced for applications deployed in multi-level structure [7], where a single computing job is partitioned into multiple tasks by a central system and is assigned to different worker nodes that are placed on servers with limited amount of physical resources. Further, these limited resources were utilized efficiently [8] ensuring QoS. Further studies in Ref. [9] worked on proposing scheduling task which accounted in efficient functioning of application with increased performances by proper placement of worker nodes to execute certain amount of task.

Further, suggestions were made on the previous studies to divide task with large computing requirements to complete them within similar time. Authors in Ref. [7] proposed scheduling strategies to distribute resources fairly among different jobs. With introduction to containers, many container-based problems were studied to deploy them. Authors in Ref. [10] proposed a resource allocation technique where it dynamically switches the container status between its sleep and awake states based on a threshold defined previously, to minimize the energy consumption in resource allocation to containers.

3.3.2 Container-Based Scheduling

Monolithic design was incapable for concurrent deployment due to its overhead in storage footprint, so with light containers, concurrent deployment was possible utilizing the cloud resources efficiently. Studies in direction went through many phases starting with a mathematical model for container scaling to utilize computing resources [11], to various container placement algorithms which were proposed to minimize the deployment cost. These algorithms were ant colony-based algorithm [12] for placement of optimal containers (selected based on resource availability) that was based on the idea of movement of colonies of ant where the trails are detected using pheromones. Authors in Ref. [13] proposed an artificial fish swarm-based algorithm as a monitoring mechanism to find an optimal loaded PM for deploying certain container. Evolutionary computation algorithm proposed by the authors in Ref. [14] is for the green-network aware placement of VMs to help conserve power dissipation from servers. The authors in Ref. [15] proposed the resource provisioning strategy to find an optimal container based on genetic algorithm in Kubernetes platform.

3.3.3 Advancement in Optimization Methods

Over the years, optimization techniques have been advanced, for the purpose to solve various practical problems. Optimization strategies have become one of the most encouraging techniques in engineering applications, with advancement in the area of computing. This helped in narrowing down the gap between optimization method and engineering practices. This section deals with the detailed advancement in optimization strategies and its applications.

One of the optimization problems in dynamic bin packing is to store items in multiple bins where number of bins needs to be minimized based on the arrival time and number of items arriving. In "Garey and Johnson list, the bin-packing problem", the bin-packing technique is adapted in two steps, for every item.

- If item gets fit into one of the currently opened bins, fit it.
- If item doesn't get fit in the opened bins, open a new bin and place it in the new one.

This adaption of items can have different criteria to choose the bins, which introduced different approximation algorithms:

- **Next fit**: Keeps only one bin open to allow to fit item; once the bin capacity is full, next bin is opened.
- **First fit**: Allows to fit items in the any bins that encountered first as is this, all the bins are kept open.
- **Best fit**: Allows to minimize the number of bins to store items by placing the item into the bin with maximum capacity.
- **Worst fit**: Attempts to store item into the bin with minimum holding capacity, which in turn uses maximum number of bins to store them.

From classical bin packing, this technique evolved over years. Advancement in different segments of the algorithm led to a number of versions of bin-packing problems. Here, number of bins and size of bins are constant while item sizes can very. "Maximum resource bin-packing and bin-covering problem" [16] maximize the number of bins based on threshold values. Although these classical approaches sought to solve some of the optimization problems but with advancement in the computing studies, limitation arose, involving the time of arrival of items as such techniques dealt with items arriving over time and have to be packed at the time of their arrival. To overcome the limitation of classical bin packing with repacking of items, studies in [17] introduced "relaxed online bin-packing model" which allowed repacking an already-packed item. But, this variant further encountered a restriction over the item departure as it does not consider the departing of items over time. The authors in Ref. [18] eliminated the limitation of "relaxed online bin-packing model" by introducing another extension to classical bin packing, that is, "dynamic bin packing". It additionally keeps track of departing of items over time. Currently, dynamic bin-packing algorithm is the most efficient variant of bin packing in use.

Another well-known optimization technique being mostly used is PSO. It emerged as a requirement of being applied to continuous as well as discrete optimization problems using local and global models. Initially, standard PSO uses evolutionary mechanism to optimize problems, but carries a limitation of slow convergence. To eradicate the limitation, several extensions to standard PSO were introduced comprising better convergence rate. PSO itself is an extension on the base work of "swarm intelligence" which is defined as "the collective behavior of decentralized and self-organized swarms". Ant colony optimization introduced by Marco Dorigo in 1992 was the first optimization algorithm based on swarm intelligence which aims to search for an optimal path in a graph. It is based on the behavior of ants seeking a path between their colony and a source of food", but this strategy was not widely used due to its low convergence rate, and it was underperformed in large search space due to slower performance. The evolution of same continued to artificial bee colony proposed by the authors in Ref. [19] which divided the swarm into three groups each having separate task to perform in swarm; later, a well-established technique, namely PSO, introduced by "James Kennedy and Russell Eberhart" in Ref. [6], proposed "methods for optimization of continuous non-linear functions". Further, PSO was categorized into two types based on search space for swarm, and they are as follows:

- Discrete PSO—On discrete-valued search space, combinatorial and discrete optimization problems apply discrete PSO algorithm. It comprises position and velocity generating real/discrete values after every iteration.
- Binary particle swarm optimization (BPSO)— It is a variant of the PSO algorithm that is specifically designed for solving binary optimization problems. In binary optimization, the goal is to find the optimal combination of binary values (0 or 1) that satisfy a set of constraints or minimize a given objective function. Binary PSO works by maintaining a population of particles, where each particle represents a potential solution encoded as a binary string [6]. The velocity and position updates in binary PSO are designed to preserve the binary nature of the solutions, ensuring that the particles always maintain valid binary values. Like regular PSO, binary PSO uses a combination of individual and global best solutions to guide the movement of particles toward the optimal solution. Binary PSO has been applied to a wide range of optimization problems, including combinatorial optimization, feature selection, and machine learning, with good results. A modified binary version of PSO [20] has been applied to optimize the task in this chapter, and proposed BPSO is compared with other strategy available.

3.4 THE PROPOSAL

This section presents system model, depicting various stages processed in cloud responsible for microservice deployment on physical machine using containers. Proposed models for microservice-to-container and container-to-server mapping are also discussed with their detailed explanation.

3.4.1 System Model

This section introduces application model for deployment of microservices in cloud environment as shown in Figure 3.5. Cloud manager is responsible for various interactivities before deployment of microservices to PMs using containers. It has various phases involved comprising specific function, where admission controller checks availability of resources in PMs for incoming microservices. It is followed by admission scheduler, which stores them in a queue and deploys microservices to the resource provisioner via suitable scheduling mechanisms. The SLA manager checks the QoS of each application and pauses the microservice during any violation. Resource provisioner selects best-fit container for each microservice based on resource requirement. It is also responsible for finding warm containers in PMs. Auto-scalar monitors current status of resources for scaling out/down the PMs. Finally, resource manager deploys containers on to suitable PMs.

3.4.2 Problem Statement and Formulation

This work explores the novel approach of assigning and integrating containers into application deployment while minimizing the computing resources. Researchers have used various techniques to deal with efficient utilization of computing

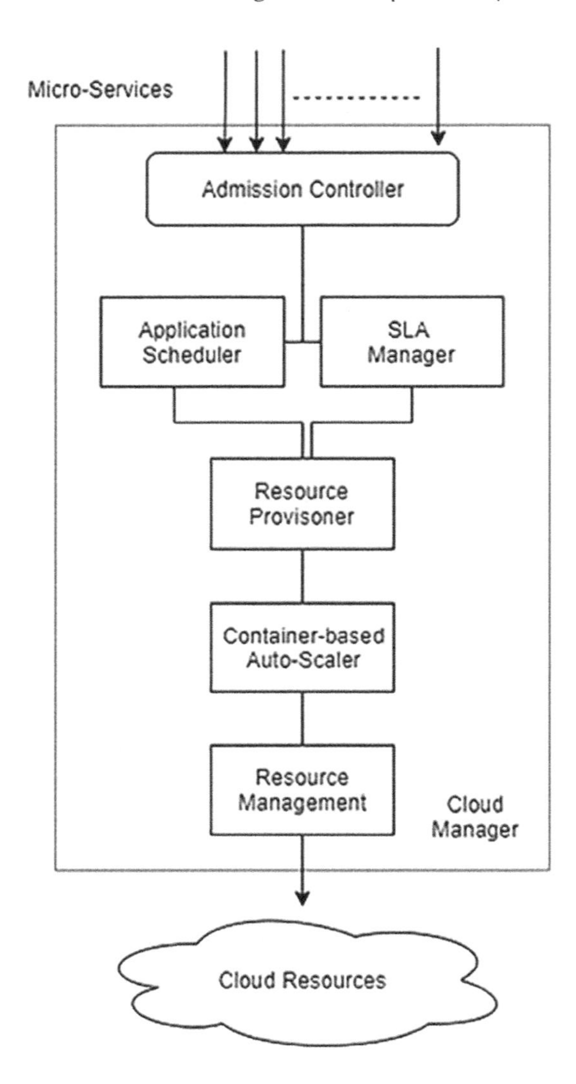

FIGURE 3.5 Microservice-based application deployment architecture [4].

resources, and still many research works are in progress to make it feasible. Therefore, this work addresses following challenges in deployment of application using containers.

How to integrate containers to map microservices on physical machine to build an efficient scheduling mechanism for application deployment?

This challenge brings two research aspects that need to be addressed, including:

- Placement of microservices over containers in order to minimize resource wastage.
- Placement of containers over servers to minimize the power consumption and resource utilization.

This work attempted to design new scheduling mechanism for processing microservices in the cloud. This work's contribution of the proposed strategy is proposed in three stages. These contributions are mentioned later. Firstly, we have designed different heuristics to efficiently fit container to each microservices using resource requirements of microservices. These algorithms are best-fit approach considering minimum residual-load and best-fit approach considering minimum CPU capacity of containers. Then, we have designed container-to-physical machine mapping based on availability of resources. Containers are deployed on to suitable PMs/VMs using two optimization techniques for minimizing the number of PMs and resource wastage. The optimization techniques involve dynamic bin packing (worst case), dynamic bin packing (best case), and BPSO. Finally, each strategy is compared with one another to find the best technique for efficient utilization of computing resources while deploying microservices over physical machine. Variations are observed and analyzed via plotted graph.

3.4.2.1 Application Deployment Cost Formulation

Resources are abbreviated as CPU and MM for CPU and memory, respectively. An application is represented by set of loosely coupled microservices $M = \{M_1, M_2, \ldots, M_k\}$, k here represents total number of microservices. A microservice l: $l \in M$ is set of tuples represented by $\left(\mathrm{CP}^{\mathrm{CPU}}, \mathrm{CP}^{\mathrm{MM}}, \mathrm{BC}_l\right)$, where $\mathrm{CP}^{\mathrm{CPU}}$ denotes amount of requested CPU, $\mathrm{CP}^{\mathrm{MM}}$ denotes amount of requested memory, and BC_l denotes budget constraint of microservice l. The values of $\mathrm{CP}^{\mathrm{CPU}}$ and $\mathrm{CP}^{\mathrm{MM}}$ can be calculated as shown in Equations (3.1) and (3.2).

$$\mathrm{CP}^{\mathrm{CPU}} = |\mathrm{CR}_i| \times \text{size of}(\mathrm{CR}) \tag{3.1}$$

$$\mathrm{CP}^{\mathrm{MM}} = |M_i| \times \text{size of}(M) \tag{3.2}$$

Equations (3.1) and (3.2) denote CPU capacity ($\mathrm{CP}_i^{\mathrm{CPU}}$) and memory ($\mathrm{CP}_i^{\mathrm{MM}}$) capacity of container I, while CR denotes one core of CPU and M as one unit of memory. Now, processing time PT_{li} of microservice l is calculated as shown in Equation (3.3).

$$\mathrm{PT}_{li} = \frac{\mathrm{SZ}_l}{\mathrm{CP}_i^{\mathrm{CPU}}} \tag{3.3}$$

where SZ_l is denoted as size of microservice. Now, processing cost (PT_{li}) of container i for processing microservice l for CPU usage is denoted by $\mathrm{CO}^{\mathrm{CPU}}$ in unit interval T_1. It can be calculated as shown in Equation (3.4):

$$\mathrm{CO}_{il}^{\mathrm{CPU}} = \mathrm{CP}_l^{\mathrm{CPU}} \times \frac{\mathrm{PT}_{li} \times \mathrm{CO}_i^{\mathrm{CPU}}}{T_1} \tag{3.4}$$

Similarly, for memory usage, cost is

$$\mathrm{CO}_{il}^{\mathrm{MM}} = \mathrm{CP}_l^{\mathrm{MM}} \times \frac{\mathrm{PT}_{li} \times \mathrm{CO}_i^{\mathrm{MM}}}{T_1} \tag{3.5}$$

Total processing cost of microservice l is

$$\mathrm{CO}_{il} = \mathrm{CO}_{il}^{\mathrm{CPU}} + \mathrm{CO}_{il}^{\mathrm{MM}} \tag{3.6}$$

and cost of deploying each container is

$$\mathrm{CDT}_i = \frac{\mathrm{CO}_i^{\mathrm{DT}}}{T_2} \tag{3.7}$$

Here, $\mathrm{CO}_i^{\mathrm{DT}}$ represents the cost of deploying container i for unit interval of time T_2.

We minimize the total cost of deploying an application as denoted by CO_a:

$$\mathrm{CO}_a = \sum_{l=1}^{z} \mathrm{CO}_{il} + \sum_{i=1}^{k} \mathrm{CDT}_i \tag{3.8}$$

3.4.2.2 Resource Wastage Formulation

To fully utilize the multidimensional resources, the wasted resources based on remaining resources available on each physical machine can be calculated as shown in Equations (3.9) and (3.10) [21].

$$L_l^{\mathrm{CPU}} = \frac{\mathrm{CP}_i^{\mathrm{CPU}} - \mathrm{CP}_l^{\mathrm{CPU}}}{\mathrm{CP}_i^{\mathrm{CPU}}} \tag{3.9}$$

$$L_l^{\mathrm{MM}} = \frac{\mathrm{CP}_i^{\mathrm{MM}} - \mathrm{CP}_l^{\mathrm{MM}}}{\mathrm{CP}_i^{\mathrm{MM}}} \tag{3.10}$$

where L_l^{CPU} and L_l^{MM} are normalized CPU and memory resources. Equations (3.9) and (3.10) are used in Equation (3.11) to compute the resource wasted denoted by W_i; U_i^{CPU} and U_i^{MM} are normalized CPU usage and memory usage of containers. Resource wasted is calculated as given in Equation (3.11):

$$W_i = \frac{\left| L_i^{\mathrm{CPU}} - L_i^{\mathrm{MM}} \right| + \in}{U_i^{\mathrm{CPU}} + U_i^{\mathrm{MM}}} \tag{3.11}$$

where W_i is the resources wasted, U_i^{CPU} is normalized CPU usage, and U_i^{MM} is normalized memory usage of containers.

3.4.2.3 Power Consumption Modeling

Power consumption of PM is described as linear relationship between the power consumption and the CPU utilization (U_j^{CPU}) [22]. Power consumption for jth PM is defined as given in Equation (3.12):

$$P_j = \left(P_j^{\mathrm{busy}} - P_j^{\mathrm{idle}} \right) \times U_j^{\mathrm{CPU}} + P_j^{\mathrm{idle}} \tag{3.12}$$

where P_j^{busy} and P_j^{idle} are considered average power values when jth PM is busy and idle, respectively.

This work aims to minimize power consumption and resources wastage which can be formulated as given using Equations (3.13)–(3.17):

$$\text{MIN}\sum_{j=1}^{k} P_j = \sum_{j=1}^{k}\left[y_j \times \left(\left(P_j^{\text{busy}} - P_j^{\text{idle}}\right)\left(\sum_{i=1}^{m} x_{ij} \times U_i^{\text{CPU}}\right) + P_j^{\text{idle}} \right)\right] \tag{3.13}$$

$$\text{MIN}\sum_{j=1}^{k} W_j = \sum_{j=1}^{k}\left[y_j \times \frac{\left|\left(\left(1-\sum_{i=1}^{m} x_{ij} \times U_i^{\text{CPU}}\right) - \left(1-\sum_{i=1}^{m} x_{ij} \times U_i^{\text{MM}}\right)\right)\right| + \in}{\sum_{i=1}^{m} x_{ij} \times U_i^{\text{CPU}} + \sum_{i=1}^{m} x_{ij} \times U_i^{\text{MM}}} \right] \tag{3.14}$$

such that

$$\sum_{j=1}^{k} x_{ij} = 1 \quad \forall\ i \in I \tag{3.15}$$

$$\sum_{i=1}^{m} x_{ij} \times U_i^{\text{CPU}} < y_j \quad \forall\ j \in J \tag{3.16}$$

$$\sum_{i=1}^{m} x_{ij} \times U_i^{\text{MM}} < y_j \quad \forall\ j \in J \tag{3.17}$$

Considering Equations (3.13) and (3.14) as separate objective functions, these are used in binary PSO as an objective to minimize power/energy consumption (P_j) and (W_j) resource wastage, respectively. The constraint given in Equation (3.15) ensures the placement of one container on one and only one physical machine, whereas Equations (3.16) and (3.17) ensure that resources of each physical machine should not exceed the overall resources of all the containers placed on it. In the above given formulation, binary search space in the BPSO will be required with required dimensions and decision variables to minimize objective function. A mapping is needed that will satisfy the mentioned constraints and minimizes the objective functions.

3.4.2.4 Binary PSO Formulation

Binary PSO is a variant of PSO. The procedure follows by giving an objective function $f\left(\vec{D}\right)$ where $\vec{D}$ is the decision variable. The objective is to search for $\vec{D}$ which further will minimize function f with iterations. A three-vector set $\vec{D}, \vec{P_i}, \vec{V_i}$ sent each particle in the population with d dimensional vector which is the dimension of the search space. Here, $\vec{D_i}$ represents the present position of each particle in a search space, as a solution, $\vec{V_i}$ denotes the velocity of every particle and $\vec{P_i}$ denotes the previously found best position of particle i. In BPSO, velocity of the particle is indicated as number of bits flipped per iteration. A stationary particle will have zero number of bits flipped, whereas a particle which can move furthest will have all its bits flipped. Every particle constitutes position and velocity associated with it, which gets updated based on previous values of Pbest (particle best), Gbest (global best), and velocity, after each iteration. After *i*th iteration, the velocity of the particle is updated using Equation (3.18):

$$v_{id}^{i+1} = w \cdot v_{id}^{i} + c_1 \cdot r_1 \left(\text{Pbest}_{id}^{i} - x_{id}^{i}\right) + c_2 \cdot r_2 \left(\text{Gbest}_{id}^{i} - x_{id}^{i}\right) \tag{3.18}$$

In Equation (3.18), c_1 and c_2 are considered as "cognition learning" and "social learning" rates, respectively. Maximum steps required to take toward Pbest and Gbest are decided by these rates. Inertia weight, that is w, controls the velocity. r_1 and r_2 are random numbers between [0, 1] and are real.

Given the velocity, the update in the particle position is done using Equation (3.19):

$$x_{id}^{i+1} = \begin{cases} 1, & r < S\left(v_{id}^{i+1}\right) \\ 0, & \text{otherwise} \end{cases} \tag{3.19}$$

Here, S is considered as sigmoid function and r is a random number uniformly distributed in the range [0, 1]. This function is given in Equation (3.20):

$$S\left(v_{id}^{i+1}\right) = \frac{1}{\left(1 + \exp\left(-v_{id}^{i+1}\right)\right)} \tag{3.20}$$

In the above-mentioned equations, x_{id}, Gbest_{id}^{i} and Pbest_{id}^{i} have binary digits as there value.

3.4.3 The Proposed Methods

This section deals with strategies used to provision microservices to physical machine. The proposed strategy involves three stages, that is, provisioning of microservices to containers which include three forms of mapping. Secondly, strategy for containers to PM mapping using dynamic bin packing and finally container-to-PM mapping using binary PSO, for generation of resource utilization and power consumption values for the process. These strategies are briefly discussed below.

3.4.3.1 Microservices-to-Container Mapping

This stage deals with the mapping of microservices to suitable container using three different techniques to minimize the resource wastage of the containers. The first method named as best fit represented by the flowchart depicted in Figure 3.6 does the mapping based on the minimum residual load first, here residual load CM_{ij} is calculated as load of container i to the requested load of the microservices. Every microservice is deployed on one container considering the fact that resource requirements of microservices do not exceed the resource capacity of selected container. Best fit selects the container with minimum residual load. Worst fit selects the containers with maximum residual load with threshold of 0.5. Resource wastage policy selects the containers with minimum resource wastage.

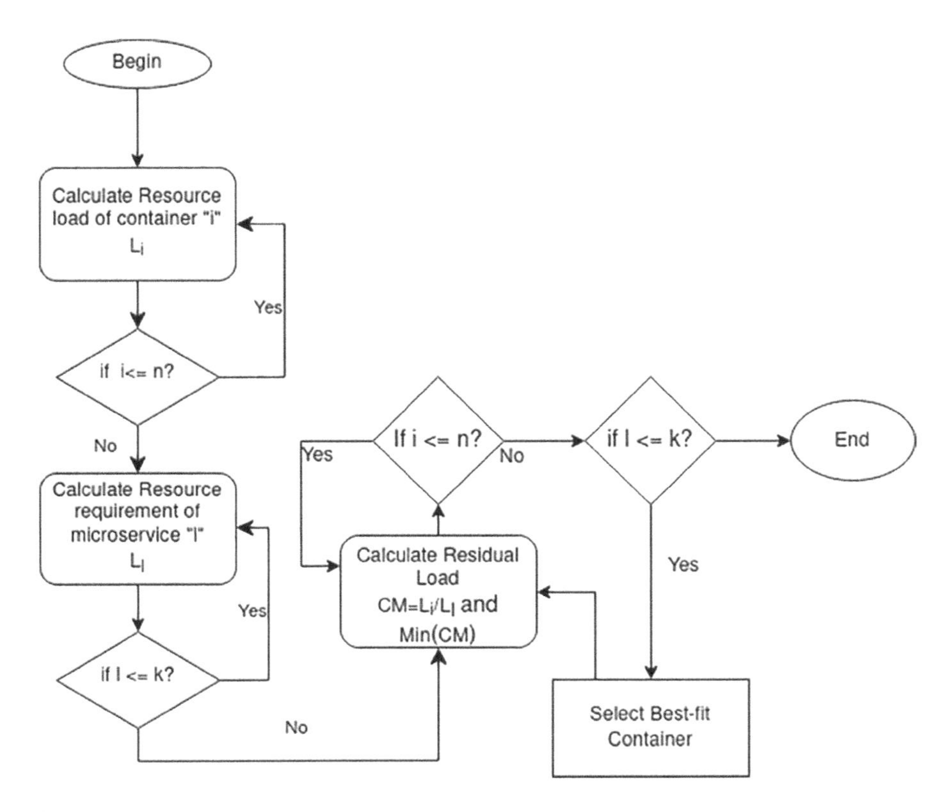

FIGURE 3.6 Best-fit approach based on residual load.

3.4.3.2 Container-to-PM Mapping

This stage is responsible for deploying containers comprising tasks, on active physical machines. This stage also has two different modes of deployment: using dynamic bin-packing strategy and using binary PSO strategy. The containers that are selected to process microservices wait in the global queue of resource manager. The dynamic bin-packing method is used to minimize the number of active PMs required to process the microservices in order to minimize the resource wastage. This objective is adopted by designing dynamic bin packing to find a suitable PM for containers based on the requirements of their computing resources. Considering active PMs as bins with their weights as CPU and memory capacity, PMs are sorted based on weights and from the pool of containers. One container is selected and packed in one PM with least weight. After deployment of the container, the available resources are updated based on the container capacity that are deployed. Once the threshold of the bin/PM is reached, another minimum weighted bin/PM is selected for packing. ACP_{CPU} and ACP_{MM} are the currently available CPU and memory capacity of PM, which is initialized to CP_{CPU} and CP_{MM}. These are updated after every assignment of container, by deducting CPU and memory capacity from available CPU and memory capacity of PM. If the available capacity of the selected PM is reached, then it activates another PM from the sorted list, and the same process is continued until all the containers are assigned to PMs.

Another method used for mapping containers to PM is binary PSO strategy. In binary PSO, *i*th container is mapped to *j*th physical machine, and this mapping is represented by the decision variable x_{ij}. A binary matrix of $n \times m$ size represents each such solution of container allocation with the constraint as sum of the row is always one. It means no two or more same containers be assigned to physical machines. Initially, initial feasible solution is generated. To initialize the solution matrix, a random value is generated between $[1, m]$ randomly, and the x_{ij} value is set to 1 according to the value of j. This technique helps to satisfy the conditions given in Equation (3.16). Still, after update in the particle position using Equations (3.19) and (3.20), there can be a possibility where updated particles do not satisfy all constraints. To eradicate this limitation, for each particle position update, an array is maintained which keeps track of the *i*th row element which has value 1. If after updating, the value of another element of the same *i*th row is trying to be 1, the array detects it and does not allow updating particular position. This helps to improve the effectiveness of the binary PSO for further computing. Figure 3.7 shows the various steps of PSO.

Binary PSO takes the output from the initialization phase where an initial swarm of particles with their position matrices is generated in binary search space. For a limiting number of iterations, either of the objective functions is computed using Equation (3.13) or (3.14), and for each iteration, position and velocity of each particle

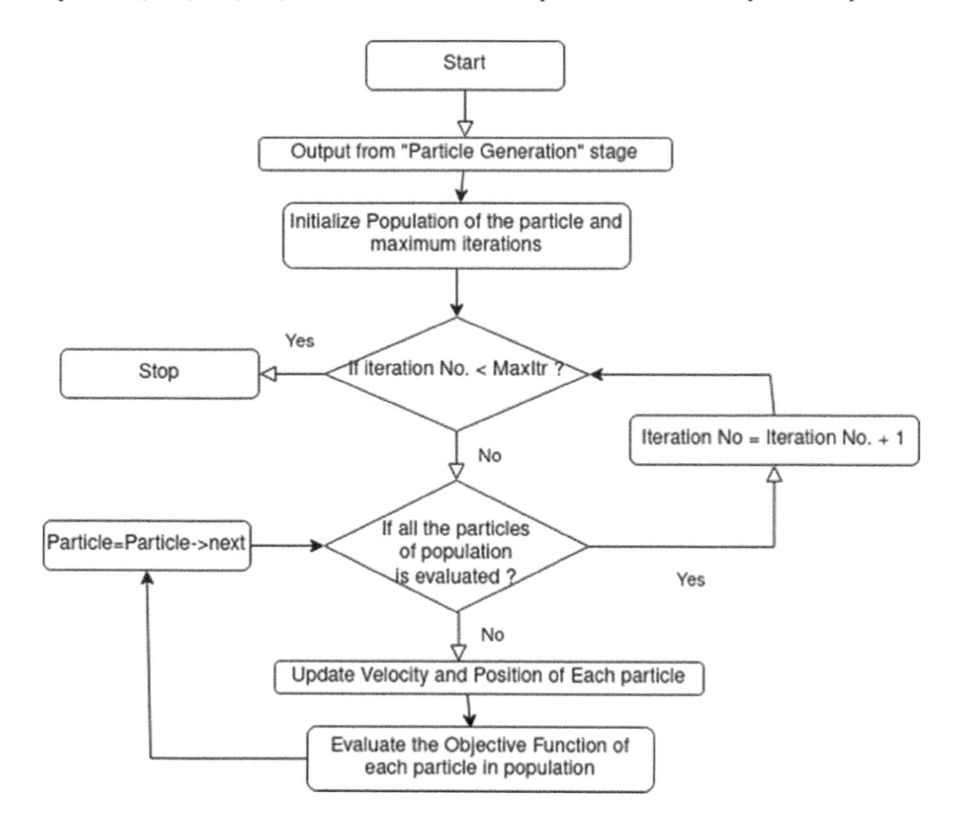

FIGURE 3.7 Flow of instruction in PSO technique.

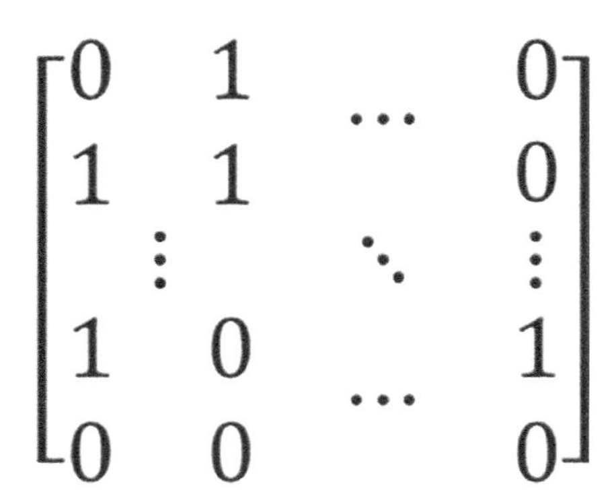

FIGURE 3.8 An example of a generated solution.

in the swarm are updated based on the previous particle best and global best values of the computed objective function. Finally, the particle with minimum global best value is selected and is compared with the computed values of dynamic bin-packing approach. Here, the solution for the best position of the particle contains the information of container allocation over PMs in the matrix form. The procedure of applying BPSO in this work is similar to general PSO with few tweaks. Here, each particle is in the form of binary matrix of dimensions $n \times m$, where 1 in the matrix denotes the deployment of services to particular server else 0 is assigned. An $n \times m$ position matrix, denoted by x_{ij}, is generated and is initially flooded with random binary value taking constraints into account. This creates a feasible particle depicting services placement on physical server. At every iteration, the position matrix for particles is altered on the basis of velocity vector. Velocity vector is denoted by v_{ij} in BPSO and is a $n \times m$ binary matrix that provides the current velocity of the particle. The position x_i represents a solution, or a point represented by the matrix as shown in Figure 3.8. In the search space, particle moves in the search space by flipping determined by number of bits flipped per iteration. If none of the bits are changed, particle is considered to be stationary and moves further with the increment in number of bits that are flipped. Equation (3.18) represents the update in velocity of a particle after *k*th iterations, and update in position is represented by Equation (3.19). Each solution is represented by a binary matrix as in Figure 3.8. It represents whether *i*th container is assigned to *j*th server or not, with constraint that row sum always equals to 1.

3.5 RESULT AND ANALYSIS

In this section, the experiment results are given along with the simulation settings.

3.5.1 Simulation Settings

For the dynamic bin-packing approach, the datasets are generated from the Google cluster trace with tasks comprising computing resource requirements. Load is calculated for each task after normalizing CPU and memory resource capacity to analyze

TABLE 3.1
Simulation Parameters

Simulation Parameters	Values
Dataset-1	50 microservices, 10 containers
Dataset-2	100 microservices, 20 containers
Dataset-2	150 microservices, 30 containers
Number of PMs	20
Capacity of CPU cores in PMs	$[100-800]\times 10{,}000$ MIPS
Capacity of CPU cores in containers	$[20-160]\times 10{,}000$ MIPS
Capacity of memory blocks in PMs	$[120-1600]\times 2$ MB
Capacity of memory blocks in containers	$[12-320]\times 2$ MB

the result. Each dataset is used for every heuristic to observe the change in parameters with change in dataset. Twenty heterogeneous physical machines with different computing resource capacities are included on which containers need to be mapped. The graph is plotted for the three cases for mapping of microservices to containers, and three different datasets and the results are analyzed, accordingly. The simulation parameters for evaluation of proposed methods over existing state-of-art algorithms are represented in Table 3.1.

Binary PSO strategy for experimentation is maintained with a setup that constitutes a swarm/population consisting of a set of 40 particles and is iterated for a maximum of 30 iterations throughout. The rate of cognition learning c_1 and rate of social learning c_2 are tuned to 1.997, so that the solution is not caught up in local optima. Inertia weight, that is w, is chosen between [0.5, 1], randomly. To avoid sigmoid function to reach its saturation, v_k value 1 is made limited between (–10.0 and +10.0). Fifty containers instances are generated comprising computing resources for CPU and memory. Servers are kept heterogeneous in terms of resource capacity and counts to 20 in number. Same CPU and memory capacity is selected in the dynamic bin-packing approach for comparison. To generate the resource capacity of containers randomly, a linear correlation between memory and CPU is considered. The resource capacity of containers and resource requirements of microservices are kept persistent for all heuristics.

3.5.2 Analysis and Results

This section provides the result obtained and discusses the variations in performance of computing resources for two algorithms for placement of microservices to containers, that is, based on min-residual-load vs best-fit approach on CPU capacity and three heuristics for placement of containers to servers, that is best-fit dynamic bin packing, worst-fit dynamic bin packing, and PSO. All the experiments consist of same number of heterogeneous servers. Experiments are performed on three different datasets for each heuristic for observing the variation in computing metrics.

3.5.2.1 Comparison of Resource Wastage

3.5.2.1.1 For Microservices-to-Containers Placement Based upon Residual Load and CPU Capacity of Container

On the three different metrics, the variation of different algorithms on three different datasets is plotted in the three sub-graphs of Figure 3.9. Figure 3.9 shows the optimization performed by binary PSO to minimize resource wastage. Figure 3.9a and b depicts the variation of PSO on microservice-to-container placement based on minimum residual load and min-CPU capacity of containers, respectively. Similarly, Figure 3.9c–f shows the minimization of resource wastage with different number of iterations.

Figure 3.10 shows the comparison of different proposed algorithms on different datasets. In the graph shown in Figure 3.10a, it can be observed that for a dataset comprising 10 different types of containers for deploying 50 microservices, for all the three different heuristics, the microservice-to-container placement based on

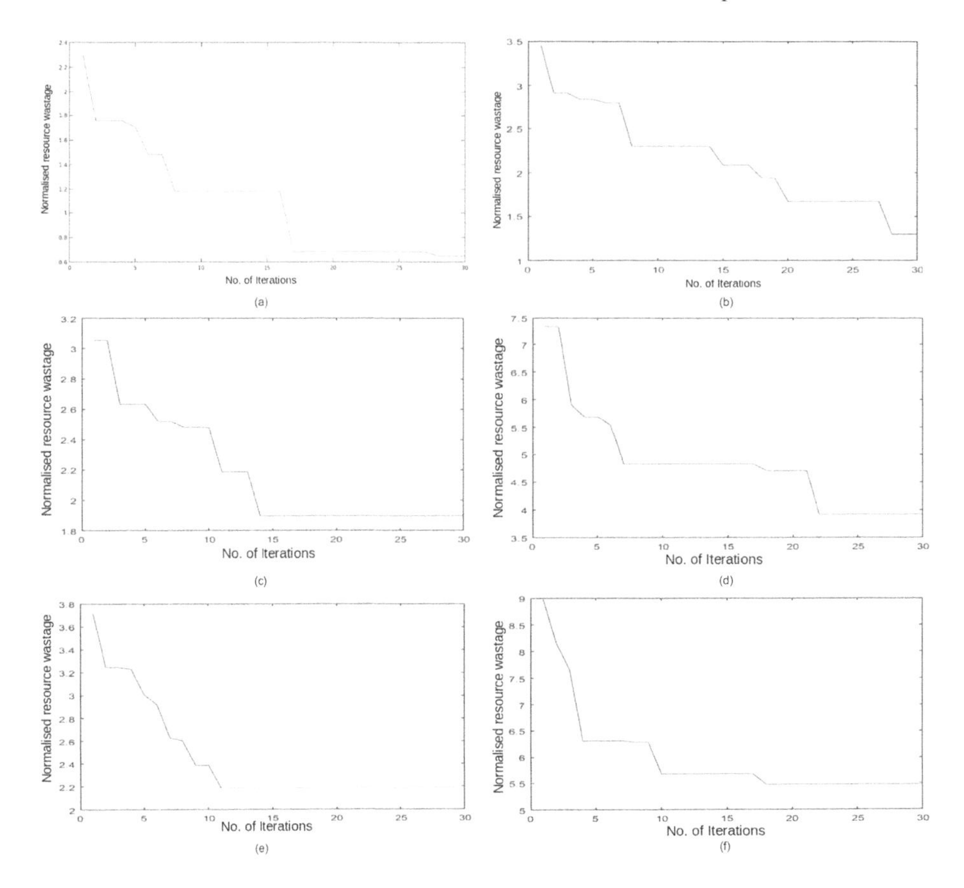

FIGURE 3.9 Convergence of PSO in different settings. (a) Residual load (50 services), (b) CPU capacity (50 services), (c) residual load (100 services), (d) CPU capacity (100 services), (e) residual load (150 services), and (f) CPU capacity (150 services).

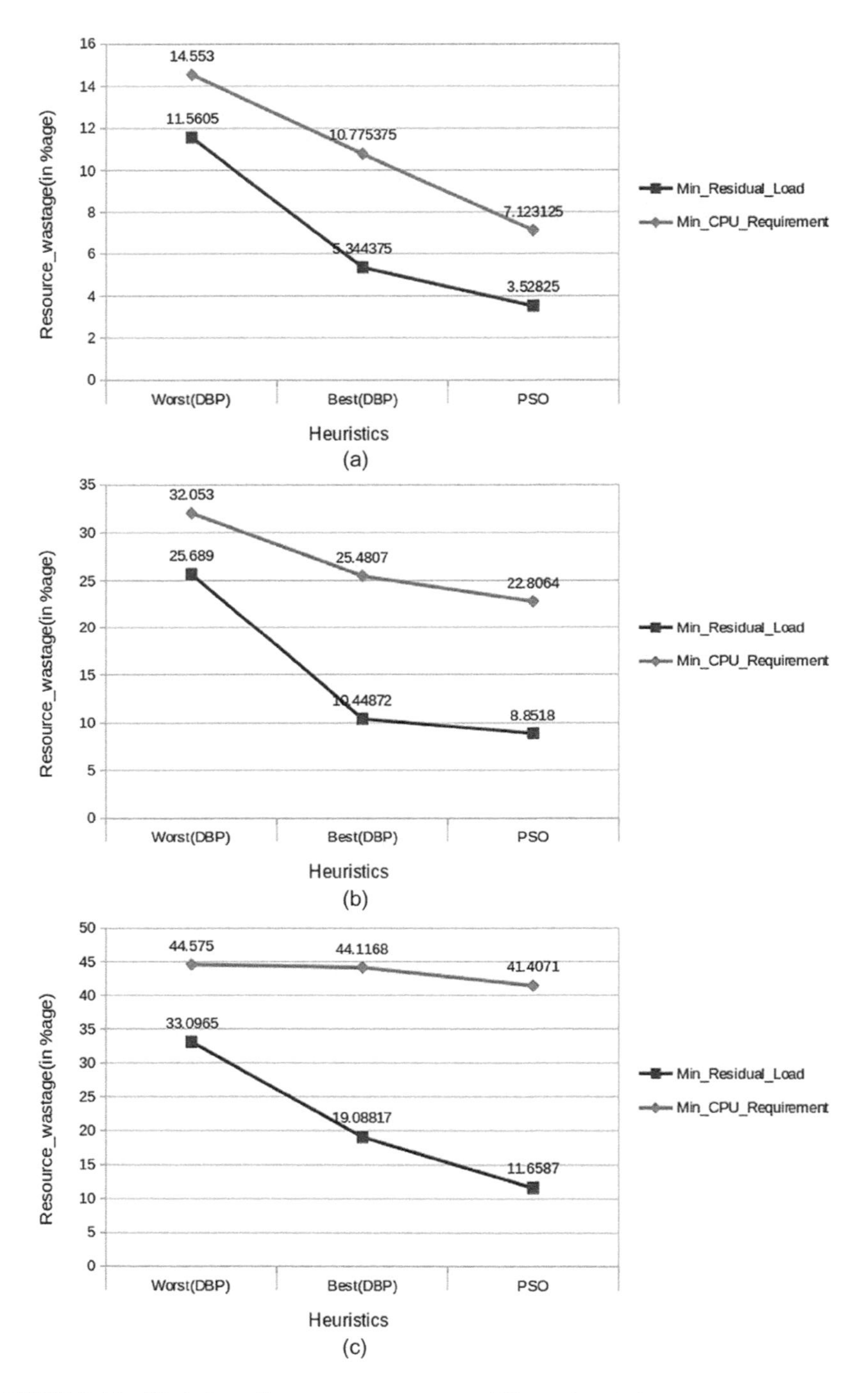

FIGURE 3.10 Variation of resource wastage on different heuristics. (a) For 10 container-types and 50 microservices, (b) for 20 container-types and 100 microservices, and (c) for 30 container-types and 150 microservices.

minimum residual load outperforms the placement based on minimum CPU capacity of containers as observed in the graph; for all the heuristics, the resource wastage is lower for residual-load placement than placement based on CPU capacity. Similarly, Figure 3.10b and c displays the variation of percentage of resource wastage for different algorithms on datasets consisting of 20 types of containers to deploy 100 microservices and 30 types of containers to deploy 150 microservices, respectively. The variation in the resource wastage parameter is observed to be similar as for Figure 3.10a. Hence, based on above three graphs in Figure 3.10, it can be concluded that for all the three heuristics, microservice-to-container placement based on min-residual load always performs better than that based on min-CPU capacity of containers, that is, best fit.

3.5.2.1.2 For Container-to-Server Placement Based on Three Heuristics

Figure 3.11 displays the variation of different heuristics with different datasets. Microservice-to-container placement is based on residual load. It can be observed that even with the increase in number of microservices and containers, the pattern for variation in resource wastage percentage remains same for different datasets for all the three algorithms, that is, dynamic bin packing (worst-case scenario), dynamic bin packing (best case scenario), and PSO technique; for each dataset, the maximum percentage of resource wastage is observed in dynamic bin packing (worst case), which got improved while applying best-case dynamic bin packing, and finally, the best result is obtained in the case of PSO technique. This pattern is observed all over the experiment for all the different sets.

Similarly, in the case of placement based on CPU capacity as plotted in the graph depicted in Figure 3.12, all the three heuristics show the similar characteristic as that for microservice-to-container placement for these heuristics. Hence, we can conclude that PSO technique optimizes the resource wastage better than dynamic bin packing in all the scenarios.

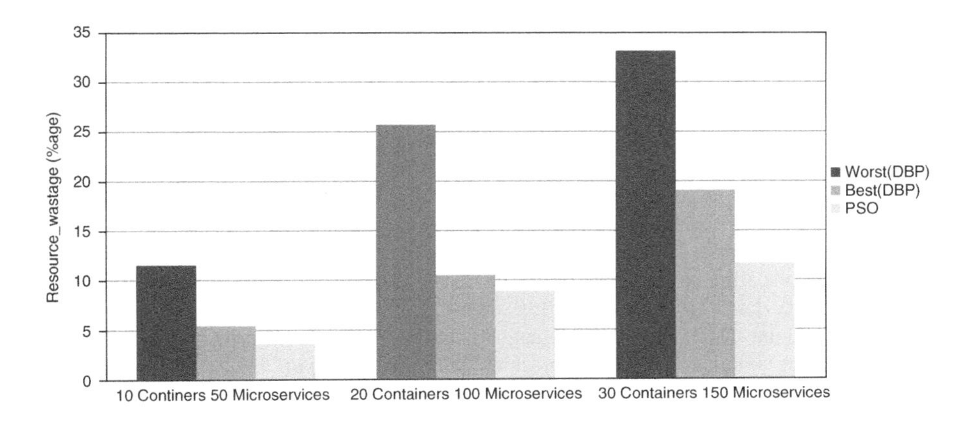

FIGURE 3.11 Resource wastage (residual load) on different heuristics for different datasets.

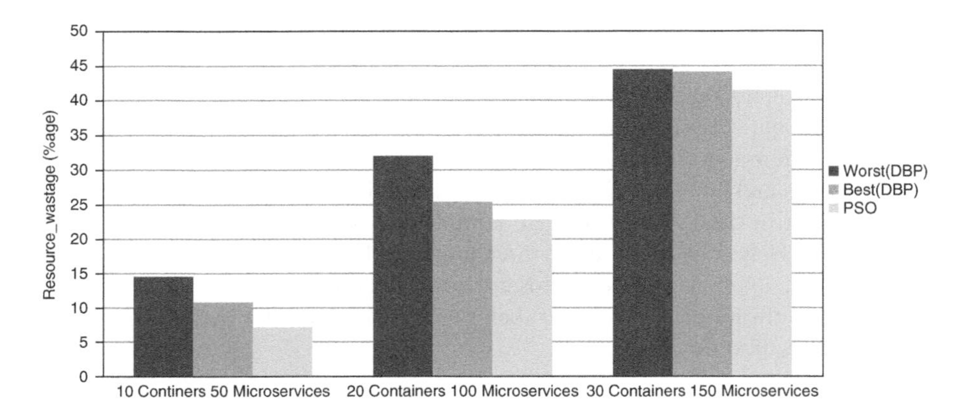

FIGURE 3.12 Resource wastage variation (CPU capacity) on different heuristics for different datasets.

3.6 CONCLUSION

In this work, microservice-container and container-server placement strategies are proposed for processing of microservices in order to minimize the overall resource wastage. The contribution of proposed work is accompanied by a two-phase process. First phase uses two different algorithms for microservice-container mapping to minimize the resource wastage, and the second phase comprises three different algorithms for container-to-server placement in order to minimize the overall computing resource wastage. The obtained results verify the effectiveness of the proposed work. The proposed BPSO algorithm results in lower resource wastage as compared to dynamic bin-packing method. This in turn minimizes the overall resource usage cost. Further, for future work, multi-objective evolutionary methods, for example, MOPSO, NSGA, etc., can be applied for optimizing more than one objective simultaneously.

REFERENCES

1. P. Mell and T. Grance, "The NIST definition of cloud computing," *Cloud Comput. Gov. Background Benefits Risks*, vol. 145, pp. 171–173, 2011, doi: 10.1016/b978-0-12-804018-8.15003-x.
2. D. Mourtzis and E. Vlachou, "Cloud-based cyber-physical systems and quality of services," *TQM J.*, vol. 28, no. 5, 2016, doi: 10.1108/TQM-10-2015-0133.
3. L. Qi, Y. Chen, Y. Yuan, S. Fu, X. Zhang, and X. Xu, "A QoS-aware virtual machine scheduling method for energy conservation in cloud-based cyber-physical systems," *World Wide Web*, vol. 23, no. 2, 2020, doi: 10.1007/s11280-019-00684-y.
4. S. N. Srirama, M. Adhikari, and S. Paul, "Application deployment using containers with auto-scaling for microservices in cloud environment," *J. Netw. Comput. Appl.*, vol. 160, 2020, doi: 10.1016/j.jnca.2020.102629.
5. A. M. Potdar, D. G. Narayan, S. Kengond, and M. M. Mulla, "Performance evaluation of Docker container and virtual machine," *Procedia Comput. Sci.*, vol. 171, 2020, doi: 10.1016/j.procs.2020.04.152.
6. J. Kennedy and R. Eberhart, "Particle swarm optimization," in *IEEE International Conference on Neural Networks - Conference Proceedings*, vol. 4, pp. 1942–1948, 1995. doi: 10.4018/ijmfmp.2015010104.

7. K. Kc and K. Anyanwu, "Scheduling hadoop jobs to meet deadlines," in *Proceedings -2nd IEEE International Conference on Cloud Computing Technology and Science, CloudCom 2010*, pp. 388–392, 2010, doi: 10.1109/CloudCom.2010.97.
8. M. Alicherry and T. V. Lakshman, "Optimizing data access latencies in cloud systems by intelligent virtual machine placement," in *Proceedings - IEEE INFOCOM*, pp. 647–655, 2013, doi: 10.1109/INFCOM.2013.6566850.
9. Y. C. Kwon, M. Balazinska, B. Howe, and J. Rolia, "Skew-resistant parallel processing of feature-extracting scientific user-defined functions," in *Proceedings of the 1st ACM Symposium on Cloud Computing, SoCC'10*, pp. 75–86, 2010, doi: 10.1145/1807128.1807140.
10. K. Govindarajan, V. S. Kumar, and T. S. Somasundaram, "A distributed cloud resource management framework for High-Performance Computing (HPC) applications," in *2016 8th International Conference on Advanced Computing, ICoAC 2016*, pp. 1–6, 2017, doi: 10.1109/ICoAC.2017.7951735.
11. F. Zhang, X. Tang, X. Li, S. U. Khan, and Z. Li, "Quantifying cloud elasticity with container-based autoscaling," *Futur. Gener. Comput. Syst.*, vol. 98, pp. 672–681, 2019, doi: 10.1016/j.future.2018.09.009.
12. C. Kaewkasi and K. Chuenmuneewong, "Improvement of container scheduling for Docker using ant colony optimization," in *2017 9th International Conference on Knowledge and Smart Technology: Crunching Information of Everything, KST 2017*, pp. 254–259, 2017, doi: 10.1109/KST.2017.7886112.
13. W. Li, A. Kanso, and A. Gherbi, "Leveraging Linux containers to achieve high availability for cloud services," in *Proceedings -2015 IEEE International Conference on Cloud Engineering, IC2E 2015*, pp. 76–83, 2015. doi: 10.1109/IC2E.2015.17.
14. A. P. M. De La Fuente Vigliotti and D. M. Batista, "A green network-aware VMs placement mechanism," in *2014 IEEE Global Communications Conference, GLOBECOM 2014*, pp. 2530–2535, 2014, doi: 10.1109/GLOCOM.2014.7037188.
15. C. Guerrero, I. Lera, and C. Juiz, "Genetic algorithm for multi-objective optimization of container allocation in cloud architecture," *J. Grid Comput.*, vol. 16, no. 1, pp. 113–135, 2017, doi: 10.1007/s10723-017-9419-x.
16. J. Boyar et al., "The maximum resource bin packing problem," *Theor. Comput. Sci.*, vol. 362, no. 1–3, 2006, doi: 10.1016/j.tcs.2006.06.001.
17. G. Gambosi, A. Postiglione, and M. Talamo, "Algorithms for the relaxed online bin-packing model," *SIAM J. Comput.*, vol. 30, no. 5, 2000, doi: 10.1137/S0097539799180408.
18. E. G. Coffman, Jr., M. R. Garey, and D. S. Johnson, "Dynamic bin packing," *SIAM J. Comput.*, vol. 12, no. 2, pp. 227–258, 1983, doi: 10.1137/0212014.
19. D. Karaboga and B. Basturk, "Artificial Bee Colony (ABC) optimization algorithm for solving constrained optimization problems," in P. Melin, O. Castillo, L. T. Aguilar, J. Kacprzyk, and W. Pedrycz (Eds.), *Foundations of Fuzzy Logic and Soft Computing*, Lecture Notes in Artificial Intelligence (LNAI), vol. 4529, pp. 789–798. Berlin/Heidelberg: Springer Science & Business Media, 2007, doi: 10.1007/978-3-540-72950-1_77/COVER.
20. A. Tripathi, I. Pathak, and D. P. Vidyarthi, "Energy efficient VM placement for effective resource utilization using modified binary PSO," *Comput. J.*, vol. 61, no. 6, 2018, doi: 10.1093/comjnl/bxx096.
21. D. Kumar and Z. Raza, "A PSO based VM resource scheduling model for cloud computing," in *Proceedings -2015 IEEE International Conference on Computational Intelligence and Communication Technology, CICT 2015*, pp. 213–219, February 2015. doi: 10.1109/CICT.2015.35.
22. X. Fan, W.-D. Weber, and L. A. Barroso, "Power provisioning for a warehouse-sized computer," *ACM SIGARCH Comput. Archit. News*, vol. 35, no. 2, 2007, doi: 10.1145/1273440.1250665.

4 RSS-Based Smart Device Localization Using Few-Shot Learning in IoT Networks

Mohammad Zeeshan, Ankur Pandey, and Sudhir Kumar

CONTENTS

4.1 INTRODUCTION

Localization and tracking of smart devices in a connected Internet of Things (IoT) network is a key service and challenge due to the significant advancements of IoT devices and increasing demands of smart environment applications. Traditionally, the Global Positioning System (GPS) has been used for large-scale outdoor localization and navigation; however, it is known that GPS does not provide higher location accuracies in indoor, underground, and environments with sophisticated non-line-of-sight conditions [1]. The localization accuracy of GPS-based systems in such environments is in general ≥ 5 m and at times even in the 10 m range which is not convenient for indoor positioning and navigation systems in smart IoT environments. Additionally, various upcoming technologies like efficient indoor area mapping and vision-based navigation solutions for big buildings require 3-D localization with high

DOI: 10.1201/9781003438588-4

localization accuracy [2]. This is particularly difficult to achieve with GPS-based systems, as they struggle to localize accurately even in 2-D indoor environments. Therefore, to achieve accurate and reliable localization in complex indoor and fading environments, we need to explore non-GPS localization alternatives.

4.1.1 Motivation and Literature Review

In the literature, we find methods that exploit location-dependent signals and/or signal parameters from cellphones including 3G, 4G, and 5G, Bluetooth Low Energy mesh, and wireless local area network (WLAN) channel state information [3–5]. Specifically, WLAN-based localization is promising and popular due to the pervasive nature of Wi-Fi signals, the widespread availability of already existing Wi-Fi access points (APs) throughout a large number of buildings of interest such as offices, malls, and airports. This translates to a lower cost of deployment due to the already existing infrastructure. Further, it is much simpler to obtain Wi-Fi signal parameters such as received signal strength (RSS) even on edge devices such as smartphones and low-cost micro-controllers. Recently, machine learning (ML)- and deep learning (DL)-based approaches have demonstrated state-of-the-art performances [1], albeit the performance of these methods is hindered due to the following phenomena:

- **Device Heterogeneity:** Smart devices of different makes and manufacturers provide varying RSS values due to the variation in chipset and antenna sizes. It is observed that two devices from different manufacturers, when placed at the exact same distance from an AP, register different RSS values [2]. This difference can be as high as 10 – 20 dBm for different devices. Thus, the same fingerprinting method will provide different location estimates to users of different smart devices even at the same location.
- **RSS Temporal Variations:** Wi-Fi signals experience scattering, multipath fading, and other such effects in a typical indoor environment. When the layout of this environment changes over time due to the change in furniture placement, opening and closing of doors, or change in the APs, the RSS values of each location are perturbed significantly [3].

Thus, the fingerprinting methods' performance degrades significantly to the above-mentioned phenomena. In this chapter, we demonstrate in detail the performance of various ML and DL methods along with codes so that users can understand the localization process in detail and show that the few-shot learning approach best addresses the above-mentioned bottlenecks of fingerprinting-based localization.

In the previous works that attempt to solve the above-mentioned challenges such as Siamese embedding location estimator (SELE) [1], the learnt embeddings are such that the Euclidean distance between them is representative of the actual physical distance between the pair of inputs fed to the SELE network. Additionally, the network is 'fine-tuned' on a small number of RSS samples of data from later unseen months to achieve high performance. In TransLoc [3], cross-domain mapping is learnt which helps transform the data from one domain to another. Although both these approaches have demonstrated good performance across the months on the common public dataset, the issue with them is that they do need at least a few samples from later months based on which they generalize over the months and improve the localization performances. In contrast to this, our proposed method only requires very few data points from the later months as a 'reference', and the model itself is not trained on these RSS samples from later months as per the few-shot learning approach.

In the upcoming sections, we will demonstrate with working examples some of the popular ML-/DL-based approaches that are widely used for performing fingerprinting-based localization using Wi-Fi RSS data. Eventually, we also discuss our few-shot learning-based approach using Siamese network in detail.

4.2 LOCALIZATION METHODOLOGY

4.2.1 Fingerprinting-Based Localization

Amongst the various available methods, fingerprinting-based methods are the most popular due to low cost and computational time with edge device computing. In fingerprinting, the location is estimated in offline and online phases:

- **Offline Phase:** In this stage, RSS values are scanned and recorded at multiple locations in the deployment area for all the APs that are present in the IoT environment. These recorded values are then stored as the 'offline dataset'. Training and testing of localization models are performed on the offline dataset. Hence, it is imperative that the data collection process is truly representative of the scenarios where the system has to be finally deployed.
- **Online Phase:** In this stage, the localization method developed in the previous stage is deployed to perform localization on the real-time data from different smart IoT devices. The results of this phase truly present the accuracy of the proposed localization algorithm as in real world device heterogeneity, device orientation, temporal variations, and device removal occur.

As explained earlier, we now present a few of the most popular ML-based localization methods and then propose a few-shot approach for localization. Here, we utilize the RSS dataset named 'long-term Wi-Fi fingerprinting dataset and supporting material' from Universitat Jaume I (UJI) [6]. The dataset can be obtained from this link. This dataset consists of RSS readings from 620 Wi-Fi APs. There are separate RSS values for 25 months with each month consisting of its own training and testing data. Due to this month-wise separation in the dataset, we can clearly visualize the two important issues discussed in the previous section, namely temporal variation over longer periods of time and device heterogeneity as the last month consists of data from different smart devices.

The first step in fingerprinting-based localization is loading the RSS values for the model. We demonstrate how to open and load a comma-separated values (CSV) data file from a dataset as a numpy array. Python is an open-source language that is widely used for building an ML-based model; hence, we utilize Python throughout this chapter for the learning of the readers. The following code snippet demonstrates a custom Python function `csv _ opener` which opens a CSV file and returns the contents in the `numpy.ndarray` format.

Listing 1: Opening Data Files

```
1       #importing pandas library
2       import pandas as pd
3
4       #function for returning CSV's content as a numpy array
5       def csv_opener(fname):
6       df = pd.read_csv(fname)
7       read_value = df.to_numpy()
8
9       return read_value
10
11      #storing CSV file's content in 'rss_data' variable
12      rss_data = csv_opener('FULL_PATH_TO_CSV_FILE/trn01rss.
        csv')
13
14      #storing labels in variable 'label data'
15      label_data = csv_opener('FULL_PATH_TO_CSV_FILE/
        trn01crd.csv')
16
17      #printing data type and shape of data to make sure its
        as expected
18      print(type(rss_data))
19      print(rss_data.shape)
20      print(label_data.shape)
```

The above snippet returns

```
1       <class  'numpy.ndarray'>
2       (576, 620)
3       (576, 3)
```

This means that the contents of the CSV file are loaded into a `numpy.ndarray` and that its shape is (576, 620) indicating that in this particular data file, 576 data points are there and each of them has 620 features. Note that each of these features denotes the RSS value for the available 620 Wi-Fi APs. Also, the label's shape indicates that for each data point, the label is a 3-D coordinate whose fields are the x and y coordinates and the floor number of the location where that particular data is loaded.

4.2.2 k-Nearest Neighbours (k-NN)

The k-nearest neighbour (k-NN) algorithm [7] is a popular supervised learning algorithm with widespread applications in varied fields such as medical data analysis, business data analysis, and localization as well, amongst many others [8]. It essentially uses the features and labels of the training data and estimates the categorization of unlabelled data. Although k-NN is more popular for performing classification tasks, it can efficiently perform regression tasks as well. k-NN performs regression such that amongst the k-NN of the raw input data the mean of the outputs of the nearest neighbours is considered. This mean is assigned to be the output of the raw input data. Note that if we take the weighted mean instead of the simple mean, the algorithm is termed as weighted k-NN [9].

In the context of performing localization using the given RSS dataset, the k-NN in effect attempts to find k-RSS samples from the training dataset which are the 'closest' to the raw RSS input. Amongst these k-nearest samples, the mean of their label features is computed, and this mean is the model's final estimation of location for the given RSS input.

For implementing k-NN regression on the dataset for performing localization, we use the Scikit-learn library [10]. The following code snippet demonstrates the entire localization method.

Listing 2: k-NN Implementation

```
1     #importing relevant module from scikitlearn library
2     from sklearn import neighbors
3
4     #using the dataset obtained in code-snippet 1
5     rss_data = ...
6     rss_label = ...
7
8     #instantiating, training, and calling the kNN method of
      sklearn
9     knn  =  neighbors.KNeighborsRegressor(7,
      weights='uniform')
10    predictions  =  knn.fit(rss_data,  rss_label).
      predict([rss_data[0]])
11
12    #printing the prediction, and expected output for a
      datapoint
13    print("kNN regressors estimation: ", predictions)
14    print("Actual coordinate value: ", rss_label[0])
```

In the above code snippet, we use the `rss _ data` and `label _ data` which are obtained in code snippet #1. The k-NN regressor is trained on this data. Note that we have set the parameter k as 7 and weights as 'uniform'. The choice of these parameters is user dependent, and it is up to the user to decide the suitable one which works best with the given RSS data. We demonstrate testing the model on the first data point of the RSS value at hand, which is `rss _ data[0]` and its label is `label _ data[0]`. Following is the output obtained:

```
1	kNN regressors estimation:  [[10.40554331 27.17299647
	3.       ]]
2	Actual coordinate value:  [12.91385188 29.21654402  3.]
```

It is observed that both the estimated and expected coordinates are printed. A quick calculation of the Euclidean distance (of only x and y coordinates that are the first two dimensions) between these gives us a value of 3.23 m. This means that the estimated location for the data sample `rss _ data[0]` is 3.23 m off of the actual position for this particular data point.

4.2.3 Decision Tree (DT)

Decision Tree (DT) is another popular ML algorithm that essentially builds a tree of binary classifiers based on various 'splits' in the input data features. This method is prominently used in data-driven decision-making for businesses and is popular for performing IoT-based tasks such as localization [11]. Again, it should be noted that although originally DTs have been used for performing classification, they can also be used for performing regression which is demonstrated in this chapter. For implementing the DT regressor on the dataset, we use the same library as before, the Scikit-learn library. The following is the complete implementation of DT:

```
1	#importing relevant module from scikitlearn library
2	from sklearn.tree import DecisionTreeRegressor
3	
4	#using the dataset obtained in code-snippet 1
5	rss_data = ...
6	rss_label = ...
7	
8	#instantiating, training, and calling the DT method of
	sklearn
9	DT = DecisionTreeRegressor(max_depth=3)
10	predictions = DT.fit(rss_data, rss_label).
	predict([rss_data[0]])
11	
12	#printing the prediction, and expected output for a
	datapoint
13	print("DT regressor's estimation: ", predictions)
14	print("Actual coordinate value: ", rss_label[0])
```

In the above code snippet, we are again using the `rss _ data` and `label _ data` which are obtained in code snippet #1. Here, we have set the parameter `max _ depth` as 3. And once again, the choice of parameters is completely dependent on the type of data and user's requirements. Also, we run the model on the first data point of the dataset, similar to what we performed for k-NN, and the following is the output obtained:

```
1       DT regressors estimation:   [[ 9.98749189 27.47428874
        4.94871795]]
2       Actual coordinate value:   [12.91385188 29.21654402  3.]
```

Again, the Euclidean distance between our estimated and the actual coordinate is 3.41 m, and in this case, the floor estimated by DT is 4.948 which is closer to 5, thereby indicating that the model's regression output for the floor is quite erroneous for this particular data point.

4.2.4 Multi-Layer Perceptron (MLP)

Multi-layer perceptron (MLP) is one of the most popular ML models which is used extensively in various fields. MLPs are essentially a nonlinear mathematical model with multiple learnable parameters that can be adjusted, or 'learnt', by training the model on a given dataset. Due to the presence of fully connected layers and nonlinear activation functions, MLP is able to unravel all possible linear and nonlinear relations between the inputs and outputs [12].

In the context of localization, the MLP model makes an effort to minimize the discrepancy between the model's predicted coordinate and the real position corresponding to a certain set of input RSS data. A loss function, in our instance the mean squared error (MSE), provides this difference. The accurate locations are estimated by tuning the learning parameters of the model to minimize the MSE to a minimum value, as it learns across the training split of the dataset.

TensorFlow is a well-known open-source DL framework that we utilize to create the required MLP in Python [13]. With TensorFlow, users have complete control over the design of fully customizable DL models. For example, users can choose the number of hidden layers, neurons per layer, activation functions for each of the layer, regularization, loss, and other parameters. The design of a five-layered MLP with dropout regularization and batch normalization is presented in the following code snippet.

```
1       #importing TensorFlow modules relevant to our MLP
2       from tensorflow.keras.models import Sequential
3       from tensorflow.keras.layers import Dropout, Dense,
        BatchNormalization
4       from tensorflow.keras.optimizers  import  Adam
5       import pandas as pd
6       import numpy as np
7
8
9
10      #using the dataset obtained in code-snippet 1
11      rss_data = ...
12      rss_label = ...
```

```
13
14    #function for designing a 5-layered MLP using
      TensorFlow Sequential API
15    def build_mlp_model():
16    model = Sequential()
17
18    model.add(Dense(256, input_dim=620, activation='relu'))
19
20    model.add(Dense(256, activation='relu'))
21    model.add(Dropout(0.2))
22
23    model.add(Dense(128, activation='relu'))
24    model.add(Dropout(0.2))
25
26    model.add(Dense(64, activation='relu'))
27    model.add(BatchNormalization())
28     model.add(Dropout(0.2))
29
30    model.add(Dense(3))
31    model.summary()
32    return model
33
34    #calling 'build_mlp_model' function to build the model
35    mlp = build_mlp_model()
36
37    #defining optimiser, loss function, and training the MLP
38    opz = Adam(learning_rate = 1e-2)
39    mlp.compile(optimizer=opz, loss= 'mse',
      metrics=['mae'])
40    history = mlp.fit(rss_data, rss_label, epochs=100)
41    predictions = mlp.predict(np.array([rss_data[0]]))
42
43    #printing the prediction, and expected output for a
      datapoint
44    print("MLP regressors estimation: ", predictions)
45    print("Actual coordinate value: ", rss_label[0])
```

1. In the above code snippet, a five-layered MLP is designed and trained on our dataset. Specifically, there are 256, 256, 128, 64, and 3 neurons in each of the five layers, and the activation function used for the first four dense layers is the rectified linear unit (ReLU) whereas the last layer is provided with a linear activation because it needs to perform regression. Additionally, dropout regularization is used in the second, third, and fourth dense layers, along with batch normalization after the fourth layer. We select MSE loss function for the training, and Adam optimizer is utilized to execute the loss function optimization for determining the learnable weights. Again, it should be noted that the user is in charge of selecting the model hyperparameters. Different model hyperparameters should be selected according to

the dimensionality of the RSS data and its number of samples. The following output is obtained after running code snippet #4.

```
1     Model: "sequential"
2
3     Layer (type) Output Shape Param #
4     ==================================================
5     dense (Dense) (None, 256)   158976
6
7     dense_1 (Dense)      (None, 256)   65792
8
9     dropout (Dropout)    (None,  256)  0
10
11    dense_2 (Dense)      (None, 128)   32896
12
13    dropout_1 (Dropout) (None, 128)   0
14
15    dense_3 (Dense)      (None, 64)    8256
16
17    batch_normalization (BatchNo) (None, 64)       256
18
19    dropout_2 (Dropout) (None, 64)    0
20
21    dense_4 (Dense)      (None, 3)     195
22    ==================================================
23    Total params: 266,371
24    Trainable params: 266,243
25    Non-trainable params: 128
26
27
28    Epoch 1/100
29    18/18 [===========] - 1s 3ms/step - loss: 198.7981 -
      mae: 11.2517
30    .
31    .
32    .
33    Epoch 100/100
34    18/18 [===========] - 0s 5ms/step - loss: 4.1288 -
      mae: 1.6101
35    MLP regressors estimation:  [[12.102449  29.613571
      3.9933233]]
36    Actual coordinate value:  [12.91385188 29.21654402  3.]
```

In the output log, we observe that firstly a model summary is printed due to the fact that we put a `model.summary()` command in the `build _ mlp _ model` function. This summary contains the details of each of the layers that are used, along with the count of trainable and untrainable parameters associated with these layers. After that, the actual training of the model starts wherein optimization of the loss function is performed using the selected optimizer. We clearly observe that the loss value is decreasing as the training epochs increase. At the end of 100 epochs, an MSE of 4.4068 m and a mean absolute error (MAE) of 1.61 m are obtained with the selected hyperparameters in the presented code. Additionally, we again test the model's output for the first data point of the dataset, whose results are also shown in the logs. The Euclidean distance between the actual and estimated coordinate is 0.9 m. Also, the model estimated the floor to be 3.99, which with a decision boundary at $z=4.0$ means that the final estimated floor is 3 (as $3.99<4.0$).

It is observed that MLP with the chosen hyperparameters is performing better than the k-NN and DT methods. As discussed before, MLP exploits all possible interconnections between predictor variables and complicated nonlinear relationships between independent and dependent variables, thereby providing better performance, although it should be kept in mind that both the k-NN and DT's performance can also be improved, given that proper hyperparameters are selected for those models.

4.2.5 Siamese Network-Based Few-Shot Approach for Localization

Siamese network is a specialized deep-learning architecture that is extensively used for several applications which involve a relatively smaller dataset size and unseen classes of data [14]. This architecture is recently used successfully for a wide variety of applications including localization [1,15], leaf classification [16], and chromosome classification [17], to name a few. The overall architecture of Siamese network is as follows.

Few-shot learning task for an ML model refers to the task of the model correctly classifying data points that belong to those classes which the model has not seen at all during its training [18,19]. This means that the model must be capable of generalizing well over unseen classes of data. For the task at hand, the RSS data from later months are essentially 'unseen' classes for our model which has been primarily trained on the first month's data. The temporal variation of RSS over the months effectively renders the RSS fingerprints of later months significantly different from those of the initial months that the data from later months is treated as unseen classes. The system for localization is a two-stage architecture with a Siamese network-based approach and comprises training and testing phases, in accordance with the fingerprinting-based method's methodology as mentioned in Section 4.2. The network consists of two sub-networks, namely embedding functions, which in our case is selected as MLP. It should be noted that these two MLPs have the exact same weights as shown in Figure 4.1 such that the same transformation is performed onto the pair of data that has been fed to the model. The embedding function returns a fixed length vector which we call embeddings. The obtained embeddings are then stacked together, i.e. concatenated, and are passed on to another small three-layered MLP which outputs the predicted label for the pair of RSS which are being input initially into the model. Specifically, the MLP that we are using as the embedding functions in our case is a 4-layered MLP with an input, two hidden, and an output layer of sizes 620, 256, 128, and 64 units, respectively. The leaky rectified unit activation function [20] is used by two hidden layers, and the output layer uses the sigmoid activation [21]. The small MLP which acts upon the concatenated embeddings to produce the final predicted label has 3 layers, of which one is the input layer, one is the hidden layer, and one is

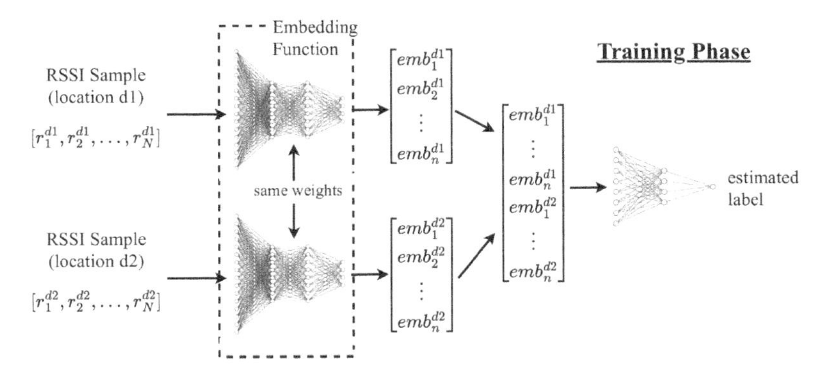

FIGURE 4.1 Siamese architecture for few-shot approach: training phase.

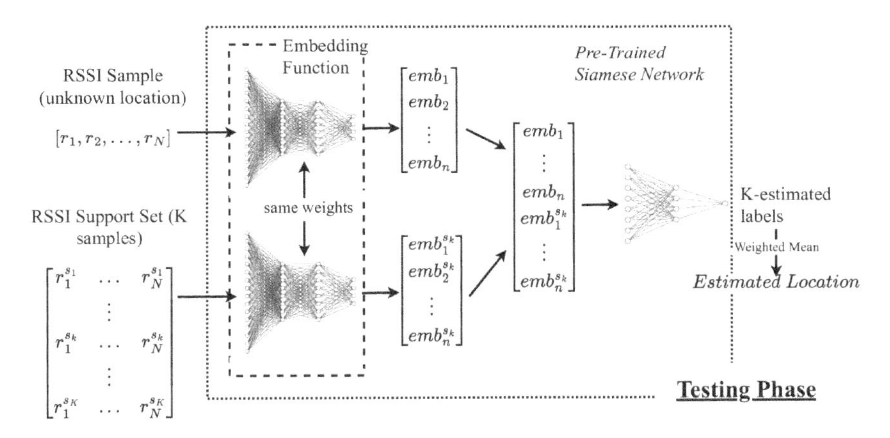

Figure 4.2 Siamese architecture for few-shot approach: testing phase.

the output layer. They have 64, 32, and 1 unit, respectively. The first two are using the ReLU activation function [20], and the output layer uses the sigmoid activation function as shown in Figure 4.2.

In the training phase, we construct a so-called 'pair-wise' dataset from the original dataset, and the Siamese network is trained on these pair-wise samples. Additionally, the labelling of these pair-wise samples is done based on the distance between the two points. Algorithm 4.1 shows the steps to form the pairs.

Data: Input dataset, each of whose elements is of the form (Ri, yi). Ri is the ith RSS sample and yi is the corresponding location coordinate. Dataset contains a total of T samples.

Result: Paired dataset, each of whose elements is of the form (ri,0, ri,1, li) where (ri,0, ri,1) are the two elements of the ith element of the pair-wise dataset and li is the corresponding distance-based labelling.

- Form all possible pairs from the T total data present in the dataset.
- Note that such pairs are possible.

- For each of the pairs, find the corresponding distance-based label l_i as $l_i = \frac{1}{1+\|y_{i,0} - y_{i,1}\|_2}$.
- Hence, the pair-wise dataset has been created, whose ith sample is of the form $(\mathbf{r}_{i,0}, \mathbf{r}_{i,1}, l_i)$.

Algorithm 4.1: Pair-Wise Dataset Creation

Once the pair-wise dataset is created, we move on to the training of the Siamese network. The labelling scheme mentioned in Algorithm 4.1 is such that for the RSS samples whose corresponding locations are close to each other are labelled closer to 1, whereas the samples which are far away from each other are labelled closer to 0. Note that in the extreme case of the samples belonging to the same location, the label is 1.0, whereas for points that are infinitely far away, the label is 0.0.

The Siamese network is trained on the pair-wise dataset, and a modified contrastive loss is used to optimize the weights of the network [22]. For a data sample $(\mathbf{r}_1, \mathbf{r}_2)$ whose label is l_{actual}, let the Siamese network's predicted label be $l_{predicted}$. The loss function and the labelling scheme we have used are slightly different from the original proposed in Ref. [22]. Here, we are applying the loss on the actual and predicted labels $(l_{actual}, l_{predicted})$ instead of the distance between the embeddings. Additionally, our labelling scheme is dissimilar as we label the pair of data that are closer to each other towards 1.0 and those farther away towards 0.0. In Ref. [22], they have chosen the opposite labelling scheme.

In the testing phase, we take the help of a 'support set' for performing the regression for localization. The support set consists of RSS samples, one each from all possible locations selected randomly. We then pair the unknown RSS sample with each of the elements of support set one by one and pass them through the trained Siamese network.

Data: Trained Siamese Network, Support Set $\left[\left(\mathbf{r}^{S_1}, y^{S_1}\right), \ldots \left(\mathbf{r}^{S_K}, y^{S_K}\right)\right]$.

Input: Input RSS data *r* whose location is to be found.

Result: Estimated coordinates for the sample, *y*.

- Pair the unknown RSS sample **r** with the elements of support set one by one.
- Each of these pairs is given to the Siamese network sequentially, and their predicted labels are stored as $\left[l^{S_1}, \ldots, l^{S_i}, \ldots, l^{S_K}\right]$ where l^{S_i} is the label for the paired data $\left(\mathbf{r}, \mathbf{r}^{S_i}\right)$.
- The estimated coordinate of the unknown sample **r** is found by taking the weighted mean of all the coordinated elements of the support set, with the estimated labels being the weights, i.e. $y = \sum_{i=1}^{K} l^{S_i} y^{S_i}$

Algorithm 4.2: Testing Phase Operation of the Siamese Network

Hence, we obtain the estimated location of the unknown RSS sample **r** with the help of a support set and the trained Siamese network.

4.2.5.1 Few-Shot Approach with the Siamese Network

It was observed in the previous section that the Siamese network utilizes a support set for performing the final coordinate estimation. We utilize this support set for performing the few-shot location regression when needed. For data samples coming from a new distribution, we re-form the support set. Here, also we select one RSS sample corresponding to all possible locations in consideration, but the difference here is that these samples are taken from the new distribution of data. It should be noted that we are not training or fine-tuning the model on the data from the new distribution.

We are merely using them in the support set for the Siamese network to make a better estimation. This simple technique helps attain much better results than other methods which have been presented so far, as is shown in the subsequent section.

4.3 RESULTS

In the previous section, we demonstrated the ML approaches along with their implementation in Python. In each of those implementations, we tested the learnt models as well on a single data point from the training data itself. In this section, we perform thorough testing of the previously mentioned models, and a comparison of their performances is presented.

It should be noted that the dataset contains RSS data for 25 separate months. The data for each month contains samples of both the training and testing data. But a closer inspection of the data for each month reveals that although a healthy amount of training and testing data have been provided for the 1st month, the remaining 24 months have very few training data instances and relatively more testing data. This makes the dataset very competitive and difficult to obtain accurate results. Hence, the testing is performed on two different cases namely **Case-I** and **Case-II**.

(1) **Case-I Tests:** The models are trained only on the Training data of the first month, and these trained models are tested on Testing data from all the 25 months. This test essentially *demonstrates how the temporal variation in RSS fingerprints over a long period of time adversely affects the localization performance.* This test also serves as the baseline for observing the changes in the performance of Case-II tests.
(2) **Case-II Tests:** The model is trained on the Training data of the first month. For each of the subsequent 24 months, the models are trained on the (relatively smaller amount of) Training data of that particular month, on top of its training on the first month's data. This additional training on the new month's data serves to 'fine-tune' the model for the particular month.

To quantify the performance of a localization algorithm, we define a mean localization error (MLE) parameter which is given as:

$$\text{MLE} = \frac{1}{N}\sum_{i=1}^{N}\sqrt{\left(x_i - \widehat{x_i}\right)^2 + \left(y_i - \widehat{y_i}\right)^2} \tag{4.1}$$

where (x_i, y_i) is the true location of ith sample, $\left(\widehat{x_i}, \widehat{y_i}\right)$ is the estimated location of ith sample, and N is the total number of data samples in consideration. Note that we do not perform the Case-II tests with the Siamese network because we want to demonstrate the few-shot capability of the Siamese network.

Table 4.1 demonstrates that MLE for various methods ranges from 2.9 to 4.39 m for Case-I tests and from 2.66 to 3.49 m for Case-II tests.

More specifically, MLE for the presented methods for each of the 25 months present in the **UJI** data is presented for Case-I in Figure 4.3 and for Case-II in Figure 4.4. The key observations are as follows:

- The performance of the Siamese network-based approach is the best amongst all the presented methods. Even after the 11th month, MLE is 2.5 m only. It is to be noted that a significant portion of the APs is changed after the 11th month and, hence, the localization performances of other state-of-the-art methods drastically reduce; however, the performance of the proposed few-shot method is not affected by such a change as the embeddings are learnt with much lower MSE loss compared to other methods.
- When the localization models which have already been trained on initial RSS data are trained (fine-tuned) further on a few samples from new RSS data, the performance of the models improves significantly.
- As can be seen in Figure 4.4, MLE does not worsen drastically after the 10- to 11th-month mark when the models are fine-tuned on data from newer months, whereas it does worsen significantly when the models are not calibrated for the changed environment.

Therefore, we realize that with changes in the environment over time, the Wi-Fi RSS distribution changes. This change reduces the performance of localization models

TABLE 4.1
Mean Localization Error for Different Methods in Consideration

Method	MLE (m) Case-I	MLE (m) Case-II
k-NN	3.10	2.87
DT	4.39	3.49
MLP	2.90	2.66
PCA + MLP	3.06	2.97
Siamese few-shot	**2.43**	-

Note that Case-II results are always comparatively better than Case-I

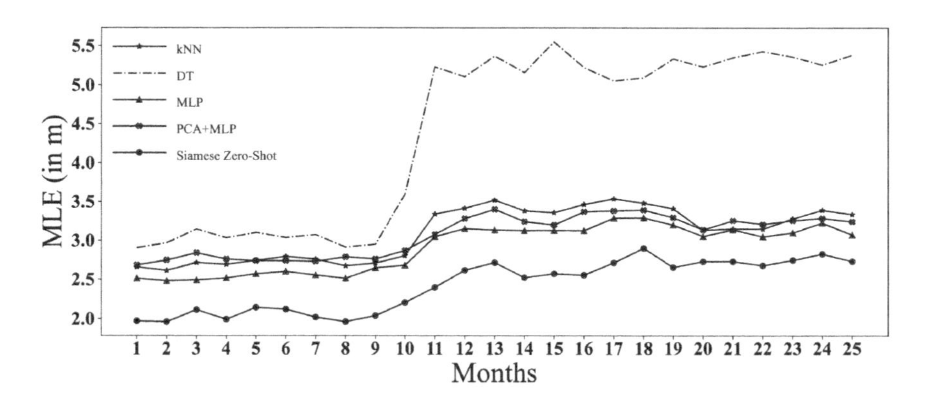

FIGURE 4.3 Test results for Case-I testing. Mean Localization Error (MLE) comparison over 25 months.

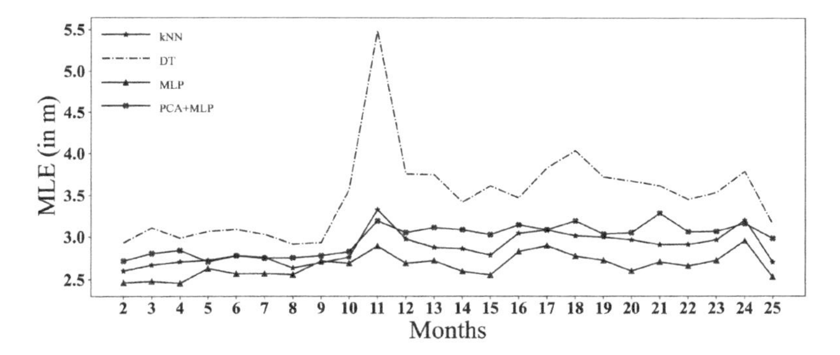

Figure 4.4 Test results for Case-II testing. Mean Localization Error (MLE) comparison over 24 fine tuned months.

which are trained on RSS data before the environment changes (11th month). In general, to mitigate this, the simplest way is to further train the already trained model on samples from the data of the new environment.

Further, if we observe the change in RSS values by plotting the RSS values from an initial month to a later month, the following piece of code will be used to accomplish this:

```
1     %config  InlineBackend.figure_format  =  'svg'
2
3     #importing MatplotLib - the standard library
4     #used for visualizations in Python
5     from matplotlib import pyplot as plt
6     #defining Global fontsize for the plots
7     plt.rcParams["font.size"]  =  20
8
9     #function for plotting 2 RSS samples at once
10    def rss_plotter(signal1, signal2):
11
```

```
12      #subplot with 2 rows 1 column
13      f, axarr = plt.subplots(2, 1)
14
15      #defining height & width of each subplot
16      f.set_figheight(6)
17      f.set_figwidth(12)
18
19      #plotting the first signal signal1
20      #np.arange() used to create vector of 620 numbers for
        the x-axis
21      axarr[0].plot(np.arange(signal1.shape[0]),   signal1)
22      axarr[0].title.set_text("Wi-Fi rss of all 620 APs")
23      axarr[0].set_ylabel("rss")
24      axarr[1].plot(np.arange(signal2.shape[0]),   signal2)
25      axarr[1].set_ylabel("rss")
26      axarr[1].set_xlabel("Wi-Fi  AP  number")
27      #f.savefig("rss.pdf",  bbox_inches  =  'tight')
28
29      #calling the function on 1000th and 3000th rss samples
30      rss_plotter(x[0],  x[1])
```

The above code snippet returns the two plots in Figure 4.5 for the first and second RSS samples in the training split of Month-#1 data, and the two plots in Figure 4.6 are again for the first and second RSS samples in the training split of Month-#24 data. Also, all four of these RSS samples are corresponding to the same location $y=(12.913852, 29.216543, 3.0)$.

It is observed from Figure 4.5 that all the APs beyond AP number 60 provide −100dB RSS values, that is they are not being detected at all. The APs till AP numbered 50–60 are effectively providing all the information for performing localization. However, in Figure 4.6, it is observed that along with APs numbered up to 100, a few APs belonging to the 300–500 range are also sensed, and hence, they should also be used for localization.

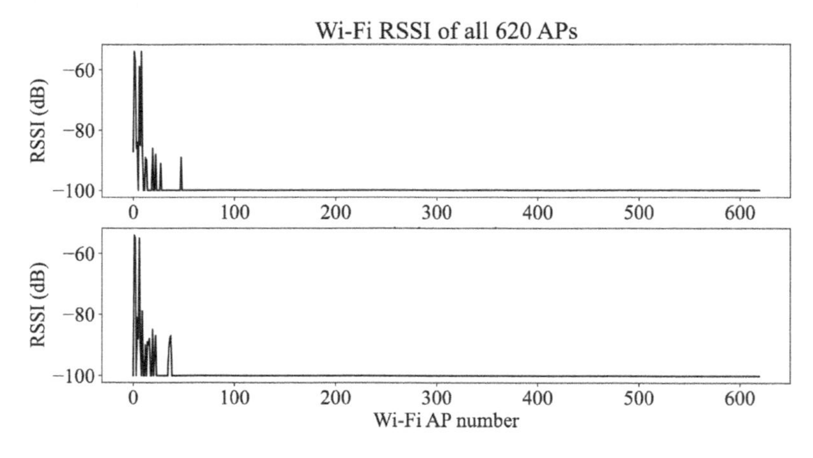

FIGURE 4.5 RSS values for month-#1 data. Location for both samples y = (12.913852, 29.216543, 3.0).

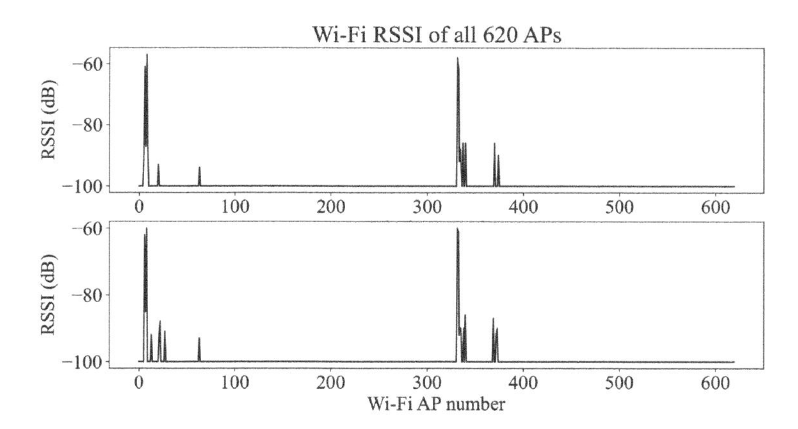

FIGURE 4.6 RSS values for month-#24 data. Location for both samples y = (12.913852, 29.216543, 3.0).

Due to this difference in distribution with respect to APs, the models which have only been trained on the first month data (Case-I tests) fail to perform well on the later month's RSS data.

4.4 CONCLUSIONS

In this chapter, we demonstrated multiple Wi-Fi RSS fingerprinting-based localization techniques and proposed few-shot learning techniques using Siamese network for localization. This chapter detailed k-NN, DT, MLP, and Siamese network-based few-shot approach as ML methods for performing RSS-based localization. The proposed Siamese network provided an MLE of 2.43 m for Case-I, which is significantly better than the other presented methods highlighting the advantages of few-shot learning for addressing the temporal variations and device heterogeneity challenges in the localization based on fingerprinting approaches. In the non-few-shot-based methods, with MLE of 2.9 and 2.66 m, respectively, the presented MLP offered the best MLE for both Cases I and II. Additionally, the Case-II testing demonstrated how the pretrained models can be fine-tuned using a few data samples from the months of concern in order to solve the issue of temporal variations in RSS across lengthy periods of time. This chapter also presents detailed codes for the readers that can be used to implement localization techniques in the Python programming language.

REFERENCES

1. A. Pandey, R. Sequeira, and S. Kumar, "SELE: RSS-based siamese embedding location estimator for a dynamic IoT environment," *IEEE Internet of Things Journal*, vol. 9, no. 5, pp. 3672–3683, 2022.
2. G. Lui, T. Gallagher, B. Li, A. G. Dempster, and C. Rizos, "Differences in RSSI readings made by different Wi-Fi chipsets: A limitation of WLAN localization," in *2011 International Conference on Localization and GNSS (ICL-GNSS)*, IEEE, Tampere, Finland, 2011, pp. 53–57.

3. L. Li, X. Guo, M. Zhao, H. Li, and N. Ansari, "TransLoc: A heterogeneous knowledge transfer framework for fingerprint-based indoor localization," *IEEE Transactions on Wireless* Communications, vol. 20, no. 6, pp. 3628–3642, 2021.
4. A. Pandey, R. Vamsi, and S. Kumar, "Handling device heterogeneity and orientation using multistage regression for GMM based localization in IoT networks," *IEEE Access*, vol. 7, pp. 144354–144365, 2019.
5. M. Zeeshan, A. Pandey, and S. Kumar, "CSI-based device-free joint activity recognition and localization using Siamese networks," in *2022 14th International Conference on COMmunication Systems & NETworkS (COMSNETS)*, IEEE, Bangalore, India, 2022, pp. 260–264.
6. G. M. Mendoza-Silva, P. Richter, J. Torres-Sospedra, E. S. Lohan, and J. Huerta, "Long-term Wi-Fi fingerprinting dataset for research on robust indoor positioning," *Data*, vol. 3, no. 1, p. 3, 2018.
7. O. Kramer, *K-Nearest Neighbors*. Berlin, Heidelberg: Springer, 2013, pp. 13–23. [Online]. Available: https://doi.org/10.1007/978-3-642-38652-72
8. Z. Guowei, X. Zhan, and L. Dan, "Research and improvement on indoor localization based on RSSI fingerprint database and K-nearest neighbor points," in *2013 International Conference on Communications, Circuits and Systems (ICCCAS)*, vol. 2, IEEE, Chengdu, China, 2013, pp. 68–71.
9. H. Yigit, "A weighting approach for KNN classifier," in *2013 International Conference on Electronics, Computer and Computation (ICECCO)*, IEEE, Ankara, Turkey, 2013, pp. 228–231.
10. F. Pedregosa, G. Varoquaux, A. Gramfort, V. Michel, B. Thirion, O. Grisel, M. Blondel, P. Prettenhofer, R. Weiss, V. Dubourg et al., "Scikit-learn: Machine learning in Python," *The Journal of Machine Learning Research*, vol. 12, pp. 2825–2830, 2011.
11. A. H. Salamah, M. Tamazin, M. A. Sharkas, and M. Khedr, "An enhanced WiFi indoor localization system based on machine learning," in *2016 International Conference on Indoor Positioning and Indoor Navigation (IPIN)*, IEEE, Alcala de Henares, Spain, 2016, pp. 1–8.
12. H. Alla, L. Moumoun, and Y. Balouki, "A multilayer perceptron neural network with selective-data training for flight arrival delay prediction," *Scientific Programming*, vol. 2021, 2021.
13. M. Abadi, A. Agarwal, P. Barham, E. Brevdo, Z. Chen, C. Citro, G. S. Corrado, A. Davis, J. Dean, M. Devin et al., "TensorFlow: Large-scale machine learning on heterogeneous distributed systems," arXiv preprint arXiv:1603.04467, 2016.
14. G. Koch, R. Zemel, R. Salakhutdinov et al., "Siamese neural networks for one-shot image recognition," in *ICML Deep Learning Workshop*, Lille, France, vol. 2, Lille, 2015.
15. Q. Li, X. Liao, M. Liu, and S. Valaee, "Indoor localization based on CSI fingerprint by Siamese convolution neural network," *IEEE Transactions on Vehicular Technology*, vol. 70, no. 11, pp. 12168–12173, 2021.
16. B. Wang and D. Wang, "Plant leaves classification: A few-shot learning method based on Siamese network," *IEEE Access*, vol. 7, pp. 151754–151763, 2019.
17. S. Jindal, G. Gupta, M. Yadav, M. Sharma, and L. Vig, "Siamese networks for chromosome classification," in *Proceedings of the IEEE International Conference on Computer Vision Workshops*, Venice, Italy, 2017, pp. 72–81.
18. F. Sung, Y. Yang, L. Zhang, T. Xiang, P. H. Torr, and T. M. Hospedales, "Learning to compare: Relation network for few-shot learning," *in Proceedings of the IEEE Conference on Computer Vision and Pattern Recognition*, Salt Lake City, UT, USA, 2018, pp. 1199–1208.

19. O. Vinyals, C. Blundell, T. Lillicrap, D. Wierstra et al., "Matching networks for one shot learning," in *Advances in Neural Information Processing Systems*, Barcelona, Spain, vol. 29, 2016.
20. Y.-D. Zhang, X.-X. Hou, Y. Chen, H. Chen, M. Yang, J. Yang, and S.-H. Wang, "Voxelwise detection of cerebral microbleed in CADASIL patients by leaky rectified linear unit and early stopping," *Multimedia Tools and Applications*, vol. 77, no. 17, pp. 21825–21845, 2018.
21. A. C. Marreiros, J. Daunizeau, S. J. Kiebel, and K. J. Friston, "Population dynamics: Variance and the sigmoid activation function," *NeuroImage*, vol. 42, no. 1, pp. 147–157, 2008. [Online]. Available: https://www.sciencedirect.com/science/article/pii/S1053811908005132.
22. R. Hadsell, S. Chopra, and Y. LeCun, "Dimensionality reduction by learning an invariant mapping," in *2006 IEEE Computer Society Conference on Computer Vision and Pattern Recognition (CVPR'06)*, vol. 2, IEEE, New York, NY, USA, 2006, pp. 1735–1742.

5 Data-Driven Risk Modelling of Cyber-Physical Systems

Shakeel Ahamad, Ratneshwer Gupta, and Mohammad Sajid

CONTENTS

5.1 INTRODUCTION

A computer system is extensively used in almost all business domains. The use of computer systems extends from the primary utility task to the advanced safety-critical Functionality. The domains where the system performs a safety-critical task are the nuclear power plant, Air control system, Surgical robotic devices, and mission-critical systems and so on. The failure of these systems leads to a colossal loss to the environment, business, and human beings, which is unacceptable. Most of the domain using the safety-critical system (SCS) has now moved to the cloud-based system. These systems are vast and complex in dealing with the safety and risk analysis of these SCSs manually or program based. These systems require automatic safety and risk analysis. Computer systems work in various critical environments, e.g. nuclear power plants, surgical machines, financial software, missile, etc. The failure and insufficient quality of the system lead to unacceptable consequences [1]. Critical systems should have high safety and low risk [2–5]. So, the study of risk is significant in obtaining an optimised system's quality [6]. Artificial intelligence (AI) is the leading and innovative technology of the current age. AI technology can imitate human behaviours and introduce automatic and intelligent systems. AI algorithms can help in the automatic, innovative method to analyse,

DOI: 10.1201/9781003438588-5

predict, and estimate the risk of SCSs according to today's needs [5]. However, creating an effective AI model is difficult because real-world problems and data are dynamic and variable [7]. In software engineering, a significant limitation of the system is enhancing the system's quality using less time as the size and complexity of the software system are rapidly growing. So, the software developer is looking forward to a better method to predict the risk of cyber-physical systems (CPS) at a low cost. Intelligent systems are a solution for predicting the risk of CPS [8]. The theoretical and practical features of prediction methods using AI and machine learning in creating CPS are explored in the newly published research work Prediction for CPS [9].

CPS, such as airplanes and medical equipment, get more complex and sophisticated, and designing them is becoming more difficult. It is preferable to employ automation to supplement engineers' efforts rather than completely replace them because many of a system's qualities are not readily quantifiable [10]. CPS require high availability to reduce the risk. To promote the deterministic execution of CPS, this work explores the idea of upgrading the current CPS paradigm by adding a risk management layer. The author has proposed an approach for modelling the risk of CPS using reinforcement learning (RL) to improve trust in the system based on data. Further, the potential source of the risk is also identified, and the author proposed the best way to reduce them.

The rest of the chapter is summarised as follows: in Section 5.2 procedure for implementing the reinforcement technique for the risk modelling is given. Section 5.3 illustrates the procedure using the case study of a nuclear medicine hospital for guiding the patient to a safe state. The significance of the approach and challenges to implementing the approach are given in Sections 5.4 and 5.5, respectively. Finally, the summary of chapter is given in Section 5.6.

5.2 PROCEDURE FOR RISK MODELLING

The methodology for risk modelling consists of various steps. The significant steps of the analysis are given in Figure 5.1.

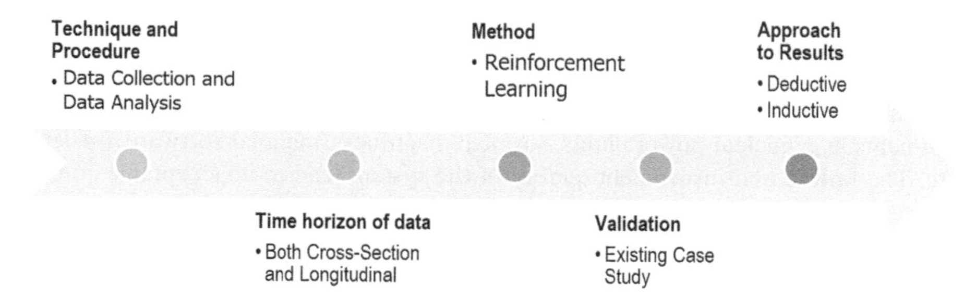

FIGURE 5.1 Risk analysis of the cyber-physical systems using reinforcement learning.

5.2.1 Reinforcement Learning Technique

The results of this study intend to present novel approaches for designing and deploying verified control software for SCSs, such as autonomous vehicles, robotic surgical equipment, air traffic control, shipping and warehousing, rail networks, and robotic surgery [11]. The authors propose using correct-by-construction control software algorithmically for safety-critical applications using techniques with mathematical underpinnings (formal control methods).

The author has already used the Markov model with reward (stochastic model) in previous research to predict safety and risk in a combined manner [12]. The stochastic model is extensively used to predict the dependability and performability of the system. There is a relationship between the stochastic models and RL. The relationship is shown in Figure 5.2.

RL is one of the learning techniques (i.e. Q-learning) where an agent interacts with a dynamic environment and learns through repetitive trial and error. An agent can make important decisions based on the accumulated reward using the reinforcing learning approach without being explicitly trained to complete the goal [13]. The reward can be anything, e.g. cost or loss. The accumulated reward process is defined based on quality parameters [14] (Figure 5.3).

Results: The outcome of this study would be a RL algorithm for safety and risk analysis of SCSs.

Validation: The result would be validated based on the available safety and risk modelling approach. Depending on the methods, statistical analysis or theoretical perspectives can be used to validate our explanation of the observation.

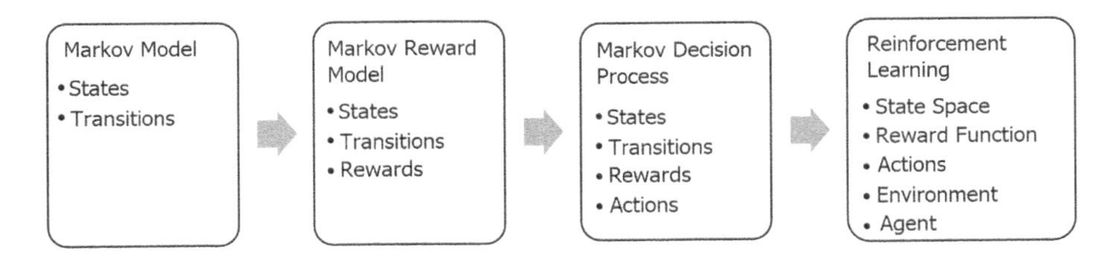

FIGURE 5.2 Relationship between the stochastic models and reinforcement learning.

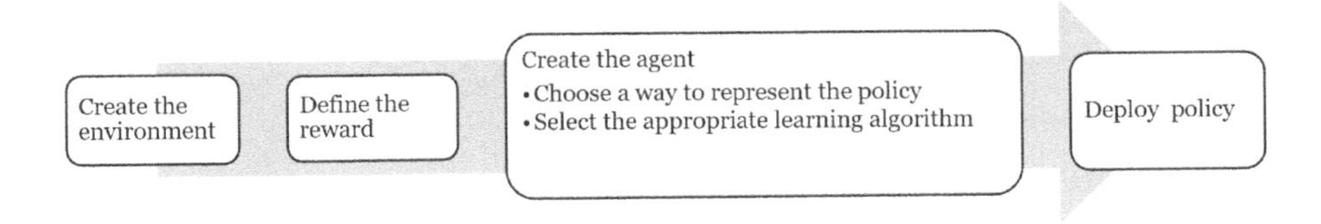

FIGURE 5.3 Steps for training an agent using reinforcement learning.

5.3 CASE STUDY

Modern medical devices use software systems, which are used to perform critical tasks and are subject to risk. So, implementing strict risk-reduction measures to reduce the likelihood of dangerous incidents is extremely necessary [15]. So, for implementing our approach (RL-based approach), Medical software systems are taken for risk management. When patients encounter potentially hazardous or harmful environments, the aim of the learning agent is to keep them out of hazardous states. The agent must concurrently be able to leave or get to a safe location quickly. Risk in medical devices depends on various factors like design, vendor management, and how the system is controlled by the user.

RL aims to teach a software agent to gather up the behaviour through frequent interactions with its environment. The RL agent gradually begins to operate at its best in a given environment as it continuously improves its learning based on experience [16].

Let's make the analogy RL agent and a novice patient. Patients may not be fully aware of their surroundings and may find themselves in a perilous situation when interacting with a medical environment for a medical checkup. They enter the world without prior knowledge, much like an RL agent. Consequently, we must train the agent to respond safely and effectively [17]. To implement RL, possible states of the systems should be determined. Some of them are unacceptable or undesirable. So, this state should be avoided. This objective can be accomplished using RL agents, guiding a patient to achieve a safe condition and avoid dangerous states.

An explanation of a neural network-based medical decision support system (MDSS) designed for patient safety from the risk of radioactive radiation [18]. The MDSS consists of a database, a computer-aided predictor, and a computer-aided diagnosis.

For a case study, an automatised system is considered in the nuclear medicine (NM) department to navigate the patient to the diagnostic room to reduce the risk of the radioactive material, as shown in Figure 5.4.

The states in the environment represent all potential sites within the hospital for diagnosis. Figure 5.4 (dark black square) depicts certain dangerous spots in the

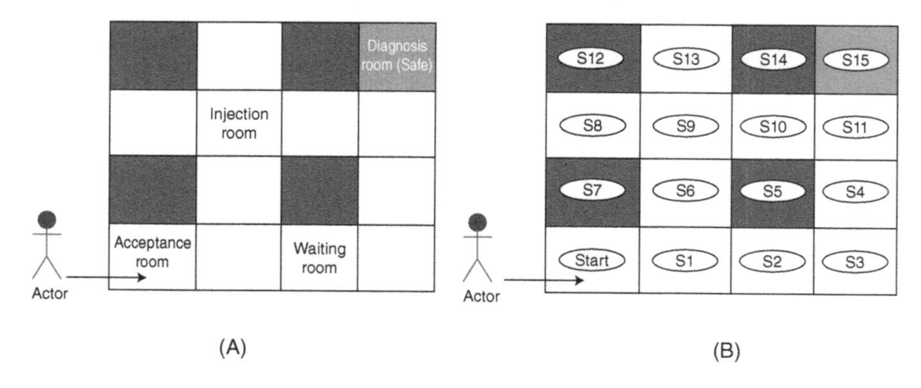

FIGURE 5.4 Abstract risk scenario of the nuclear medicine department. (a) Position of the different rooms and (b) states of the environment.

hospital, but at other places, the patient can roam around the building using the aisles (white squares). The light black square shows the area that the patient must reach with the least amount of travel time and without placing oneself at risk. The squares in light black represent terminal states.

The authors assume that an ideal NM department would consist of five places: Patients are welcomed into the department in the Acceptance Room (AR). At the same time, they wait in the Waiting Room (WR) for the radiopharmaceutical injection and are administered the appropriate injection in the Injection Room (IR). Tests are conducted in the Diagnostic Room (DR), and patients wait for the examination in the Hot Waiting Room (HWR) after receiving injections until the radiation level is within the acceptable range. A short-range radiofrequency identifier (RFID) is installed in the final four rooms so that the system can follow patients. For treatment, the patient follows the correct path (i.e. AR → WR → IR → HWR → DR). To receive notifications, the patient has a personal digital assistant (PDA). The biggest concern is that the patient can incorrectly perceive the proper procedure and enter the wrong room, forcing the physicians to act fast and exposing them to radioactive substances.

Figure 5.4 illustrates potential risky states. These scenarios are characterised as hazardous states. All those instances show potential risks that need to be eliminated. An intelligent agent has been trained to inform the patient through PDA to guide the correct path and avoid mistakes. The tracking system uses RFID. The Risk Management System is an RL agent trained to help patients in risky situations. The Mapping Activity component transforms the agent's recommendations into actions.

Figure 5.4 shows one of the possible system configurations. So, the agent does not know anything in advance. So, the agent should train to reach the DR safely. Q-learning is at the heart of all RL. The Q-learning algorithm is one of the RL approaches to achieve the goal. Q-learning does not require prior knowledge of the environment in which it works. RL helps in learning the agent to select the appropriate action in a particular situation. In Q-learning, the complete environment is divided into a small portion called grid. Q-learning trains the agent to reach the safe state with minimal time. The Q-learning algorithm has a set of actions, a set of possible states, and reward functions. There is one state, as given in Figure 5.4. The grid world is defined based on possible states.

Reward is the function of the state and actions for the state. We need to define rewards as the final element of the environment. Every state (position) in the hospital has a different reward value, which aids the AI agent's learning. Regardless of where the agent starts, it will always aim to maximise its overall rewards. All states besides the goal are rewarded negatively (i.e. punished). This encourages the AI to choose the quickest route to the objective by reducing its penalties!

$$\text{Reward function } R(S,A) = \begin{cases} 1, & \text{if patient reaches safe state} \\ 0, & \text{Aisles position} \\ -1, & \text{Risky state} \end{cases}$$

Depending on the choices made by the AI agent, the robot can be pushed into one of four directions: up, right, down, or left. Of course, the AI agent needs to learn to steer clear of dangerous areas when driving.

To maximise cumulative rewards and minimise cumulative punishments, the AI agent must find the shortest paths between the safe region (light black) and all other places (white) of the NM facility where the robot is allowed to go. The agent must also learn to avoid dangerous regions (dark black). The next objective is to develop a Q-learning model to train the agent about its surroundings.

The steps of the learning process are as follows:

1. To start this new episode, pick a random, non-terminal state (white square) for the agent.
2. Choose one of the possible actions from the action set based on the greedy epsilon algorithm.
3. Carry out the selected action and move to the next location.
4. Calculate the temporal difference and update the new state's accumulated reward (Q-value).
5. Proceed to #1 if the new (current) state is terminal; otherwise, go to #2.

Initialize $Q(s, a)$ arbitrarily
Repeat (for each episode)
Initialize s
Repeat (for each step of episode):
Choose a for s using policy derived from Q
Take action a, observe r, s'
Update

$$Q(s,a) \leftarrow Q(s,a) + \alpha\left[r + \gamma \max_{a'} Q(s',a') - Q(s,a)\right]$$

$s \leftarrow s'$;
Until s is terminal

The initial Q-values for each state are zero. For 1,000 episodes, the entire procedure will be repeated. This will give the AI agent enough time to figure out the quickest routes between the safe state and every other place in the warehouse where it is permitted to move, all while avoiding running into any risky places. The table displays the Q-table for our example. It will have 16 rows representing all potential states for this environment and four columns indicating various actions (U for up, D for down, L for left, and R for right). At the beginning of training, the Q-table is initialised with zeros.

To get the Q-values for the table scenario of Q-learning, use the equation below. By making the following remark, it somehow expresses the learning process.

$$\text{Updated} \quad Q(s,\, a) = Q(s,\, a) + \propto \left[R(s,\, a) + \gamma \max Q^{*}\left(s^{*}, a^{*}\right) - Q(s,\, a)\right]$$

Here,

$Q(s,a)$ is the current Q-value

γ discount factor

$\propto$ learning rate

The learning rate ($\propto$) can range from 0 to 1, with a value of 1 denoting that the agent strives to learn everything (exploring without taking into account prior knowledge) and a value of 0 denoting that the agent learns nothing (exploitation of prior knowledge). The discount factor (γ) value varies from 0 to 1, with 0 representing solely consideration of short-term advantages and 1 representing only consideration of long-term benefits. The authors will employ the epsilon-greedy approach, in which our agent will decide randomly whether to go exploring or take advantage of the environment by flipping a coin. The epsilon value can range from 0 to 1, with 0 denoting that an agent will engage in 100% exploitation and one denoting complete exploration.

Trained Q-values:

```
[[2.1468149e-01 5.0288569e-03 5.7640146e-03 5.0499504e-03]
[2.0392979e-03 1.9301618e-03 5.6613097e-04 1.9842336e-01]
[1.4280579e-03 1.5164332e-03 1.6588701e-03 2.2259131e-01]
[7.5076125e-04 3.4668643e-05 1.3483281e-04 8.6149670e-02]
[4.9718606e-01 1.7597415e-03 2.2053448e-03 1.7449846e-03]
[0.0000000e+00 0.0000000e+00 0.0000000e+00 0.0000000e+00]
[1.9222178e-01 1.3820914e-05 7.4349045e-06 4.6949990e-06]
[0.0000000e+00 0.0000000e+00 0.0000000e+00 0.0000000e+00]
[2.0858478e-03 2.6829327e-03 2.6783263e-03 5.7745880e-01]
[2.8261850e-03 7.8306305e-01 7.1616378e-04 8.6327153e-04]
[5.7096976e-01 8.4089604e-04 3.4205476e-04 8.6625316e-04]
[0.0000000e+00 0.0000000e+00 0.0000000e+00 0.0000000e+00]
[0.0000000e+00 0.0000000e+00 0.0000000e+00 0.0000000e+00]
[8.7613255e-02 2.6037581e-02 9.1628128e-01 2.4912380e-02]
[1.3661268e-01 9.9033320e-01 9.7716846e-02 1.5331402e-01]
[0.0000000e+00 0.0000000e+00 0.0000000e+00 0.0000000e+00]]
```

5.4 SIGNIFICANCE OF THE APPROACH

The importance of the work that has been suggested can be summed up as follows:

Reduced Labour: The traditional tools and techniques for safety and risk analysis are time-consuming and require a lot of labour. If you have a repetitive task, you can use AI as a digital assistant to reduce your workload. Looking at

Optimum Decision-Making to Maximise Safety with Low Risk: AI systems trained via RL surpass human competitors in video games and board games like Go and Chess.

Ability to Analyse the Safety and Risk of the Whole System: SCS has various components connected to accomplish the task, e.g. an autonomous vehicle makes steering wheel decisions by examining data from numerous sensors, such as camera frames and lidar measurements [19]. Without neural networks, the issue would be divided into smaller components: a module that examines the camera input to find valuable features, another module that filters the lidar measurements, and perhaps one component that aims to combine the sensor outputs to create a complete picture of the environment around the vehicle, a 'driver' module, and so on [20].

Ease the Traditional Complex Problem: The key benefit of RL is that it can encode complicated behaviours, making it possible to apply it to problems that would be extremely difficult to solve with conventional algorithms.

Scheduling: Scheduling issues can arise in various situations, such as managing resources on the production floor and controlling traffic lights. To resolve these combinatorial optimisation issues, RL provides a fantastic alternative to evolutionary techniques [21].

5.5 ISSUES IN THE IMPLEMENTATION OF REINFORCEMENT LEARNING IN RISK MODELLING

The risk analysis of CPS has various challenges. The current safety standards do not account for the technical idiosyncrasies of AI systems. They are seldom ever appropriate to these systems. Developing secure CPS using AI techniques helps reduce the risk of hazardous situations. This is significant because risk management for AI-using systems needs to be adequately addressed to consider the new problems posed by the technology. But, there is a need to put forth assurance cases in support of AI applications' certification and quality control. A comprehensive list of criteria based on the source of risk for CPS can be examined and evaluated during risk analysis to determine the most effective risk mitigation strategies.

The critical issues and challenges needed to implement RL in the safety and risk analysis of the critical safety system demanding careful addressing, and to be carefully considered in this work, have been listed below:

- To achieve an acceptable safety and risk level, extensive training is necessary.
- Setting up the issue appropriately can be challenging.
- A trained deep neural network policy is a 'black box', which refers to a network with an internal structure that is so complicated (sometimes made up of millions of parameters) that it is complicated to comprehend, justify, and assess the choices made. Because of this, it is challenging to develop formal performance guarantees for neural network policies.

5.6 SUMMARY OF THE CHAPTER

When training data or sufficient domain knowledge are lacking, RL might be helpful. You have a general idea of what you want but don't know how to get there. This study presents a RL-based risk management approach for the healthcare industry. If a patient mistakes in following the correct procedure during a nuclear examination, a software agent is trained to assist the patient in achieving a safe state. The training phase has been carried out using the Q-learning algorithm. The trial outcomes demonstrate the agent's capacity to offer suggestions for helping the patient. Future studies will apply the suggested strategy to more complicated environments with various potentially hazardous scenarios and compare it to other RL methods.

REFERENCES

1. A. Pereira and C. Thomas, "Challenges of machine learning applied to safety-critical cyber-physical systems," *Machine Learning and Knowledge Extraction*, vol. 2, no. 4, pp. 579–602, 2020, doi: 10.3390/make2040031.
2. A. Vani, M. Naved, A. H. Fakih, A. N. Venkatesh, P. Vijayakumar, and P. R. Kshirsagar, "Supervise the data security and performance in cloud using artificial intelligence," *AIP Conference Proceedings*, vol. 2393, 2022, doi: 10.1063/5.0074225.
3. S. Dey and S. W. Lee, "Multilayered review of safety approaches for machine learning-based systems in the days of AI," *Journal of Systems and Software*, vol. 176, p. 110941, 2021, doi: 10.1016/j.jss.2021.110941.
4. S. Mohseni, M. Pitale, V. Singh, and Z. Wang, "Practical solutions for machine learning safety in autonomous vehicles," *CEUR Workshop Proceedings*, vol. 2560, pp. 162–169, 2020.
5. F. Tambon et al., "How to certify machine learning based safety-critical systems? A systematic literature review," *Automated Software Engineering*, vol. 29, no. 2, pp. 1–90, 2022, doi: 10.1007/s10515-022-00337-x.
6. G. MacHer, M. Seidl, M. Dzambic, and J. Dobaj, "Architectural patterns for integrating AI technology into safety-critical systems," *ACM International Conference Proceeding Series*, 2021, doi: 10.1145/3489449.3490014.
7. A. Steimers and M. Schneider, "Sources of risk of AI systems," *International Journal of Environmental Research and Public Health*, vol. 19, no. 6, 2022, doi: 10.3390/ijerph19063641.
8. A. Aksjonov and V. Kyrki, "A safety-critical decision making and control framework combining machine learning and rule-based algorithms," arXiv:2201012819v1, pp. 1–11, 2022, [Online]. Available: http://arxiv.org/abs/2201.12819.
9. J. Hegde and B. Rokseth, "Applications of machine learning methods for engineering risk assessment: A review," *Safety Science*, vol. 122, p. 104492, 2020, doi: 10.1016/j.ssci.2019.09.015.
10. A. Chehri, I. Fofana, and X. Yang, "Security risk modeling in smart grid critical infrastructures in the era of big data and artificial intelligence," *Sustainability (Switzerland)*, vol. 13, no. 6, p. 3196, 2021, doi: 10.3390/su13063196.
11. K. Sadam, M. Naved, S. Kavitha, A. Bora, K. Bhavana Raj, and B. R. Nadh Singh, "An internet of things for data security in cloud using artificial intelligence," *International Journal of Grid and Distributed Computing*, vol. 14, no. 1, pp. 1257–1275, 2021, [Online]. Available: https://www.researchgate.net/publication/352169764.
12. P. Radanliev, D. de Roure, M. van Kleek, O. Santos, and U. Ani, "Artificial intelligence in cyber physical systems," *Artificial Intelligence SoC*, vol. 36, no. 3, pp. 783–796, 2021, doi: 10.1007/s00146-020-01049-0.
13. H. Elahi, G. Wang, Y. Xu, A. Castiglione, Q. Yan, and M. N. Shehzad, "On the characterisation and risk assessment of AI-powered mobile cloud applications," *Comput Stand Interfaces*, vol. 78, 2021, doi: 10.1016/j.csi.2021.103538.

14. X. Yuan, P. He, Q. Zhu, and X. Li, "Adversarial examples: Attacks and defenses for deep learning," *IEEE Transactions on Neural Networks and Learning Systems*, vol. 30, no. 9, pp. 2805–2824, 2019, doi: 10.1109/TNNLS.2018.2886017.
15. J. A. Yaacoub, O. Salman, H. N. Noura, N. Kaaniche, A. Chenab, and M. Malli, "Cyber-physical systems security: Limitations, issues and future trends," *Microprocessors and Microsystems*, vol. 77, pp. 1–33, 2020, [Online]. https://doi.org/10.1016/j.micpro.2020.103201
16. M. Rabe, S. Milz, and P. Mader, "Development methodologies for safety critical machine learning applications in the automotive domain: A survey," *IEEE Computer Society Conference on Computer Vision and Pattern Recognition Workshops,* pp. 129–141, 2021, doi: 10.1109/CVPRW53098.2021.00023.
17. A. Sharma and U. K. Singh, "Modelling of smart risk assessment approach for cloud computing environment using AI & supervised machine learning algorithms," *Global Transitions Proceedings*, vol. 3, no. 1, pp. 243–250, 2022, doi: 10.1016/J.GLTP.2022.03.030.
18. A. Coronato, G. Paragliola, M. Naeem, and G. de Pietro, "A reinforcement learning-based approach for the risk management of e-health environments: A case study," *Proceedings - 14th International Conference on Signal Image Technology and Internet Based Systems, SITIS 2018*, pp. 711–716, 2018, doi: 10.1109/SITIS.2018.00114.
19. U. A. Butt et al., "A review of machine learning algorithms for cloud computing security," *Electronics (Switzerland)*, vol. 9, no. 9, pp. 1–25, 2020, doi: 10.3390/electronics9091379.
20. A. Chehri, I. Fofana, and X. Yang, "Security risk modeling in smart grid critical infrastructures in the era of big data and artificial intelligence," *Sustainability (Switzerland)*, vol. 13, p. 3196, 2021.
21. N. Shevchenko, B. R. Frye, and C. Woody, "Threat modeling for cyber-physical system-of-systems: Methods evaluation," Software Engineering Institute, Carnegie Mellon University, 2018.

6 Automation of the Process of Analysis of Information Security Threats in Cyber-Physical Systems

Elena Basan, Olga Peskova, Maria Lapina, and Mohammad Sajid

CONTENTS

6.1 INTRODUCTION

In recent years, technologies associated with cyber-physical systems (CPS) and the Internet of things (IoT) have been actively developing. There is a huge range of software and hardware for creating system control mechanisms, various operating systems, and a lot of application software, and solutions can be open or proprietary. And as a result, conducting a security analysis of such a system really becomes a serious problem. Analyzing the risks of such systems is a critical task. However,

DOI: 10.1201/9781003438588-6

there is currently no standard that describes this process. The problems of standardization of information on threats to CPS, as well as attempts to determine measures to minimize risks, is an urgent problem and is considered by both the Russian and the world community. In general, in the CPS, the operation of classifying and minimizing information security issues is associated with the following problems:

- Lack of current legislation in the field of security for CPS.
- Lack of up-to-date databases of threats to CPS.
- Lack of a standardized description of the structural and functional characteristics (SFChs) of CPS.
- The lack of mechanisms, tools, or software solutions to automate the process of determining information security risks and selecting requirements for protecting CPS, which leads to the need for expert assessment, which requires large labor costs and is subjective, does not exclude errors in identifying risks, and this, in turn, leads to inadequate choice of protective measures.
- Constant change in the range of products, which requires constant updating of the methodological support of the CPS.

The purpose of this study is to develop a technique for organizing the analysis of information threats of information security in CPS, which is constructed on the learning of the system construction, risk analysis of the realization of threats, together with the competences of the attacker.

The article has the following structure.

In the "Analysis of publications" section, an analysis of publications is carried out, in which an attempt is made to analyze known threats at various levels of the architecture of the IoT. In the "Analysis of the architecture of cyber-physical systems" section, the SFChs of CPS were carried out and their main problems from the viewpoint of information security were emphasized.

Further, in the "Development of a methodology for ensuring the safety of CPS" section, the authors determined the list of indicators influencing their security and identified the level of initial security, based on the presence of certain factors. New CPS threats related to the features described in the previous sections have been identified: agility, usage of vulnerable radio networks for communication, being external of the perimeter of controlled zones. The main result was the original author's method for ensuring the security of the CPS, the novelty of which lies in the development of a set of factors that determine the likely violator for the CPS. The studies carried out in the framework of risk assessment, identification of current threats, and development of effective recommendations for ensuring the security of the CPS have made it probable to develop a knowledge base of CPS vulnerability category, intrusions, threats, which is presented in the "Cyber-physical system threat database development" section.

6.2 RELATED WORKS

The authors analyzed publications that attempt to analyze known threats at various levels of the architecture of CPS (most often we are talking about IoT systems).

The authors of the article [1] are working on the collection and description of many threats at various levels of the IoT architecture, with an accent on the use of malware. The authors have developed and described a procedure for realizing attacks on the IoT, in addition to a situation for a DDoS attack through a botnet, on the basis of which actions to enhance information security are proposed.

In the paper [2], a special document has been developed on the expansion of methodology for IoT security proved on the best applies that are used in global industries. The document contains sections describing approaches to risk assessment, security software that should be applied in order to protect the privacy, integrity, and accessibility of information. In addition, the method for calculating the impact of identified risks is described in detail. However, the information provided can only be used by an expert to independently assess risks, threats, and the choice of protection measures, and this work will require a lot of his time and attention. The lack of automation of the process in this case is a serious drawback. In addition, when assessing risks, objects that are directly "managed" by the IoT are not considered [3].

In Ref. [4], three levels of the IoT structure are distinguished: perceptive, network, application layers. The main risks and challenges associated with IoT information security are listed here, but they are presented in a disparate set that does not make it possible to formally assess these threats.

Article [5] describes a variant of the IoT architecture, divided into four levels: sense level, communication level, middleware level, and application level. The most relevant threats and attacks are highlighted for each level. In particular, threats associated with equipment (sensors, actuators, and other components) are presented at the level of perception. The network layer is characterized by various attacks, most often associated with data analysis (routing attacks, DoS attacks, DDoS attacks, phishing, and so on). The middleware layer provides the link between the network and application layers and includes a lot of physical mechanisms for instance data storage, queuing systems, intelligent systems, and so on. This layer is vital for an IoT application to be secure from various points of view, but at the same time it is itself subject to many different attacks (SQL injection, signature attacks, MitM attacks, and others). The main goal of the application layer is to guarantee the confidentiality of user data. The article discusses various hardware and software components of the IoT, formulates the requirements for security systems in the Internet of things, but does not spell out a detailed methodology for analyzing and assessing risks, which reduces the applicability of the work.

Article [6] analyzes security schemes based on known IoT threats and attacks. These attacks can be divided into two main categories: system architecture attacks and data privacy attacks. The first type of attack relies on the IoT environment, in particular its behavior and routing. These include denial of service attacks, jamming,

TCP overflow, etc. Attacks on data privacy can be launched directly during the collection of data, and some of them are related to obtaining unauthorized access to user data (e.g., cryptographic attacks). This representation seems not very successful and leads to the difficulty of assigning specific threats and attacks to certain classes due to the implicit boundaries of these classes.

In Ref. [7], despite its title, the classification of threats is given much less space than the consideration of existing malware, as well as various machine learning algorithms for detecting attacks. However, the original threat structure is presented, based on five different levels: Application, Communication, Device, Network, and Transport. In turn, a more complex scheme is presented directly for the classification of threats: Security and privacy threats are distinguished at the top level. Security threats, in turn, are divided into two classes: Availability (DOS, Physical) and Integrity (MiM, Malware). For privacy threats, only confidentiality threats are considered in detail. The IoT security application area scheme is also proposed.

Authors [8] list security procedures only in three categories based on three layers of IoT architecture:

- sensor layer (physical attacks, wireless attacks—IP-Sec Security Channel),
- communication layer (passive monitoring, traffic analysis, eavesdropping—Routing security),
- application layer (weak standard/global trust policies—Firewalls, Anti-virus, Anti-spyware).

The devices which were made long years ago are not compatible with today's security policies and terms and conditions. In this paper, authors are going to highlight some of the issues which may arise due to this.

From the point of view of methodological support for the analysis of various components and structures of IoT, the work [9], which considers IoT ontologies, can be useful. Forty-five IoT ontologies have been described, although not widely used. A thorough analysis of these ontologies allows us to conclude that they can be used both in the development of a new IoT ontology and in achieving the three main requirements for security and reliability: confidentiality, integrity, and interoperability. When analyzing ontologies, issues related to security requirements are considered in detail, but only threats to validity are considered in detail as threats. In this article, threats are divided into four types: external, internal, constructive, and inference validity.

The highlighted disadvantages complicate the use of the considered solutions. It is also worth considering regulatory documents to advance the security of cyber-physical and similar systems. The National Institute of Standards and Technology (NIST) has developed a special standard for enhancing the cybersecurity of critical information systems "Framework for Improving Critical Infrastructure Cybersecurity" [10], which consists of three sections: basic core, embodiment degrees, and standardized cross-section. The kernel is a set of cybersecurity activities and guidelines that are used in critical infrastructure. It contains guidance on developing individual profiles.

The implementation levels describe how to analyze the individualities of a management policy to risk administration, which is important for prioritizing the achievement of cybersecurity goals [11–14].

Thus, these documents deal with the issues of structuring and categorization of various information systems and propose mechanisms for increasing the level of their security. At the same time, they poorly consider the peculiarities of technological processes, which makes it difficult to use them for CPS.

6.3 EXAMINATION OF THE ARCHITECTURE OF CYBER-PHYSICAL SYSTEMS

If we analyze the architecture of CPS, we can say that this system is multi-component and can include thousands of different sensors that exchange information with each other, accumulate it, and transmit it to the control device [15].

Figure 6.1 shows a diagram of the CPS SFChs, which reveals new aspects in the architecture of the CPS and allows you to identify more accurately those or other threats that can be implemented on the CPS, and in this study, the component of intelligent control is highlighted [16,17].

Each of the presented system elements has its own vulnerabilities that an attacker can exploit. Intelligent technologies, which can be exposed to new types of vulnerabilities, are also an important component. For example, today the problem of analyzing the validity and authenticity of data for training neural networks is becoming urgent. This is because generating new data is a time-consuming process and often the data is taken from open sources. In this case, an attacker can substitute training data, as a result of which the network will be trained incorrectly, and, as a result, will make erroneous decisions.

CPS is a new area of research, since, in contrast to the usual computer and information systems, CPS has a number of features [18]:

1. The CPS has a mixed structure, consisting not only of information aspects that control technological processes, but also physical components.
2. The CPS has a communication environment in which there is a continuous exchange of large amounts of data inside and outside the system.
3. The CPS components can be mobile and located outside the controlled area.
4. The CPS depends on information, on external information influences directed at the control subsystem, on the interface of interaction between the operator and the system, on devices that are part of the CPS, on interaction protocols and network equipment.
5. Such systems can be controlled both by changing the parameters and by changing the structure of the CPS with the redistribution of functions between the components.
6. The CPS is often heterogeneous system that combine many different technologies.

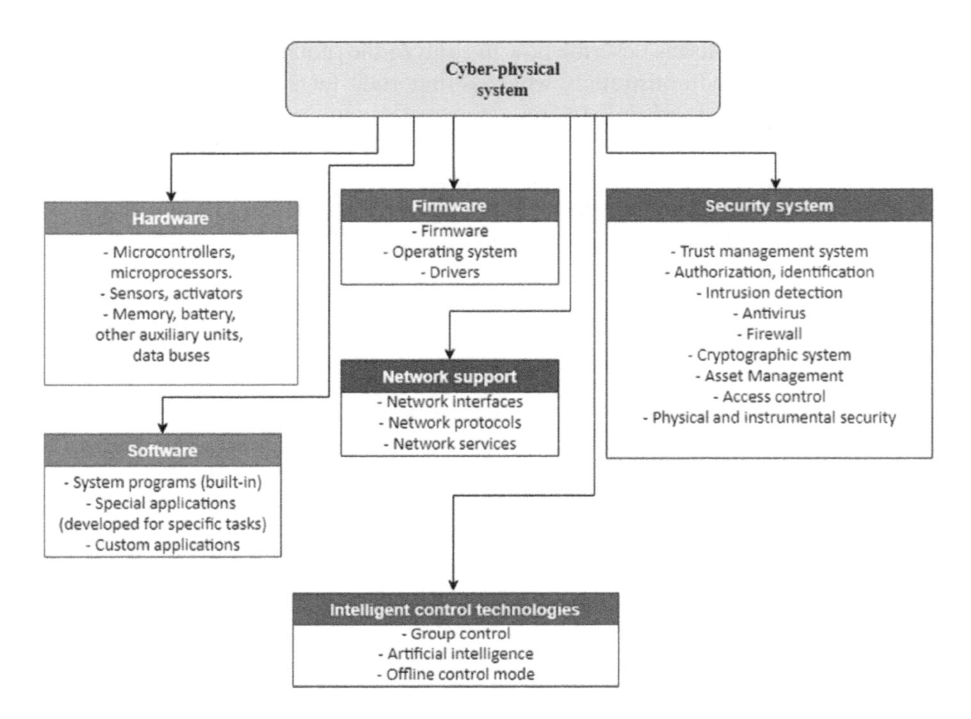

FIGURE 6.1 Components of cyber-physical systems.

6.4 DEVELOPMENT OF A METHODOLOGY FOR ENSURING THE SAFETY OF CPS

6.4.1 Assessment of the Level of Criticality

The authors have developed their own methodology for ensuring the security of the CPS, which includes several levels of calculations to ultimately determine the relevance of a particular threat.

The idea is that the owner of the CPS transmits information about his SFCh, and as a result receives a list of current threats and recommendations for the CPS. This methodology can serve as a basis for ensuring cybersecurity in an enterprise from scratch or as a mechanism to improve an existing security system.

The methodology establishes the level of criticality of the various components of CPS. The level of criticality is determined by the potential damage to the subsystem in relation to which actions are taken that endanger the security of information and modules of the CPS necessary for the system to achieve its objectives. The security categorization should also consider the vulnerability and threat information that is consistent with the CPS.

In conventional information systems, the security attributes are confidentiality, integrity, and availability of information. All negative impacts on the security system are mainly associated with the violation of one or more of these attributes. In this paper, the security attributes will be the SFChs shown in Figure 6.1.

Each attribute has subsystems, which will be estimated. The concept of a subsystem is understood as SFCh included in one or another attribute, where there will be three levels of criticality (high, low, and medium). They will depend on the consequences of an attack on a particular subsystem. The level of criticality is understood as what consequences the system will suffer when a threat is realized on one or another attribute of the CPS.

The subsystem is rated low to high.

Each level has several points: three points for a low level, five points for an average, and seven points for a high level. As an example, consider the assessment of criticality for CPS subsystems, as shown in Table 6.1.

TABLE 6.1
Distribution of Criticality Levels

Structural and Functional Characteristics	Low Level of Criticality (3)	Medium Level of Criticality (5)	High Level of Criticality (7)
	Hardware:		
• Microcontrollers, microprocessors		+	+
• Sensors, activators		+	
• Memory, battery, other auxiliary units, data buses			
	Firmware:		
• Firmware	+	+	
• Operating system		+	
• Drivers			
	Software:		
• System programs (built-in)	+	+	
• Special applications (developed for specific tasks)		+	
• Custom applications			
	Network Support:		
• Network interfaces		+	
• Network protocols		+	
• Network services		+	
	Intelligent Control Technologies:		
• Group control	+	+	
• Artificial intelligence		+	
• Offline control mode			
	Security System:		
• Trust management system	+	+	+
• Authorization, identification	+	+	
• Intrusion detection, prevention		+	
• Anti-virus		+	
• Firewall		+	
• Cryptographic system		+	
• Asset management			
• Access control			
• Physical and instrumental security			
Result	15	85	14

The criticality levels were assigned based on the importance of systems and subsystems, as well as the number of threats that are possible on a particular subsystem and were calculated using formula (6.1)

$$C = TI, \tag{6.1}$$

where *C*—level of criticality, *T* is an indicator of the weight of threats implemented in the SFCh, *I*—is the value of the subsystem.

T is calculated by formula (6.2):

$$T = \frac{\Sigma W}{Y}, \tag{6.2}$$

where *W* is the sum of the weights of threats implemented on the farm, and *Y* is the number of threats implemented in the SFCh.

The importance of the subsystem is assessed from low to high, and each level has a number of points: three points for a low level, five points for an average, and seven points for a high level.

The number of threats is considered from the threat database.

If the level of criticality exceeds 28 points, then this SFCh is assigned a high level of criticality; if it exceeds 15, then an average level of criticality is assigned; and if it does not exceed 9 points, then a low level of criticality is assigned to it.

6.4.2 Analysis of the Level of Heterogeneity of the System

When determining the list of threats to information security at the stage of producing a CPS, it is necessary to assess the possibility of imbedding a threat, especially if measures to protect information are not executed and/or their adequacy and usefulness have not been carried out. This characteristic can be determined by calculating the probability of occurrence of a specific threat to information security in the CPS with given SFCh and operating features relation to the level of enterprise safety of the CPS [19].

The degree of enterprise safety is determined by the original safety of the CPS, because of the SFCh specified in the project and the environments for its processes:

- **High:** The "High" CPS enterprise protection level will match to a variety below 50% of the calculated factors upsetting the SFCh of the UAV and the conditions for its functions.
- **Medium:** 70% of the factors affecting the SFCh of the CPS and the environments of its procedure will correspond to the level of design protection of the UAV "Medium."
- **Low:** The "Low" CPS enterprise protection level will match to the variety from 80% to 100% in terms of the factors upsetting the SFCh of the UAV and its maintenance conditions.

The results of the analysis of the CPS SFCh, its operating conditions, together with the influence of diverse peculiarities on each of the considered CPS safety levels, are shown in Table 6.2. The level of initial security of the information system is calculated

TABLE 6.2
CPS Design Safety Indicators

SFCh Operating Conditions	CPS Design Security Level		
	High (0, 3)	Average (0, 5)	Low (0, 7)
Structure of System:			
• Local data architecture,		+	+
• Distributed data architecture.			
Information System Construction—Based on:			
• Peer-to-peer network,		+	+
• Hierarchical network.			
Presence/Absence of Interconnections with Other Systems:			
• Non-interrelating,		+	+
• Interrelating.			
Presence/Absence of Interconnections to Public Communication Infrastructure:			
• Not associated,	+	+	+
• Associated via a special-purpose infrastructure,			
• Associated.			
Placement of Technical Means—Located:			
• Within unchanged proved sector,	+	+	+
• Within several proved sectors,			
• Outside the proved sector.			
Modes of Processing of Information:			
• Individual player,	+		+
• Multi-player.			
Areas of Application:			
• Industrial,		+	+
• Medical,		+	+
• Household,			+
• Social,			
• Research.			
Environment of Operation:			
• Space,	+	+	+
• Air,		+	+
• Marine,			
• Ground,			
• Underground.			
Mobility Degree:			
• Stationary,		+	+
• Mobile.			
Control Methods:			
• Autonomous control,		+	+
• Group management,		+	+
• Operator control,			
• Semi-automatic control.			

(Continued)

TABLE 6.2 (*Continued*)
CPS Design Safety Indicators

SFCh Operating Conditions	CPS Design Security Level		
	High (0, 3)	**Average (0, 5)**	**Low (0, 7)**
Operating Conditions:			
• Algoristic (defined) environments,		+	+
• Non-algoristic (undefined).			
Type of Communication:			
• City-wide wireless networks,		+	+
• Wireless wide area networks,		+	+
• Wireless personal networks,			
• Wireless local area networks.			
Type of Used Navigation:			
• Global,	+	+	+
• Local,			
• Personal.			
Type of Network Topology			
• Ad hoc network,		+	+
• Mesh network,		+	
• Star.			
Results	5	9	15

from the degree of heterogeneity and the number of points scored according to the enterprise protection level of the CPS, and this is calculated by formula (6.3):

$$P = S + G, \tag{6.3}$$

where P—is the level of the primary system security; G—the number of points scored on the enterprise protection of the information system; S—the degree of network heterogeneity, calculated by formula (6.4):

$$S = \frac{QR}{100}, \tag{6.4}$$

where Q—is constant;

R—depends on the level of heterogeneity: at low 50, on average 70, at high 90.

If S is <21.5, then the level of heterogeneity of the system is low. If S is more than 21.5, then the level of heterogeneity is average. If S is more than 38.7, then the level of heterogeneity is high.

6.4.3 The Complexity of the Attack

To correctly determine the degree of complexity of the attack process, it is necessary to assess the goals and capabilities of the attacker.

Attack tools—those software or firmware tools that an attacker will use to implement a particular attack (*A*):

- The attack is carried out by non-specialized software tools that are available on the Internet and do not require additional development and application of qualification skills (0.2);
- The attack is carried out by non-specialized software tools that are available on the Internet and require additional development and application of qualification skills (0.4);
- The attack is carried out using software and hardware available for purchase, which does not require additional development and application of qualification skills (0.3);
- The attack is carried out using software and hardware available for purchase, which require additional development and application of qualification skills (0.7);
- Complex impact on the object of attack using methods of social engineering, psychology, etc. (0.5);
- The attack is carried out using specially developed software and hardware (0.9).

The attack method is a scenario that defines the attack vector and the entry point to the system (*W*):

- The attack is carried out using remote access (0.9);
- The attack is carried out through direct physical access (0.7);
- The attack is carried out by directly affecting SFCh (0.8);
- The attack is carried out by indirectly affecting SFCh (impact on environmental parameters, etc.) (0.8);
- The attack is implemented by affecting one SFCh through another SFCh (multi-stage attack) (0.8);
- The attack is implemented by influencing the control personnel (0.6).

The complexity of the attack is calculated by formula (6.5):

$$H = WA, \tag{6.5}$$

where *H*—is the complexity of the attack,
W—the way to implement the attack,
A—tools for carrying out an attack.

6.4.4 Assessment of the Degree of Negative Consequences of the Implementation of the Threat

After that, we should determine what consequences can result from each specific attack. These types of consequences can be defined as one of the manifestations of damage. Let's define the following types of damage: economic, social, political, reputational, environmental, man-made, in the sphere of protection, state safety.

The degree of their influence on the significances of the effect from the execution of the attack will also be determined.

After analyzing cyber-attacks, it was revealed that the greatest damage is caused by the economic component. Each type of damage can be attributed to one of three degrees: low, medium, high.

6.4.5 Determining the Value of an Information Resource

Damage can be assessed based on the value of the assets possessed by the system owner [13,14].

The presented methodology highlights the following assets: information, software-and-hardware tools, software tools, information security tools, the final manufactured product, or object controlled by the CPS.

The authors propose to introduce the following objects in addition to the basic set of resources: an object managed by the CPS, or a product that is the result of labor. Adding this object is important because an attacker can change the environment settings and change the CPS reading, which ultimately can lead to incorrect actions by the operator or artificial intelligence. Conversely, the results of an attack can affect the CPS that analyzes the environment; for example, an attacker can tamper with data or block the network.

Three levels of value are also defined for each asset: high, average, short.

This characteristic is evaluated directly by the system owner. To determine what damage this or that threat will inflict on one of the above SFCh, we apply formula (6.6):

$$M = \frac{H + C + D + L}{P}, \tag{6.6}$$

where M—is the danger degree of the threat realization, H—the complexity of the attack, P—the initial level of the system security, C—level of criticality, D—type of damage, L—asset value level.

Three indicators are proposed to evaluate the danger degree of the threat realization (low, medium, and high). This assessment is characterized by the scale of the losses that may arise during the implementation of the threat. This indicator is complex and considers all previous calculations.

The danger degree of the implementation of the threat:

- Low: If the parameter M scores less than seven points, then it is assigned a low level of danger, but if the implemented threats to the UAV exceed the value of 0.7 and the type of damage exceeds 40, then the level of danger is transferred to the average level.
- Medium: If parameter M scores more than eight points, then it is assigned a hazard degree of medium.
- High: If the parameter M scores more than nine points, then it is assigned a hazard degree of high.

6.5 CYBER-PHYSICAL SYSTEM THREAT DATABASE DEVELOPMENT

6.5.1 Database Architecture

The growth in the number of information security threats associated with a growth in the amount of manufacturers, types of mechanisms, and information technologies requires the creation and constant updating of a knowledge database about threats, vulnerabilities, and security necessities for information security facilities. This will simplify the process of integrating knowledge bases with expert systems, artificial intelligence, and will also solve the problem of building links between various entities of knowledge bases and procedures for updating them.

The model of organization of the knowledge base in the field of CPS security can be represented on the basis of ontological knowledge representation models. The core of the knowledge base is a conceptual model of information systems security, which will include such typical security concepts as "threat," "vulnerability," "security requirement," "attack," "recommendation," which are supposed to be supplemented with CPS concepts, and also to establish relationships between concepts (e.g., the "subset" relationship between the concepts "attack" and "threat," an attack is a deliberate threat) [20].

The main concepts are related vulnerabilities, attacks, threats, intruders, SFCh, and requirements. The rest of the entities are auxiliary and extend the descriptions of the main concepts. The central objects are catalogs of threats and vulnerabilities, and they are related to a lot of concepts. Vulnerability is associated to SFChs, as many to many, because one vulnerability can exist in different software, and one software can have several vulnerabilities. Tables of error type, CVSS, and vulnerability databases are references that are used for a more detailed description or classification of vulnerabilities. Also, vulnerability is associated with a threat, as one-to-many, because vulnerability makes it possible to identify a single threat.

Organizations suffer damage in different areas after an attack, so the relationship between these models is many-to-many. Several threats can be used to carry out one attack, and several attacks can use one threat, so their relationship is many-to-many. Different requirements or a set of requirements can be applied to mitigate a threat, and some requirements are repeated, so several threats may have the same elimination requirement. The intruder's concepts and the consequences of the threat realization also have a many-to-many relationship with the threat model for a more detailed description. One of the key features is the ability to update threat information through integration with external data banks. The script at a certain frequency, for example, once a day, downloads an up-to-date list of vulnerabilities from an external database. The list is filtered by the specified parameters to select only the vulnerabilities of interest, for example, by the list of names or status, and converted to .json format for subsequent convenient entry into the internal database.

6.5.2 Threat Catalog

During the development of the database, a catalog of threats was developed, which includes the following classes of threats:

- Use of techniques of reverse engineering to get data about the bootloader category, the control board, and the protocol of the bootloader type;
- Danger of high voltage on the control board;
- Remote resource consumption;
- Changing system components;
- Unsanctioned access to the software and hardware of the CPS nodes because of their position outside the controlled zone;
- Obtaining information about the owner of a wireless device;
- Attacker's impact on parameters of environment;
- Electromagnetic influence on the nodes of the CPS;
- Detection by electromagnetic interference;
- Signal noise.

All threats are presented in the following format:

- Description of the threat
- Object of influence
- Consequences of the threat
- At what level of SFCh the threat is realized

Examples of threat descriptions:

1. The threat of distant resource feasting.
   Description: The threat lies in the fact that an invader can inspired the energy possessions disbursed by the system nodes (the quantity of energy spent per time unit) by continuously transfer packets to the nodes and also avoiding them from going into snooze mode. In addition, an invader can affect bandwidth of network by flooding it with packets. An invader can also launch a deauthentication process, which entails supplementary energy consumption. Also to network impact, an attacker can physically affect the resources of network nodes [21].
   Object of influence: Control board of the CPS unit, sensors.
   Consequences of the threat: Violation of the confidentiality property of the transmitted information. Violation of the properties of the availability of transmitted information, as well as the availability of network nodes.
   At what level of SFCh the threat is realized: Network.
2. The threat of unsanctioned access to the software and hardware of the CPS nodes located outside the controlled area.
   Description of the Threat: CPS nodes may be in an unprotected and uncontrolled environment, where an invader can easily intercept and use any CPS node. As a rule, the CPS software is built based on Unix-like OS,

and the quantity of types of hardware and software types is restricted. An invader with knowledge of the structure of the system will be able to inject their own code and change the configuration of the system. In addition, an invader will be able to change the hardware configuration of the device, sensors, etc.

Object of Influence: Application software, system software.

Consequences of the Threat: Partial destabilization of the network due to the failure of one or more devices. The introduction of an intruder at this stage can lead to disruption of the functioning of the CPS nodes, as well as to the disruption of the process of achieving goals by an individual CPS node or a group of nodes, inflicting a destructive effect on the network, and partial failure of the system.

At what level of SFCh the threat is realized: Software and hardware.

6.6 CONCLUSION

In conclusion, we note that this work represents a development that is important for guaranteeing information security in CPS. The main result of the paper is recommendations on the methodology for examining the security of CPS, based on the study of vulnerabilities and security threats. The analysis of the SFCh of the CPS is carried out, and their main structures are highlighted. A set of factors affecting their security was also determined, and the level of initial safety was calculated based on the presence of certain factors. New threats to the CPS have been identified, which are related with their specific properties: mobility, the use of wireless networks as communication channels, being outside the perimeter of the controlled zone.

To date, the methods for determining a probable intruder assess him in terms of goals and capabilities. At the same time, if we are talking about a CPS, then an intruder with a low potential, that is, with the ability to use standard and publicly available utilities for carrying out attacks, can cause significant damage to an object that is under the control of the CPS. Therefore, it is advisable, when considering an offender, to take into account not only his impact on information and technical means, but also on an object that is under the supervision or control of the CPS. This concept implies the need to revise the model of a potential intruder and develop a methodology for identifying it. The determination of the targets and capabilities of the intruder is also carried out by experts means, and there are practically no clear recommendations. The novelty of the proposed methodology lies in the development of a set of factors that determine the likely intruder for the CPS. In particular, attacks and attack tools that an attacker can use are considered separately.

The introduction of new risk assessment methods, the identification of new current threats, and the development of actual references for CPS have made it probable to generate a knowledge database with data of CPS threats, attacks, and vulnerabilities. Due to the fact that the knowledge database is built on the basis of an ontological model, it allows you to attribute threats and attacks, vulnerabilities, and SFCh and establish interactions between them. In the future, it is planned to develop a knowledge base about threats, attacks, vulnerabilities to the level of an information, and analytical resource that could be used by owners and developers of CPS. Such a resource should allow us to automate the process of determining security requirements for CPS and measures to minimize risks.

ACKNOWLEDGMENTS

This research was funded by the Russian Science Foundation grant number 22-11-00184, https://rscf.ru/project/22-11-00184/.

REFERENCES

1. Makhdoom I., Lipman J., Ni W., Anatomy of Threats to the Internet of Things. *IEEE Communications Surveys & Tutorials* 21(2), (2019): 1636–1675.
2. Zhou W., Yu B., A Cloud-Assisted Malware Detection and Suppression Framework for Wireless Multimedia System in IoT Based on Dynamic Differential Game. *China Communications* 15(2), (2018): 209–223.
3. Carielli S., Eble M., Hirsch F., Rudina E., Zahavi R., IoT Security Maturity Model. Practitioner's Guide, Version 1.0, 2019. URL: https://www.iiconsortium.org/pdf/IoT_SMM_Practitioner_Guide_2020-05-05.pdf.
4. Niu M., Dai H., Internet of Things Information Security and Preventive Measures. *Academic Journal of Science and Technology* 4(2), (2022): 93.
5. Choudhary S., Meena G., Internet of Things: Protocols, Applications and Security Issues. *Procedia Computer Science* (2022). DOI: 10.1016/j.procs.2022.12.030, URL: https://www.researchgate.net/publication/366716818_Internet_of_Things_Protocols_Applications_and_Security_Issues.
6. Soomro F., Jamil Z., Tahira H. R., Security Threats to Internet of Things: A Survey. *International Journal of Scientific Research in Science, Engineering and Technology* 19(4), (2022): 130–135. DOI : 10.32628/IJSRSET229423.
7. Podder P., Mondal R. H., Bharati S., Paul P. K., Review on the Security Threats of Internet of Things. *2020, International Journal of Computer Applications, Foundation of Computer Science (FCS)*, New York, USA, URL: https://www.researchgate.net/publication/348487221_Review_on_the_Security_Threats_of_Internet_of_Things.
8. Sohel M., Shah T., A Comprehensive Study on Securities and Threats in the Internet of Things (IoT). *International Journal of All Research Education and Scientific Methods (IJARESM)* 10(6), (2022): 778–783.
9. Qaswar F., Rahmah M., Raza M. A., et al., Applications of Ontology in the Internet of Things: A Systematic Analysis. *Electronics* 12, (2023): 111. DOI: 10.3390/electronics12010111.
10. National Institute of Standards and Technology. *Framework for Improving Critical Infrastructure Cybersecurity*, Version 1.1 (2018). URL: https://nvlpubs.nist.gov/nistpubs/cswp/nist.cswp.04162018.pdf.
11. Sarfraz A., Mohammad M. R., Chowdhury J. N., Interoperability of Security Enabled Internet of Things. *Wireless PersCommun* 61(2011): 567–586.
12. Tao M., Zuo J., Liu Z., Castiglione A., Palmieri F., Multi-Layer Cloud Architectural Model and Ontology-Based Security Service Framework for IoT-Based Smart Homes. *Computer Systems* 78 (2018): 1040–1051.
13. Zhang X., He Y., Information Security Management Based on Risk Assessment and Analysis. *2020 7th International Conference on Information Science and Control Engineering (ICISCE)*, Changsha, China, 2020, pp. 749–752. DOI: 10.1109/ICISCE50968.2020.00159.

14. Mozzaquatro B. A., Agostinho C., Goncalves D., Martins J., Jardim-Goncalves R., An Ontology-Based Cybersecurity Framework for the Internet of Things. *Sensors* 18 (2018): 3053–3055.
15. Abbass W., Bakraouy Z., Baina A., Bellafkih M., Classifying IoT Security Risks Using Deep Learning Algorithms. In *6th International Conference on Wireless Networks and Mobile Communications (WINCOM),* Marrakesh, Morocco, 2018, pp. 1–10.
16. Fedorchenko A. V., Doynikova E. V., Kotenko I. V., Automated Detection of Assets and Calculation of Their Criticality for the Analysis of Information System Security. *Trudy SPIIRAN* 18(5) (2019): 1182–1211.
17. Threatmodeler.Com, Security Starts, 2021. URL: https://threatmodeler.com.
18. Bakhshi Z., Balador A., Mustafa J., Industrial IoT Security Threats and Concerns by Considering Cisco and Microsoft IoT Reference Models. In *Wireless Communications and Networking Conference Workshops (WCNCW).* IEEE, Barcelona, Spain, 2018, pp. 173–178.
19. Facial Recognition Implemented at Beijing Capital International Airport, 2021. URL: https://medium.com/@pandaily/facial-recognition-implemented-atbeijing-capital-inter-national-airport-8c0b7cc1b945.
20. Basan A., Basan E., Ivannikova T., Korchalovsky S., Mikhailova, V., Shulika M., The Concept of the Knowledge Base of Threats to Cyber-Physical Systems Based on the Ontological Approach. *2022 IEEE International Multi-Conference on Engineering, Computer and Information Sciences (SIBIRCON),* Yekaterinburg, Russian Federation, 2022, pp. 90–95. DOI: 10.1109/SIBIRCON56155.2022.10016783.
21. Basan A., Basan E., Gritsynin A., Analysis of the Security Problems of Robotic Systems. *2019 2nd International Conference on Intelligent Communication and Computational Techniques (ICCT),* Jaipur, India, 2019, pp. 26–31. DOI: 10.1109/ICCT46177.2019.8969055.

7 IoT in Healthcare
Glucose Tracking System

V. Manjuladevi and S. Julie Violet Joyslin

CONTENTS

DOI: 10.1201/9781003438588-7

7.1 INTRODUCTION

The healthcare industry is one sector where people need to keep their focus. Due to improper food habits and lifestyle of humans, people are becoming increasingly susceptible to a wider range of health problems. One of the most important problems that need to be solved is the early treatment of diabetes. If not, it will have an impact on those who may be executed as a result of the concerns highlighted in the series. Indeed, it was not an easy chore to continuously keep an eye on the actions and health of someone who frequently dealt with these kinds of problems (Alekya et al., 2020). However, by using tracking devices, the IoT has made this possible.

To predict how the IoT devices are beneficial in the medical field, particularly in glucose tracking and monitoring in healthcare approaches, that is carried out based on the study (Mouha, 2021). The analysis shows the graphical visualization of the concepts, which enables readers to understand the importance of technology in healthcare, predominantly in monitoring glucose level in the human body (Islam et al., 2015).

Sensor tracking is quite helpful for sending emergency notifications to the patient's family. It also provides information about the specifics and history of local hospitals. Thus, it is very supportive and helpful, and in the future, the Internet of Things (IoT) will play a major part in making people smarter and enabling them to always be active and thus change their lives. All these are made possible with the support of proper supporting sensors that are linked up using IoT technologies (Ru et al., 2021).

Figure 7.1 shows a smartwatch with a sensor, allowing for continuous data monitoring. Patients and their families can easily start viewing the patients' glucose levels and pulse rates by linking it to their mobile phones. All of the information is stored as data in cloud, and the patient's doctor will receive the collected information thanks

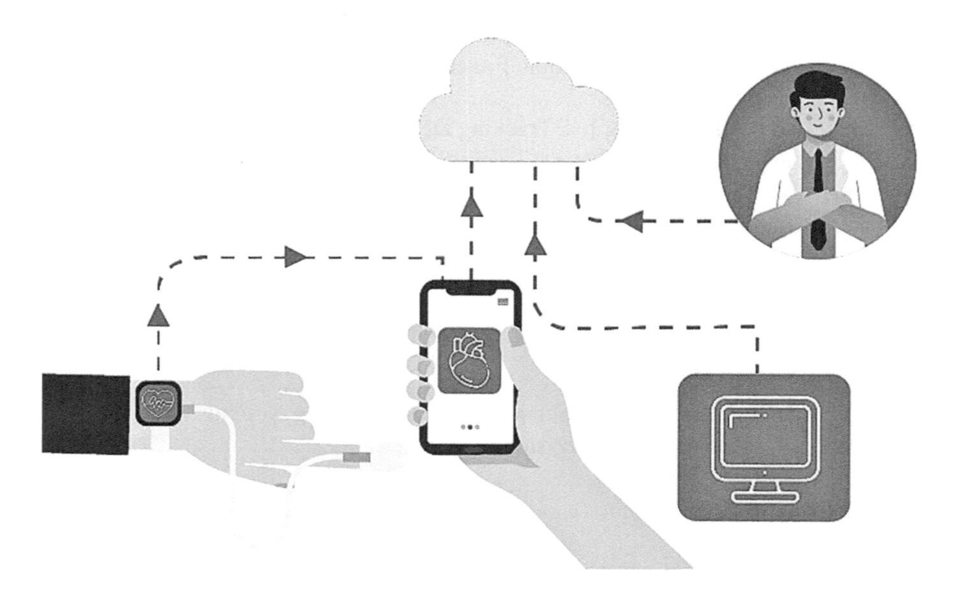

FIGURE 7.1 Tracked data is sent to doctor.

to the cloud. The patient will be asked to meet if there is an emergency or special need; otherwise, the doctor may postpone the examination. When patients or doctors are far away, these strategies may be quite useful. When the doctor is not present, the information will be transferred to a neighboring hospital with the help of a signal.

7.1.1 Organization of the Chapter

The literature review on the implications of IoT and the glucose-monitoring device is covered in Section 7.2. Section 7.3 explains about devices that connect with communication protocols including sensors and gadgets. Section 7.4 discusses the advantages of smartwatches in the glucose-monitoring system. The effective application of IoT in the healthcare sector is covered in Section 7.5. The tracking of diabetics is described in Section 7.6, and their process is described in Section 7.7. The advantages of IoT-enabled services are discussed in Section 7.8, and the monitoring process in healthcare is covered in Section 7.9. Testing blood sugar is covered in Section 7.10. At the conclusion of the chapter, all findings were thoroughly discussed.

7.2 LITERATURE STUDY

According to Abdel Rahman H. Hussein (2019), the applications utilized in the IoT have changed in many ways that have made it possible to meet a variety of needs. Without a question, the IoT has enormous potential that has improved the survival of millions of people around the globe. He described the various levels of study obstacles before concluding that the usage of IoT solutions offers users a higher level of security and privacy. This opens up the opportunity for both communal and isolated sectors to begin focusing on IoT to be implemented in health sectors, agriculture, industry, and in other areas.

Baker and Xiang (2017) explain about a variety of sensors that are deployed to keep an eye on pulse rate, pulse oximetry, and vital indicators. The benefit of using it is that it delivers accurate results with minimum data size, which is used to improve cloud storage performance. The encryption techniques that are used will be supportive for enhancing the standard that is suitable for predicting the conditions of the body and thus improving health status and saving life. It allows sharing information about the conditions of the patient in prior.

Ru et al. (2021) conducted analytical investigations using a sensor that is integrated into a system for monitoring human healthcare. Real-time tracking and monitoring are really helpful while evaluating the research process in glucose tracking. The information gathered was based on the heart rate, body temperature, physiological data, and pulse. As a result, it is utilized to issue alert signals and retain the data that is gathered for use in avoiding and managing more severe chronic conditions. Upcoming development of the approaches and concepts has the biggest potential.

Banka et al. (2018) analyzed how to track the records of patients in a place where the hospital facilities are low. By connecting the sensor and a cell phone, a network is created from which the emergency message and the patient's recorded data are delivered to the local hospitals, from where the patient receives prompt emergency care without having to wait for a diagnosis of the problem's underlying cause.

The sensor's use will also aid in warning the patient's family through SMS notifications. The assistance of IoT tracking technologies was used to accomplish this.

Alekya et al. (2020) elaborated on the guiding principles that allow for a direct connection that is efficient and useful for storing data, as well as the design principles that are employed in cloud applications and IoT systems. The wireless sensor and networking are used for system control at each level. The IoT, which is built on the healthcare system, is truly what makes the greatest wonders possible. Following completion of the deployment phases, it is combined with several various IoT services that are used to further provide a higher level of safety and security.

According to Mouha's (2021) research, the environmental component has seen the greatest transformation in the recent modern technological world thanks to IoT sensing technology. In addition to the above considerations, it is frequently employed in the medical industry. As a result, it offers the most possible benefits to people who are using these methods for human development.

In Yusuf et al.'s (2020) study, IoT for glucose monitoring is presented in smart health evaluations. It is straightforward to see how IoT technology could be used directly in smart health to produce desired results. The IoT sensors can detect the signal. However, smart automobiles and remote monitoring receive less attention. If remote monitoring with smart devices is considered, it may future deliver improved results.

Gia et al. (2017) offered a controller device powered by IoT technology, to continuously and closely watch glucose level. A complete system is implemented as part of an IoT framework, from the optical detector on the back-end server. By organization, healthcare providers and carers can quickly keep an eye on their patients from any location at any time using a smartphone app or web browser. Multiple forms of data providing values based on energy utilization can be effectively received and transferred by sensor computer nodes (such as blood sugar, vital signs, and ambient information) wirelessly. For increasing its operating time, both the detector device and the sensing device are connected with a power management unit and an energy collecting unit. With the help of a customized receiver, the patient's smartphone converts into a data-receiving connector for sensing nodes. Additionally, the deployment of the entry point offers consumers improved assistance like notification services, etc. The outcome shows that real-time remote glucose monitoring is possible and the system's energy efficiency can be improved.

Deshkar et al.'s (2017) research indicates that diabetes is a significant illness. Mobile health applications are presented in recent generations as a result of technological developments and cost reductions in wireless networks and Internet technologies. Recent advancements in wireless networks and Internet technology have resulted in the recommendation and successful implementation of increasingly advanced eHealth apps. They show the most recent applications for managing diabetes using the IoT as well as their fundamental design. We looked at the problems and difficulties that the most recent applications have. They conclude by offering potential fixes and future research trajectories for IoT.

Per Efat et al. (2020), it is a difficult task for one to keep on monitoring and tracking the records of diabetics. However, using sensors, doctors can always keep an eye on the person who is suffering from glucose issues. Health status sensor devices are connected intelligently to improve care delivery and yield improved clinical outcomes. The strategy has also decreased hospital admissions, bed days, and hospital stays. To redirect a conventional doctor's attention to a patient-friendly strategy and its primary contribution of this study develops a novel framework for a neural network and a decision support system to determine risk status. The result obtained is based on the type of the data that is collected for the research.

According to Sattar et al. (2020), decision supporting system mainly results in the prototype software that is used for increasing the end-to-end functionalities. The findings are safe, accurate, and seamless, and the server and computer are linked, creating a greater opportunity to generate therapeutic influence and increasing the comfort of the person who completely supports in protecting the patient. Thus, precaution measures are made easy and simple.

According to Al Shorman et al. (2021), patients with chronic medical disorders, especially elderly patients with diabetes, require specialist treatment and continuous medical attention. However, patients, caregivers, and the medical community can employ remote health monitoring to enhance healthcare services by utilizing cutting-edge technologies. Importantly, it is still difficult to provide effective healthcare services for those who have diabetes. It is advised in the study to examine patients with diabetes using remote health monitoring by mHealth technology. It is used to manage and control diabetes, as well as to provide better diabetes treatment and prevention assistance. The main purpose is to provide the potential developments.

According to Bansal et al.'s (2018) analysis, the number of people suffering from low glucose levels increases, and therefore proper awareness has to be given so that people can easily overcome this condition. This is made possible by recent technological developments and cost reductions in wireless networks and Internet technologies, and several electronic/mobile health (e/mHealth) applications have been proposed. Utilizing the most recent developments and cost reductions in wireless and web technology, more complex eHealth applications have recently been proposed and successfully implemented. The functionality and core architecture of the most modern IoT health applications for managing diabetes were displayed. They examined the issues and challenges these new applications must overcome. They concluded by suggesting potential IoT solutions and future research directions.

Sharmila et al. (2017), in their article, have looked at a few methods for glucose monitoring. Individual aid organizations are given more attention when being monitored. There are numerous glucometer systems on the market made using various methods. Many of these systems merely aid in glucose level self-monitoring. They recommended that a method might be proposed that is utilized for knowing basic information about nutrition, activity, medication to be taken, and precise blood glucose levels in addition to self-monitoring. Additionally, physicians should be informed of the same information for future use. The benefit of that method is that patients don't need to make an appointment in advance and can visit their doctor whenever needed. This is the strategy that will be used in the future.

Per Sattar et al. (2020), the goal of the suggested system is to provide a tool to be used for measuring glucose levels. It has been proven that a person's position is directly correlated with the amount of ketones present in their breath. In mobile phones, the graphical representation can be done for analyzing the sensitivity data that are characteristics with each other. Additionally, a web-based user interface is created to assist patients in using technology to its fullest potential.

According to Prakash et al. (2022), among other things, the healthcare business can employ predictive analytics for gaining clinical data and make better judgments. For analyzing the techniques, machine learning (ML) concepts are used to estimate upcoming accurate result gained from users. We provide a neural network classifier that accurately predicts diabetes level in the patient and helps to monitor the patient condition. A probe is used to track patient measurements and results via IoT devices. Computers receive data from smartphones and IoT devices, classify it, and then store it in the cloud. The proposed solution outperforms existing methods by 99%, according to simulation findings.

According to Longva and Haddara (2019), IoT devices are anticipated to change in the future. Let's take a closer look at future IoT-enabled devices that bring embedded intelligence health services as part of their offerings with more confidence in health facilities. Developing information technology and new IoT technology has already presented some chances for the growth of better health information systems. The use of IoT in the treatment of diabetic patients is discussed in this article along with some functions that can be enhanced and made simpler. Wearable sensors can be used to collect data on these actions and relevant information on the patient's physical and mental health. Maintaining track of patient medical records in a secure and private manner, continuous glucose monitoring is one of your possibilities. For monitoring, it is necessary to check and retrieve the most recent glucose reading, as well as other apps such as smartwatches and activity trackers.

To achieve flow control in droplets, Nandan et al.'s (2022) research provided an electronic valve with a volume control system as a scalable, advanced medical technology. The proposed paradigm of controlling the flow rate of saliva via manual switches can be implemented using IoT ideas. In this scenario, IoT is used to entirely replace the traditional system by connecting with software such as a flexible mobile application, allowing doctors to adjust flow rates in online and remote mode. IoT concept models can be implemented in rural areas such as villages. Doctors can monitor their patients from another city, building level, even their home, or anywhere in the world. The rate of process that is carried out is measured for managing the entire task. A doctor can view several patient reports on a computer or mobile app using the same IoT concepts. So, many patients can be under one doctor's care.

Per Anagha et al. (2020), the occurrence and monitoring take place simultaneously. Making use of the proposed monitoring system, the user can gain a massive set of benefits: it will be efficient and affordable, works smoothly, and continuously observes any drip in patients' condition with the help of trackers. This can be achieved by measuring the drip by comparing the measurements with the set of standard points. Here, the doctors can sit in some places and observe conditions of their patients from anywhere. When the patient's condition is clearly tracked up there the doctor can communicate along with the patience immediate and do the required needs.

According to Shinde Sayali and Phalle Vaibhavi (2017), the smart healthcare system has updated IoT features. With the use of IoT, the healthcare system has lessened the issue and complexity of patient situation monitoring and tracking. According to all poll forecasts, the entire medical system provides subpar care and underutilizes technology. Utilizing modern technology that is vastly improved solves this issue. It is possible to create smart health care equipment. It is vital to address existing hygiene. To increase awareness of glucose tracking levels and improve life quality, government programs should be improved and implemented.

According to Abdulmalek et al. (2022), IoT can enhance healthcare in countless ways including cost savings, and improved effectiveness and efficiency. IoT benefits have made it feasible to more effectively automate healthcare operations. In this approach, this work serves as a primer for professionals in the subject. They will eventually receive a whole reference guide on IoT that has been implemented at the healthcare monitoring systems. This report has carefully reviewed and analyzed new research on IoT health surveillance systems. In-depth details on their benefits and importance are covered in the study's literature evaluation. We will discuss IoT wearables in healthcare systems in this post, which includes a variety of health monitoring sensors and Quality of Service (QoS). There are also recommendations for additional research. Future analyses and evaluations will be based on connected devices, disease classification, and monitoring system devices. We also plan the following step. It is important to highlight how cutting-edge technological trends such as SDN and AI are interacting with IoT-based health monitoring system.

According to Kumbi et al.'s (2017) study, the Network of Things has improved many components of smart medical services system. The concepts of IoT have made it possible to gain insight of the process. The healthcare system is now less complicated and complex due to the IoT advancements. Our analysis of the entire medical system led us to the crucial conclusion that there are unprofitable maintenance and underused technologies. Through the application of new and improved technology, this problem can totally be resolved. Developing intelligent healthcare technology is one way to control present medical services. More people are becoming aware of intelligent illnesses, as well as welfare schemes to improve life satisfaction are being implemented.

Mittal and Navitha (2019) discussed how the integration of health monitoring and surveillance system has radically altered how patients are treated. By offering patients prompt and dependable services, the healthcare industry is transforming with high pace. It has since been identified through deep investigation. The usage of sensor-based medical devices is ongoing. As a result, the health environment is getting better. The procedure used for treating patients becomes more dependable and effective. There might be fewer patient visits to the clinic. The patients can access all of their medical information on their phone, and in an emergency, doctors can be summoned and deodorizers can also be administered.

In their proposed study, Kulkarni and Jakkan (2019) explained that the IoT has altered how we perceive those in need in regular lives. Thanks to smart objects, it has developed into a connection between vision anytime, anywhere, in any format, and in any way. Hospitals without IoT applications should be avoided by the patient as they primarily can't provide basic care to the patients. Thanks to IoT, there is an advanced healthcare system. To gather information on a mission point, smart hospital uses a variety of front-ends tools to examine the aforementioned view. Each forecast we make includes numerous techniques for gathering data in various ways, allowing the Web of Things to continuously track user activity. The problem can be resolved by fully utilizing the latest and most advanced generation of technology. Creating an intelligent health device is one way to control current hygiene. Additionally identified are recommendations to increase public awareness about sensible disorders and the implementation of government policies designed to improve one's standard of living.

In Phani Ram and Shankar's (2022) publication, the health surveillance system's executive summary is covered. The many health monitoring apps and technologies based on the IoT are covered in this article. Several implementations, apps, and services of IoT healthcare monitoring system are available. These methodologies are also described and evaluated individually for predicting the accurate set of results. Each technology has its own drawbacks and uses. A description of the study's methods and applications that were used to improve effective management is given.

7.3 HOW IoT PERFORMS ITS SENSING TASK?

IoT plays numerous roles in once life. The gadgets will be integrated into daily routine activities. The smart devices have given new dimensions through connectivity and smart gadgets like watches and sensors to interact with objects. While automated smart devices automate tasks, physical digital devices interact with physical items. This gadget (like smart wearable device) enables the automatic detection, saving, and interpretation of data from smart devices. Fusion of many technologies and communication options is the key characteristic of IoT. Other important IoT aspects include distributed intelligence for smart things, improved communication protocols, and identification and tracking systems (Longva and Haddara, 2019).

Figure 7.2 replicates the data on how the sensor is collecting and transmitting data with the help of Internet facilities, and the analysis is in the format of descriptive, diagnostic, predictive, and event-based action. The doctor reviews based on the information collected via cloud, and the information is reviewed by the concerned doctor to analyze the glucose level of the person.

7.4 USE OF SMARTWATCH IN GLUCOSE TRACKING

Smart watches offer a platform for creating IoT applications involving humans, and other devices. Smartwatches are now utilized in several IoT applications, including fitness and healthcare. The current generation of smartwatches can be used to implement several socially based Social IoT applications because they are outfitted with a variety of sensors and heterogeneous wireless protocols (S-IoT). These applications require transmitting sensor data from millions of watches via the IoT cloud (Yusuf et al., 2020).

Wearable smart devices monitor the wearer's BP, stress, period cycle, physical activity, sleep patterns, and several other activities to add to the wearer's data. Additional data can be evaluated after being connected to a mobile phone or computer to comprehend exercise levels, sleep habits, and other elements of wellness (Akkasaligar et al., 2019).

The capacity of medical staff to remotely monitor a patient's vital signs is among the main advantages of wearable technology in healthcare. This is especially helpful for Clinical Prediction that is understaffed and lacks the resources to treat patients in person for minor difficulties or those that have adopted virtual appointments. When

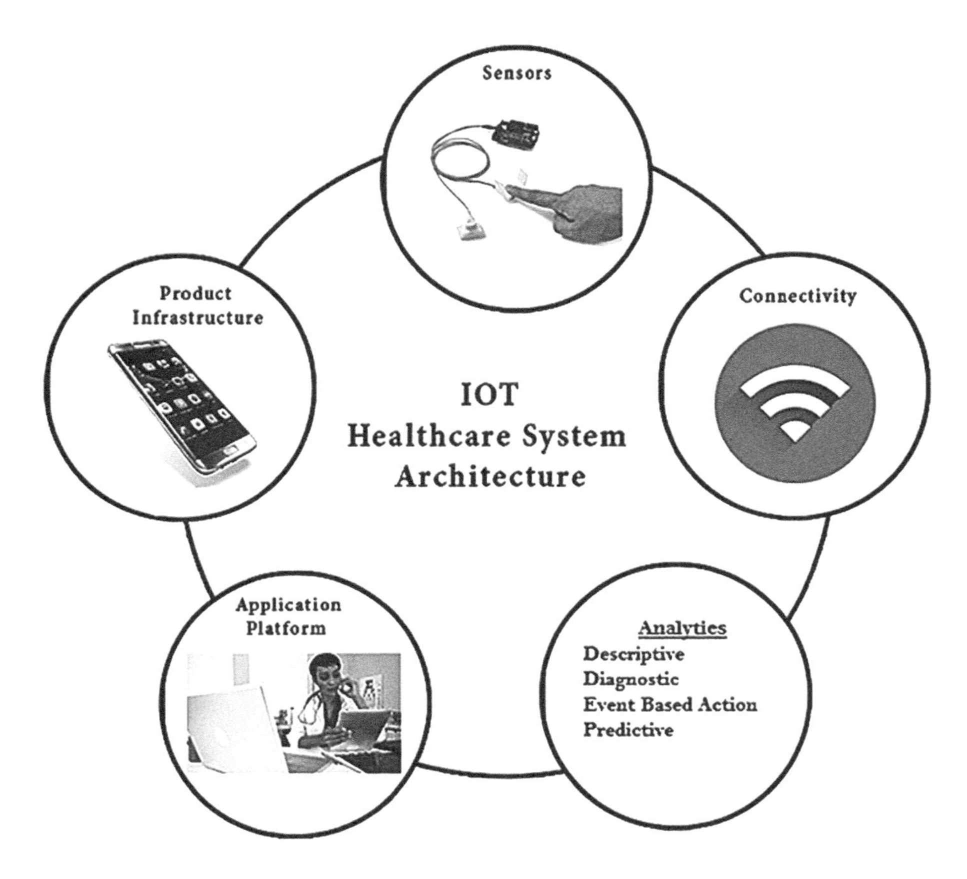

FIGURE 7.2 IoT healthcare system architecture.

questioned about their health statistics, individuals find it far simpler to promptly provide them (Hussein, 2019).

7.4.1 They Do More Than Just Keep Time

It should be obvious how to tell the time because it is a watch. Most smartwatches come up with a stopwatch as well. This is comparable to the majority of stopwatches and clocks on smartphones. It is solely intended for usage without a smartphone, though. Finally, some smartwatches offer alarms that are independent of the phone (Anagha et al., 2020), and you can set an alarm on your watch rather than your phone, and the watch will vibrate and play a sound to wake you up.

7.4.2 A Travel Companion That Is Always with You

Smartwatches may vibrate your wrist to indicate whether to turn right or left if you're following directions. You don't have to constantly check your smartphone; you may use an invisible compass instead. Instead of looking at the map, take a moment to survey your surroundings by looking up.

7.4.3 Finding a Phone, Key, or Another Item Is Even Simpler

As you are well aware, losing a phone or a set of keys is a frustrating experience. Before a significant occasion for which we cannot afford to be late, it always seems to occur (Mouha, 2021).

7.4.4 Answer Calls and Messages Right Away

You can use your watch as a substitute for pulling your phone out of your pocket. While you're moving, you can still receive and respond to messages. This is especially helpful if you are working or in another circumstance when carrying your phone is problematic (Sattar et al., 2020). Some smartwatches offer voice help, and you can communicate with someone thousands of kilometers away from you at your wrist.

7.4.5 Review Your Social Media Notifications

Who wouldn't want notifications from social media apps like Facebook, Twitter, WhatsApp, Snapchat, and others on their wrist? I always disable this option; however, some people consider it to be essential. While some watches only show your social media and message activity, others let you interact with the app (Anagha et al., 2020).

7.4.6 You Remain Connected even as You Work

Checking your calls, messages, or notifications while working out, whether you're running, cycling, swimming, or engaging in another physical activity, may be a good idea. Carrying your phone while performing those tasks is not always a smart idea, and when you do, it is uncomfortable and cumbersome. When this occurs, a smartwatch is useful. Are you submerged? It's not at all a problem! Some smartwatches have a 50-m water resistance rating. You can check your news without leaving the pool by pausing while swimming (Mouha, 2021).

7.4.7 It Gives You More Time to Be Connected Compared to Your Phone

You may be wondering why you need a wearable gadget when you have a nice smartphone. Shake it off; some wearable gadgets have batteries that are so powerful that can't be compared with a phone (Shinde Sayali and Phalle Vaibhavi, 2017). Wearable gadgets can link you for several days on a single charge.

7.4.8 You Have Access to Plenty of Entertainment

Let's say you decide to watch a YouTube video while out on a stroll and your friend keeps urging you to. You can view YouTube on your wristwatch with only a few clicks now. You can watch videos and play music on the go. The enormous screen on your phone will never be duplicated, yet it is unrivaled in terms of convenience for those few moments. Smartwatches in recent generations can also store music, which may then be performed remotely using wireless ear buds (Phani Ram and Shankar, 2022).

7.4.9 Remind You via Email

Some wearable gadgets allow you to choose the times your watch gives you reminders for various daily tasks. Small notifications throughout the day to keep track of your schedule in the back of your mind would be beneficial help you reduce stress if you know you tend to flip out in stressful situations.

7.4.10 They Function as Reliable Fitness Trackers

Many smartwatches come with a built-in feature for tracking your fitness. It will help you achieve your fitness goals. An excellent smartwatch can therefore replace a fitness tracker or pedometer if you're considering buying one (Phani Ram and Shankar, 2022). Smartwatches can track your heart rate, pulse rate steps taken, distance traveled, calories burned, and sleep. Some of them may also compute additional significant parameters. It is encouraged to wear a smartwatch with the help of Revolutions Per Minute (RPM) chip if you have a chronic condition like diabetes or hypertension since you may submit health information to your physician.

7.5 IoT IN THE HEALTHCARE

One can find that the idea of IoT creates a lot of possibilities and opportunities from the past to the present when you look there. The IoT does have a significant impact on the healthcare industry. According to a previous survey conducted by a healthcare organization that discovered that the number of individuals suffering from diabetes is continuously increasing, and when patients are not adequately monitored, their lives may be jeopardized. The IoT technology has revolutionized the industry that is largely used to save lives while also improving patients' quality of life through continuous monitoring systems (Longva and Haddara, 2019). Figure 7.3 states how the IoT techniques are used in the healthcare tracking system through sensing blood glucose levels.

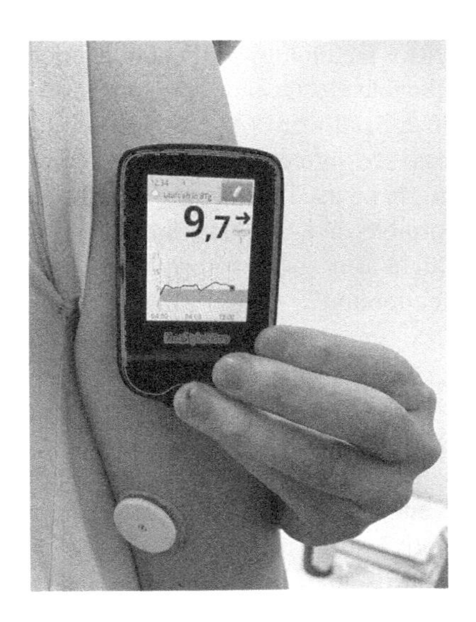

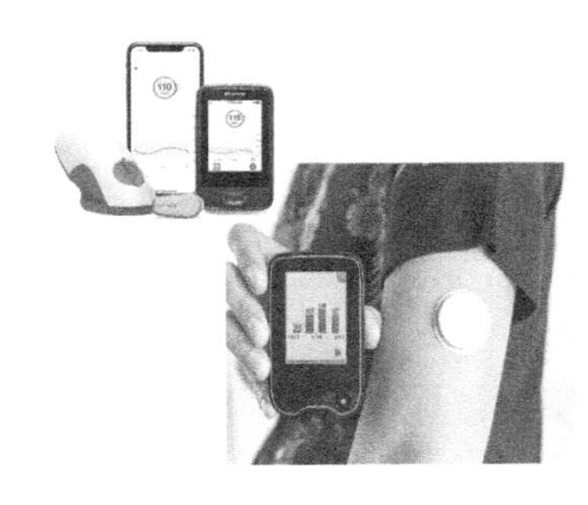

FIGURE 7.3 IoT in healthcare tracking.

7.6 DIABETICS TRACKING USING IoT TRACKING DEVICE

Diabetes is one of the most serious medical illnesses that can be discovered anywhere. This could change drastically as a result of the variation in glucose levels, which may cause critical situations to become life-threatening. The people began to worry a great deal as a result. A person who is dealing with these issues will find it difficult to get over this obstacle. The majority of doctors advice using a glucose-monitoring system, which allows for following up on the blood sugar level that is present in one's body, to avoid this. Before the development of technology, finger-pricking methods were used to measure the amount of glucose in a person's body. These methods follow tradition, but they have some serious drawbacks, especially if the patient is a youngster or an elderly person who has various types of health problems. The Wireless Sensor Network (WSN) is used to monitor without injuring or causing discomfort, and therefore, these disadvantages are avoided by using non-invasion approaches.

7.7 INTEGRATING WIRELESS SENSORS FOR TRACKING

The WSN system is always in operation, and in the event of an emergency, it will begin issuing alarm notifications in the form of brief messages, along with the live location in which the patients are staying, which may be very beneficial and supportive for discovering. Machine learning (ML) models, which are useful for forecasting a future event with major concerns, are utilized to make this type of monitoring efficiently. This paradigm is regarded as the most effective kind of application (Alekya et al., 2020).

7.8 SMART IoT-ENABLED SENSORS

Different businesses place a high value on smart or intelligent sensors. Application of smart medical sensors for automated drug administration, cancer screening, glucose monitoring, and other health screening, diagnosis, monitoring, and treatment applications has increased in recent years. Integrating sensor technology with nanotechnology, microelectromechanical systems (MEMS), flexible electronics, cellphones, and wireless sensor networks can significantly enhance healthcare (Al Shorman et al., 2021). In this presentation, unindustrialized medical sensor technologies, goods, and research will be examined along with how they might be used to enhance healthcare productivity (Gia et al., 2017). Research into portable, non-invasive, smart breath sensors—which are becoming more and more sought after for monitoring and diagnosing a variety of diseases—will receive special attention.

7.8.1 Increasing Efficiency

Performance is the outcome an organization achieves for each effort. The efficiency factor rises when there are more returns for the same effort. Reducing waste is a different strategy to boost productivity (money, time and energy). By enhancing current processes, IoT helps enterprises to increase productivity. The user is now completely aware of how well the machine is functioning thanks to smart machines and sensors.

And, understanding the machine's energy usage allows for the reduction of wasteful expenses while enhancing essential operations. Smart machines can disclose in-depth information on their performance, how long they have been operational, and the times when they have had disturbances (Mouha, 2021).

7.8.2 Enhancing Safety

Accidents at work can be reduced with sensors and fragmentary machine and human monitoring. The right actions can be made to avoid harm to workers and the machine before the crucial point by streaming data to alert operators of a malfunctioning machine or hot part (Gia et al., 2017).

The sensor senses the pulse, blood pressure, and ECG, and sends the information through the Internet to the healthcare providers as shown in Figure 7.4.

7.9 MONITORS FOR THE HEALTHCARE INDUSTRY

Healthcare professionals and patients alike have several new opportunities to monitor each other and their environments thanks to IoT devices (Efat et al., 2020). Consequently, both in terms of health care providers and their clients, the proliferation of wearable IoT devices has both advantages and disadvantages.

7.9.1 Digital Patient Surveillance

The much more common application of IoT medical devices is patient monitoring and healthcare technologies. Connected devices collect health data from patients who are not physically present in a health sector, thus reducing the necessity for

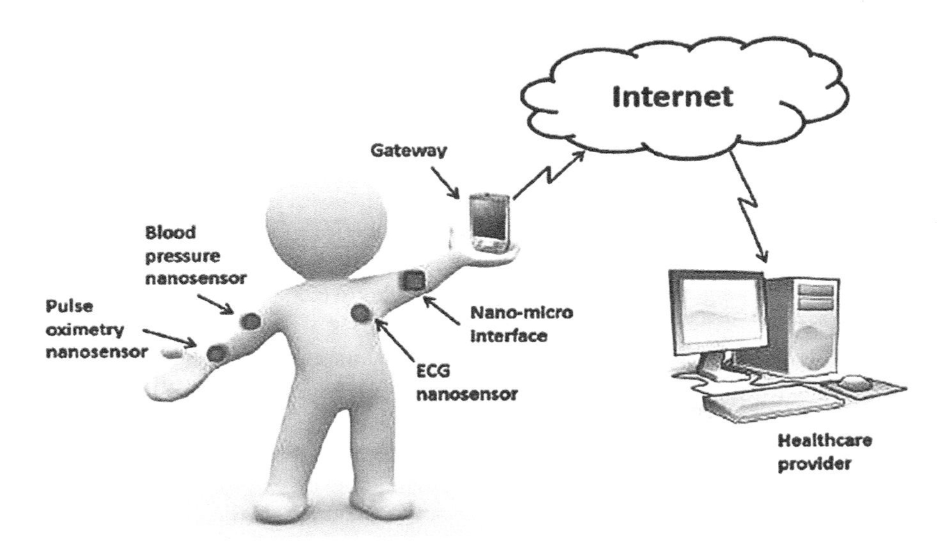

FIGURE 7.4 Tracking/sensing using sensors.

patients to visit physicians. Among these are heart muscle, hypertension, respiration, and other indicators (Deshkar et al., 2017).

Whenever an IoT gadget collects patient information, it communicates the data to a web app which the patient's health provider examines (Kumbi et al., 2017). It can be used for reviewing the data and to generate medication solutions or warnings. For example, such Internet detector may detect that a patient's pulse rate is abnormally low and transmit an alarm so that medical workers can play an active role (Efat et al., 2020).

7.10 SCREENING BLOOD SUGAR

Glucose monitoring has traditionally been difficult. Manually checking and recording glucose levels is not only time-consuming, but it also gives the patient with information on their blood sugar level during the test. The disseminated questionnaire might not be adequate to identify the issue if levels fluctuate dramatically (Sharmila et al., 2017). By continuously and automatically monitoring patients' insulin levels, sensing nodes can assist in overcoming these issues (Khan et al., 2012).The quantity of manual recording may be decreased and patients can be alerted when their blood sugar levels are troublesome by using glucose monitoring system. Some of the challenges in building an IoT glucose-monitoring device are as follows (Baker and Xiang, 2017).

A. It is discreet enough to monitor patients continuously without disturbing them.
B. It uses insufficient energy and requires frequent recharging.

However, these are not insurmountable obstacles, and there are gadgets that are used for predicting the statistics and it connects to different types of devices through which the information is shared to the healthcare industry. The movement that is processing is systematized in the Figure 7.5.

7.11 FINDINGS AND ANALYSIS

Earlier, technology was utilized in the form of sensors for monitoring heartbeat, glucose level, and blood pressure. These sensors were costly, and those who were affected had to wait for a long time to purchase each type (Chitra et al., 2016). However, as IoT in healthcare has grown, there are many different types of advantages that everyone may start choosing from based on their financial situation. Additionally, it seemed simple to adopt and monitor the patient's glucose level. As a result, it greatly promotes the growth of the human healthcare sectors (Nandan et al., 2022). It will be easy for the doctors to call their patients and consult with them about their health conditions, making it possible to start treatment right away. With the help of the tracking system, when the glucose level rises or falls, an immediate message notification will be sent to the nearby hospital along with the patient history from the beginning until the last treatment that they have undergone (Anagha et al., 2020).

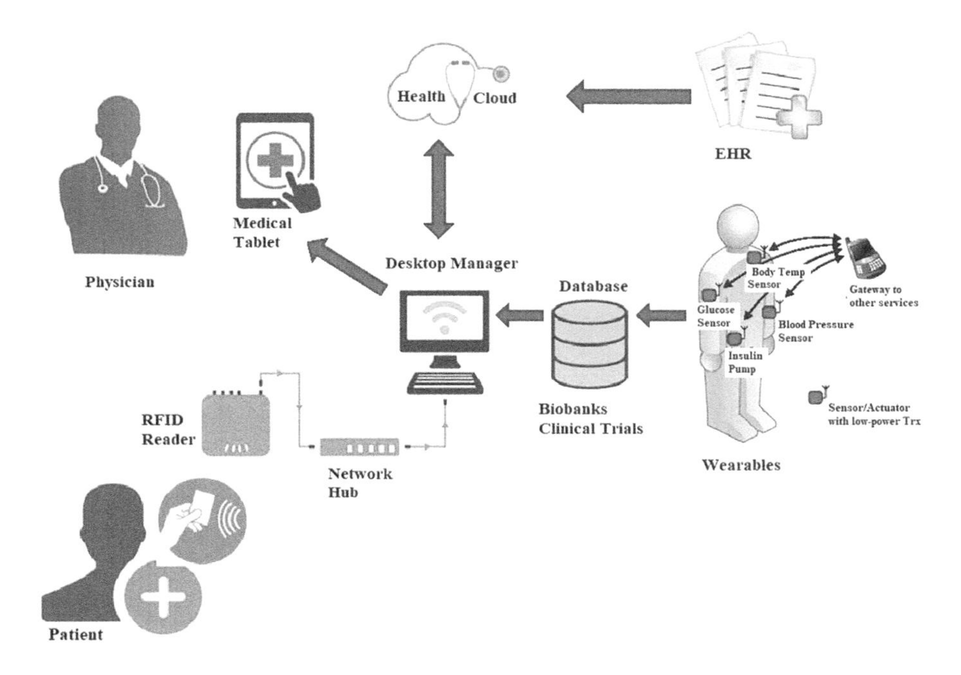

FIGURE 7.5 Patients connected with doctor.

The sheer number of people affected by the set of issues that happen due to the loss of glucose level in the body. We have implemented a real-time survey in the form of a Google form and hypothesized that inadequate life and food cycles are the primary culprits. Even though many people make serious efforts to organize everything, majority of them fail due to genetics or inheritance. The biggest challenge at this time is to designate one nurse or someone to watch after the patient's condition around the clock without taking a break. What would happen if both of it were done at the same time when the blood glucose level suddenly drops at night? The only effective approach is for people to start viewing technology as a helpful collaborator, making it feasible to track data every single second (Ru et al., 2021)

There are ever more people who are profiting from these modern technologies. However, just like a coin with two sides, the constant record tracking causes individuals to worry, even though there are many advantages. Prakash et al. (2022) verify that even though 93% of individuals can defend themselves from danger, to 5%–7% of people, it causes harm. The development of earlier detection and tracking may be very beneficial for extending people's lives (Banka et al., 2018). The analysis was separated into different categories among them. The persons who are aged above 40 are more affected than other ages, and its graphical representation is shown in Figure 7.6.

Here is a survey on the use of IoT devices in the glucose-monitoring system.

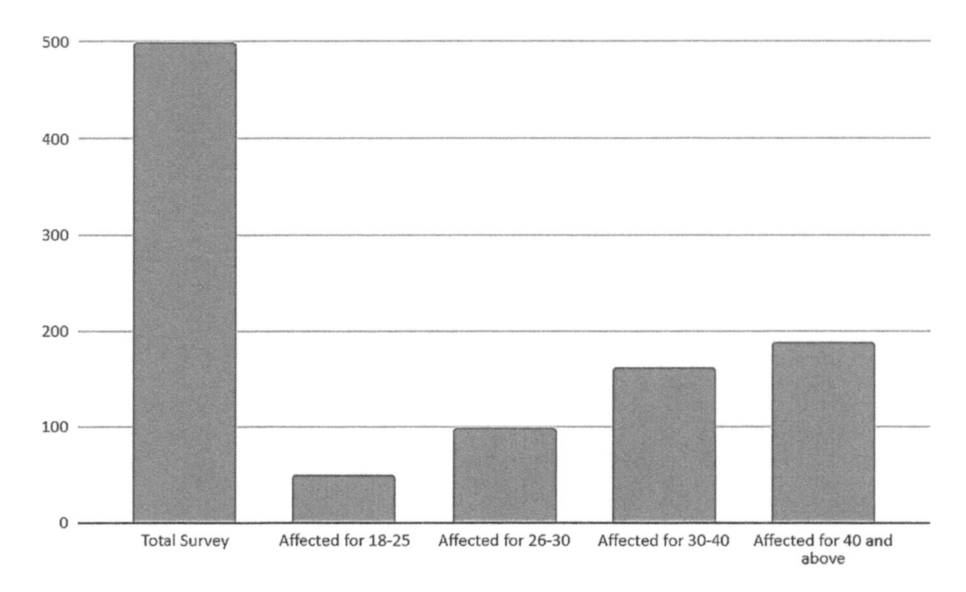

FIGURE 7.6 Analysis graph.

7.11.1 Impacts of IoT in Healthcare Glucose Tracking System

IoT works its own magic in assessing and enhancing the standard of the healthcare system and services. The goal of the research is to identify IoT technologies that have the biggest influence on evaluating glucose levels without the presence of humans. The tracking device has the potential to save the life of someone who truly needs assistance. Smart tracking sensor sends a notification message to the user, and it is then retrieved and processed by the Arduino Uno, which serves as the system's brain and allows data to be sent to the cloud via Wi-Fi modules (Abdulmalek et al., 2022).

To make this get processed, there is no need to purchase any type of monitoring devices. Instead, the user can directly start connecting it using their smartphones (Mittal and Navita, 2019). This system is used for creating the awareness among diabetic patients about the blood glucose levels. Diabetes is believed to play a vital critical role in person's life; thus, it stan ds to reason that its effects could vary greatly and eventually result in life-threatening illnesses.

Doctors recommend their patients frequently monitor their glucose level for knowing the insulin level that is segregated in one's body for taking the precaution measures in prior.

Here, using conventional methods results in distress that may also start skin diseases from scratch. Particularly when the patient is a toddler, the issue appeared to be somewhat complicated. The work proposed by Bansal et al. (2018) is based on the evaluation of data gathered from a variety of sources of age groups in relation to the use of IoT approaches for monitoring glucose levels. It is mainly done for creating the awareness of inspecting the use of such tracking system. This application is the fastest developing in the industry.

More people are becoming affected by diabetes every day. A study was conducted in the healthcare sector for monitoring glucose levels. People become more aware of their health situations as a result of the development of innovative and efficient technologies like continuous monitoring and glucose flash-based monitoring. However, the deployment of trackers equipped with inexpensive sensors and web of things technology is even more expensive (Nandan et al., 2022).

The information that is collected even from the remote locations is sent to the doctors who are situated in another location (Verma and Sood, 2018). So, with the help of those received data, the doctor can further proceed the treatment without doing any test, which will really save the life of people in danger. But the flaw that occurs over here is the lack of awareness about the use of such sensors. This makes the people confuse whether it will really work effectively (Bansal et al., 2018).

The working principles and process of IoT to measure glucose levels are not so difficult (Nandan et al., 2022). The initial step will hold the deployment of the sensors, monitor, finder, etc. These devices are mainly used for collecting accurate data. Analogue data collected by sensors and other types of external devices must be transformed into digital data before being incorporated. That particular data is used for further processing.

Once the acquired data have been translated to the digital form and has undergone pre-processing, it is transferred to a data center or the cloud. The final data that we obtain are maintained for analysis and are used to forecast the outcome. This serves as a critical component for tracking and monitoring the status and determining what should be done next (Gelogo et al., 2015). Using IoT-enabled technologies, the objective of this proposed research work is to build a chain of awareness regarding the problems with the glucose levels in the human body. To demonstrate how much better to use sophisticated tracking-enabled systems, the operating concept and procedures that are used are clearly defined (Sattar et al., 2020).

The data gathered will be helpful in determining how much effort has to be put into educating individuals about the value of using the IoT to forecast their blood sugar levels (Kulkarni and Jakkan, 2019). Additionally, it makes no difference in terms of age when people started using these technologies; everyone, from young children to older people, can gain advantages from a straightforward sensor. As a result, a tiny spark can brighten the lives of many people by prolonging their lives.

7.12 PROPOSED GLUCOSE TRACKING SYSTEM

The IoT-based healthcare glucose tracking system now allows remote monitoring using IoT and smart gadgets. The IoT technology aids in monitoring of patient's pulse, obtaining an analysis, and sending the report to surrounding hospitals. Today, diabetes is a major health problem affecting individuals of all ages. People with diabetes often have a lot of diseases, including skin diseases, bladder failure, and damage to all body parts like heart. We analyzed nearly 500 people according to their age. In particular, those with diabetes who are older than 40 are affected at a rate of 37.6%, followed by people in their 30s and 40s at a rate of 32.6%, people in their 26s and 30s at a rate of 19.8%, and people in their 18s to 25s at a rate of 10%. In this investigation, it was shown that only individuals over the age of 40 had glucose issues. Patients are embracing wearable gadgets, such as smartwatches, to provide detailed health monitoring and receive comprehensive analytical reports. Many people will be aware of their health and lives in the future, and IoT smart tracking systems will assist them to manage their day-to-day activities.

7.13 CONCLUSION

In the future, IoT smart devices will analyze the information, deliver the report to patient's family, alert them individually, and send it to the doctor directly. Sensors would also find applications in the physical industry and the healthcare sector in the future. Smart gadgets primarily use a variety of methods to analyze data from both public and private sources, such as surveillance for public day-to-day activities and also monitoring hospitals. In sequence for us to apply smart devices in the healthcare sector, data must be secure, maintained safely, and easily accessible from cloud resources. This information makes it easier to focus attention on diabetes patients, make quick decisions, and administer the appropriate treatments. Nearby hospitals have received this report, and they are prepared to care for the patients whenever it is necessary to protect their health and lives. So, when the person who is aged above 40 years are monitored with the proper sensor using IoT can be kept in a safer zone. Once proper guidance and support are given to them, they can easily be protected from these typical glucose issues.

REFERENCES

Abdulmalek, S., A. Nasir, W.A. Jabbar, M.A.M. Almuhaya, A.K. Bairagi, A.-M. Khan, and S.-H. Kee, "IoT-Based Healthcare-Monitoring System Towards Improving Quality of Life: A Review", *Healthcare*, 2022, DIO: 10.3390/healthcare10101993.

Akkasaligar, P.T., S. Potnis, and S. Tolnur, "Review of IoT Based Health Monitoring System", *International Journal of Research in Advent Technology*, 95–99, 2019.

Alekya, R., N.D. Boddeti, K.S. Monica, R. Prabha, and V. Venkatesh, "IoT Based Smart Healthcare Monitoring Systems: A Literature Review", *European Journal of Molecular & Clinical Medicine*, 2021, 7, 2020.

Al Shorman, O., M.S. Masadeh, and B. Al Shorman, "Mobile Health Monitoring-Based Studies for Diabetes Mellitus: A Review", *Bulletin of Electrical Engineering and Informatics*, 2021. DOI: 10.11591/eei.v10i3.3019.

Anagha, R., G. Keethana, S. Ashwini, and M. Monica, "Iot Based Intravenous Flow Monitoring System", *International Research Journal of Engineering and Technology*, 7(5), 2020.

Baker, S.B. and W. Xiang, "Internet of Things for Smart Healthcare: Technologies, Challenges, and Opportunities", *IEEE Access*, 99, 1–1, 2017.

Banka, S., I. Madan, and S.S. Saranya, "Smart Healthcare Monitoring Using IoT", *International Journal of Applied Engineering Research*, 13(15), 11984–11989, 2018.

Bansal, M., R. Pandey, and S. Ghosh, "A Review on IoT Based m-Health Systems for Diabetes", *Journal of Emerging Technologies and Innovative Research*, 5(8), 49–54, 2018.

Chitra, P., S. Logesh, P. Vinithkumar, A. Gowtham, and R. Jenifer, "A Certain Investigation on Health Monitoring System by Using Internetof Things [IOT]", *International Journal of Engineering Research & Technology*, 4(11), 1–5, 2016.

Deshkar, S., R.A. Thanseeh, and V.G. Menon, "A Review on IoT Based m-Health Systems for Diabetes", *International Journal of Computer Science and Telecommunications*, 8(1), 13–18, 2017.

Efat, I.A., S. Rahman, and T. Rahman, "IoT Based Smart Health Monitoring System for Diabetes Patients Using Neural Network", *Cyber Security and Computer Science*, 2020. DOI: 10.1007/978-3-030-52856-0_47.

Gelogo, Y.E., H.J. Hwang, and H.-K. Kim. "Internet of Things (IoT) Framework for u-Healthcare System", *International Journal of Smart Home*, 9(11), 323–330, 2015.

Gia, T.N., M. Ali, I.B. Dhaou, A.M. Rahmani, T.W.P. Liljeberg, and H.A.M. Rahmani, "IoT-Based Continuous Glucose Monitoring System: A Feasibility Study", *8th International Conference on Ambient Systems, Networks and Technologies*, 2017. DOI: 10.1016/j.procs.2017.05.359.

Hussein, A.R.H., "Internet of Things (IOT): Research CHALLENGES AND FUTURE APPLICATIONs", *(IJACSA) International Journal of Advanced Computer Science and Applications,* 10(6), 77–82, 2019.

Islam, M.R., D. Kwak, H. Kabir, M. Hossain, and K.-S. Kwak. "The Internet of Things for Health Care: A Comprehensive Survey", *IEEE Access*, 2015. DOI: 10.1109/ACCESS.2015.2437951.

Khan, R., S.U. Khan, R. Zaheer, and S. Khan, "Future Internet: The Internet of Things Architecture, Possible Applications and Key Challenges", *10th International Conference on Frontiers of Information Technology (FIT): Proceedings*, Islamabad, Pakistan, 2012.

Kulkarni, D.D. and D.A. Jakkan, "A Survey on Smart Health Care System Implemented Using Internet of Things", *Journal of Communication Engineering and Its Innovations*, 2019. DOI: 10.5281/zenodo.2575414.

Kumbi, A., P. Naik, K.C. Katti, and K. Kotin, "A Survey Paper on Internet of Things Based Healthcare System", *Internet of Things and Cloud Computing*, 5, 1–4, 2017. DOI: 10.11648/j.iotcc.s.2017050501.11.

Longva, A.M. and Haddara, M. "How Can IoT Improve the Life-Quality of Diabetes Patients?" *MATEC Web of Conferences*, 2019. DOI: 10.1051/matecconf/201929200C SCC2019316316.

Mittal, P. and Navita M., "A Survey on Internet of Things (IoT) Based Healthcare Monitoring System", *International Journal of Advanced Trends in Computer Science and Engineering*, 8(4), 2019. DOI: 10.30534/ijatcse/2019/90842019.

Mouha, R.A., "Internet of Things (IoT)", *Journal of Data Analysis and Information Processing*, 2021. DOI: 10.4236/jdaip.2021.92006.

Nandan, R., S. Navaneet, R. Sanjay, S. Jeevan, and H.L. Kumar, "IoT Based Electronic Valve System with Quantitative Control and Patient Monitoring", *International Journal for Research in Applied Science & Engineering Technology*, 10(1), 1–5, 2022.

Phani Ram, G.B. and T. Shankar, "A Survey on IoT Health Parameters Monitoring System", *Journal of Positive School Psychology*, 6(6), 7585–7589, 2022.

Prakash, E.P., K. Srihari, S. Karthik, M.V. Kamal, P. Dileep, S. Bharath Reddy, M.A. Mukunthan, K. Somasundaram, R. Jaikumar, N. Gayathri, and K. Sahile, "Implementation of Artificial Neural Network to Predict Diabetes with High-Quality Health System", *Hindawi Computational Intelligence and Neuroscience*, 2022. DOI: 10.1155/2022/1174173.

Ru, L., B. Zhang, J. Duan, G. Ru, A. Sharma, G. Dhiman, G.S. Gaba, E.S. Jaha, and M. Masud, "A Detailed Research on Human Health Monitoring System Based on Internet of Things", *Hindawi Wireless Communications and Mobile Computing*, 2021. DOI: 10.1155/2021/5592454.

Sattar, K.N.A., M.M. Otoom, M. Al Sadig, and N. Nandini. "Detection of Levels of Blood Sugar Using Simple IoT Based Breath Analysis", *IJCSNS International Journal of Computer Science and Network Security*, 20(8), 2020. DOI: 10.13140/RG.2.2.29326.18249.

Sharmila, G., R. Sushmitha, and R. Geetha, "A Survey on Technologies Used in Glucose Monitoring for Diabetic Patients", *International Journal of Innovative Research in Computer Science & Technology*, 2017. DOI: 10.21276/ijircst.2017.5.2.6.

Shinde Sayali, P. and N. Phalle Vaibhavi," A Survey Paper on Internet of Things based Healthcare System", *International Advanced Research Journal in Science, Engineering and Technology,* 2017. DOI: 10.17148/IARJSET/NCIARCSE.2017.38.

Verma, P. and S.K. Sood, "Cloud-Centric IoT Based Disease Diagnosis Healthcare Framework", *Journal of Parallel and Distributed Computing*, 2018. DOI: 10.1016/j.jpdc.2017.11.018.

Yusuf, N., A. Hamza, R.S. Muhammad, M.A. Suleiman, and Z.A. Abubakar, "Smart Health Internet of Thing for Continuous Glucose Monitoring: A Survey", *International Journal of Integrated Engineering*, 2020. DOI: 10.30880/ijie.2020.12.07.006.

8 Intelligent Application to Support Smart Farming Using Cloud Computing

Future Perspectives

J. Rajeswari, J.P. Josh Kumar, S. Selvaraj, and M. Sreedhar

CONTENTS

8.1 INTRODUCTION

The United Nations' Food and Agriculture Organization reported that the growing population level and the capacity for food production are vastly out of balance. Overall, there will be a need to increase food production by a huge margin of 70% to match the population's needs by the year 2050. Agricultural lands are getting reduced, and natural resources are getting depleted due to the steady rise in population. Natural resource shortages, including those of fresh water and arable land, as well as declining yield trends in a number of staple crops, considerably exacerbate the problem. Making issues worse is the workforce shifts from agriculture to different other industries, which has decreased agricultural labor in most countries.

DOI: 10.1201/9781003438588-8

The implementation of the Internet of Things (IoT) as a best practice in agriculture is necessary right now to eliminate manual labor in all areas as a result of all these problems [1]. Mobile phones are widely used today. In every area of business and education, the use of smart phones is rising. In India, more than 70% of people work in agriculture and farming. The Indian economy grows immediately when agricultural productivity rises. The growth of non-agricultural options has caused a significant change in the farming and life styles of Indian farmers. Because the majority of farmers are uneducated or because they are unaware, technical advances in agriculture are not reaching them. The majority of farmers consequently are unable to reach the maximum production rate. Over 19.1% of loss is included in the global loss growth of more than 40% [2].

Therefore, the smart farm application will be introduced in order to solve all of these issues. By maximizing the use of fertilizers to increase plant efficiency, innovative agriculture, which is based on IoT technology, is envisioned to enable producers and farmers to reduce waste and improve production. Farmers can manage their livestock, increase agricultural production, cut costs, and conserve resources thanks to it. It is a high-tech system for producing crops that are sustainable and clean for the general public. In agriculture, it refers to the use of contemporary information and communication technologies. This proposed study is suggested as "smart farming" because it covers the major issues with receiving updates from the land via smart phone and also offers the required farming operations. This application will provide a variety of information on a single system. Similar to how it contains information on crops, such as soil moisture, temperature, humidity, plant height, and meteorological data. New approaches and technological advancements are continually being put forth and put into practice in order to suit the demands of humanity today. IoT has evolved as a result and solution for all these needs [2]. IoT acts as a bridge to communicate and share information between the digital and physical worlds, which is a network of all equipment that is integrated into gadgets, sensors, machines, software, and people. So, the technique IoT and big data will be used to support smart agriculture, which is expected to enhance the productivity and efficiency. The analysis of several cloud-based IoT platforms as well as technologies for data collection and communication inside IoT systems for smart farming were discussed. Rezk et al. [3] propounded a new prediction method, which is WPART coupled with an IoT-based intelligent production system, used to establish the machine intelligence approaches to think about crop yield and dryness. Tzounis et al. [4] used a three-layered IoT architecture included the perception, network, and application. This architecture was used to identify the embedded platforms and communication technologies used in IoT solutions as well as the use cases for such solution. An efficient and intelligent culture arrangement was developed by Perales Gómez et al. [5] for crop production that comprises common data storage, low-cost IoT sensors, and dossier science of logical analysis services related to the IoT. IoT encompasses the fundamental communication infrastructure and a variety of services, such as local or distant data collecting, cloud-based intelligent information analysis and decision-making, agriculture operation automation, and user interface. With regard to our economic value chain, the agriculture industry is now one of the least efficient. These capabilities have the potential to transform this. Thombare et al. [6] proposed a research based on big data to clarify crop production projections in light of the related socioeconomic difficulties. For a specific region of the field, which agricultural methods are most effective has been decided by the K-means clustering techniques? Apriori algorithms are used to estimate production. For function discovery in massive data sets, utilizing a network, which is a Hadoop-based neural network introduced by Victoria et al. [7], seems to be comparable and decentralized in the computing paradigm. Using a Hadoop YARN artificial neural network frame, five attribute selection techniques were put into practice. In order to determine the best attribute selector for swiftly identifying acceptable qualities from a variety of and high-dimensional data sets, the Hadoop binary relational memory network is combined with stability and adaptability. Lin [8] used the Apriori data mining approach to create the MapReduce representation in Hadoop. Typically, association rules algorithms that deal with a lot of attributes compute rule scarcity. This has been overcome by the proposed technique MR-Apriori algorithm. Priya et al. [9] crop suggestion system works well. Different crops can be assessed thanks to the equipment's adaptability. The best time for planting, plant growth, and harvesting can be determined using a yield chart. To tell farmers which crop they can plant on the ground, an accurate agricultural model is provided.

Suryanarayana et al. [10] created a framework that utilizes MapReduce and Hadoop approaches to analyze historical meteorological data from a region. Weather forecasting will be helpful for many important climate-sensitive industries, including water supply, air travel, agriculture, and tourism. The branch of meteorology is weather forecasting, which gathers data on the current weather from many sources, including airstream, high temperature, drizzle, and smog. In the IoT era, the meteorological service, among other things, uses a range of sensors to measure humidity and temperature. The distributed algorithms seem to be utilized in MapReduce technology to quickly analyze weather data. Using MapReduce with Hadoop has the benefit of accelerating data collecting in a case where the quantity of data is increasing daily. A feature selection process is the sole technique to identify the crucial features in a data collection. It is quick to finish and has a high accuracy rating. The focus of a lot of literature has been on feature selection algorithms. Numerous data mining techniques have been developed recently and used to the analysis of agricultural and biological datasets, producing useful classification patterns. Crop growing conditions can be controlled with the aid of monitoring data such as soil moisture, temperature, and condition, as well as the forecasting of ordinary features such as weather and rainfall. Farmers that want to maximize productivity and cut labor expenses can use this support to plan and make irrigation verdicts. Additionally, recommendations for preventive and corrective measures against pests and illnesses in farming can be delivered utilizing big data processing technology in combination with the assembled data.

8.2 METHODS AND MATERIALS

The basic structure of an IoT device consists of actuators, which move objects via sensors, wired or wireless connections, memory, input-output interfaces, and an embedded system with a CPU, communication modules, and battery power. The typical architecture flow diagram of a smart agricultural is shown in Figure 8.1. In order to detect

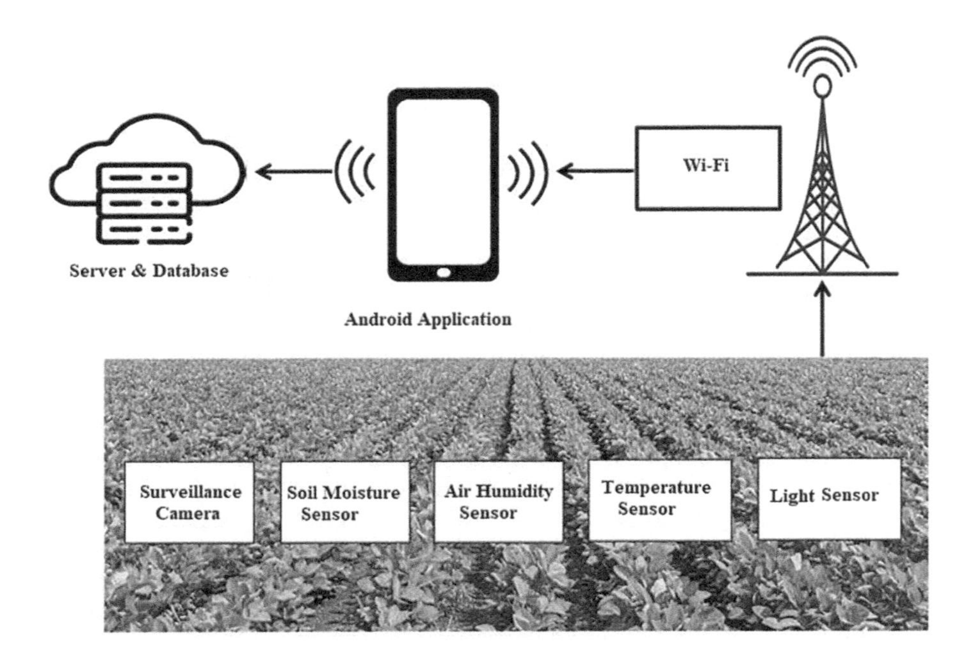

FIGURE 8.1 Architectural diagram of IoT-based smart farming.

and collect environmental parameters such as humidity, soil nutrients, and high temperatures that affect production, sensor devices are specially made to function in open spaces, especially in environment with mud, H2O, and air. Based on what has to be done, different common sensors are employed in the smart agriculture sector. Common and typical sensors, such as mechanical, optical, location, electrochemical, and air flow sensors, are used for smart farming. For the purpose of accurately gathering soil data that may then be used in a farm monitoring dashboard or smartphone application, sensors can be evenly spaced out across the length and breadth of field. Almost everything in agriculture is dependent on the weather, and farmers can greatly benefit from weather changes that can be foreseen. Through a variety of sensors, smart farming allows for the monitoring of temperature, humidity, light intensity, and soil moisture. Again, the reactive system makes advantage of them to send out alarms or automate procedures like water and air control. Intelligent pumps or automated sprinkler systems are employed in smart irrigation. To gauge the amount of moisture in agricultural land, soil moisture sensors are frequently used. These sensors delivered the details, which are useful to turn on or off the intelligent pumps. Additionally, this has been used to collect data on the wind speed, air, soil, soil moisture, leaf moisture, precipitation, air humidity, wind direction, solar radiation, and barometric pressure. Table 8.1 displays the necessary functional components for smart farming.

In this work, React Native, JSON, and JavaScript were used to construct the Android application, while Firebase Database was used for security. The sensors were utilized to measure several characteristics, including temperature, soil moisture, and humidity. Open weather API was used for collecting the weather information. The abovementioned data were processed with the help of a microcontroller, which

TABLE 8.1
Functional Requirements for Smart Farming

S. No.	Functional Requirement	Sub Requirement
1	User registration	Registration through Gmail
2	User confirmation	Confirmation via Email Confirmation via OTP
3	User login	Login with Email Id and Password
4	Forgot password	Login with Email Confirmation of OTP
5	Query form	Make a note of the problems and issues faced by user when using the application
6	Weather	Make a note of the problems and issues faced by user when using the application
7	Agro note	To list of agriculture-related information like how to plant, how much liters of water that plant need in a day, etc.
8	Sensors	To show various data from different sensors like temperature, humidity, and soil moisture
9	Database management	To show various agriculture-related data are stored
10	Exit	After user checked every information, user can exit the application

is connected to the Internet. The processed data were then updated to cloud for further process. The IoT Platform was connected with node red services, which were connected to the application. In application, user can see the parameters/data that obtained from sensors and APIs. With the help of application, user can interact with IoT devices to perform some functions such turning ON and OFF motor. Web User Interface was used for visualization of data.

8.2.1 Arduino Uno

This board has three reset catches, Wi-Fi, Ethernet, a USB connector, a micro-SD card slot, and an ATMega 32u4 MCU that is compatible with Arduino. To run Linux, the board can alternatively be connected to an Atheros AR9331 [11].

8.2.2 DHT11/DHT22 Humidity Sensor

This sensor is used to measure moisture and humidity. The land's humidity and moisture content are regularly measured using this method. Utilizing an Arduino Uno [12], this sensed data are stored in the cloud.

8.2.3 YL-69 Soil Moisture Sensor

Using this sensor, one may ascertain the soil's water content. It is widely used for applications that require precise estimations of the amount of water in the soil, such as those in agriculture, water management, greenhouses, and other research institutions. A test to determine how much muck is in the ground is joined with an electrical panel that grasps the gear to structure the device. In order for the sensor to function, a probable difference that is directly related to the dielectric permittivity of the water must be made. Voltage differences can be seen as dissimilarities in dielectric permittivity and, thus, fluctuate the water levels [13].

8.2.4 Camera

Using an IoT Arduino Uno gadget, cropped pictures are taken and then saved in the cloud.

8.2.5 Cloud Storage

The cloud is where all crop-related photographs are kept.

8.3 JSON

Web services have attracted a lot of attention in recent years as the most popular method for employing service oriented architecture [13]. In the context of web service applications, XML is the de facto standard for data interchange, a language for describing data that is platform independent and can realize interoperation between disparate systems. On the client side, an XML parser is necessary to read the XML data for web services. This brings up two major problems, such as cheaper to deploy and create DMTML components when a page reads the same domain's XML data for the security of the application and web service performance has suffered due to the inefficiency during service execution of reading and parsing XML data. In the era of human-centered Web 2.0, asynchronous JavaScript and XML (AJAX) is one of the highlights for developing web services. AJAX uses JavaScript on the client side, significantly reducing the server's effort. In order to get over the aforementioned issues with XML-based services, we study the effectiveness of using JSON in place of XML to represent the data sent in the application of web services. JSON is a compact format for exchanging key-value data. Due to its simplicity, JSON is easy for both people to read and write as well as for computers to generate and parse. JSON is independent of any programming language, although sharing several traits with the C family of programming languages, which includes C, C++, Java, JavaScript, and Perl. Because JSON is a native data type for JavaScript, processing JSON data does not require any specialized APIs or jars. JSON is a viable data exchange format for web service applications thanks to these qualities. Processing JSON data has been the subject of a great deal of research in recent years [14]. The majority of the research examined the effects of JSON-based data and XML-based data on the functionality of online services. Two elements were identified that contributed to the inefficient operation of XML-based web services. The two factors were the network communication protocol of web services and parsing and formatting XML data. It was confirmed that web service performance might be enhanced by enhancing the SOAP encoding and decoding techniques. High-performance SOAP processing was shown using a dynamic template-driven technique. The data between XML and Java are efficiently mapped by the paradigm. JSON can be parsed more effectively than XML. Researchers have previously demonstrated that JSON is capable of replacing XML as the preferred format for exchanging data in web services. This was confirmed by simulation tests on the effectiveness of XML and JSON [15].

8.4 REACT NATIVE

The mobile app's operating technique begins with the login process [16]. The React Native framework's advantages are utilized by the mobile application. A discussion is had over the dependencies used in the created modules.

React-Native-Firebase: The following features are offered by the Google product that serves as the mobile application's backend.

Authentication: Multiple modes for authentication, including social media, email, and phone, are supported. The current work makes use of phone authentication services that are offered.

Storage: It keeps all of the arranged photographs from different users because it is an image storage unit.

Real-Time Database: It functions as a storage container for the mobile application and immediately saves the account information, order information, and address details unique to each user.

React-Native-Google-Places-Auto Complete: It is a different Google product that has the ability to obtain the user's search terms and the user's current geo location. It displays location by retrieving the state and then the city. Information about the rest area is physically noted by the user.

React-Native-Vector-Icons: It is a dependency that offers custom icons together with react native elements.

React-Native-Image-Crop-Picker: The dependency has the ability to access images from a camera or picture library. Currently, the ordering option from photograph makes use of this capacity.

React-Native-Elements: As it contains all the stated components that are required, such as buttons, lists, avatars, it is a dependency that aids the developer in configuring user interface.

React-Native-Keyboard-Aware-Scroll-View: Only iOS uses this requirement to enable the virtual keyboard [17].

React-Native-Segmented-Control-Tab: A component segment is created by a dependence. This feature is currently being used to distinguish between list-based and photograph-based ordering [18].

React-Navigation: As this switches the user from one screen to another, it is the most crucial dependence of react native. This dependency also offers a tab navigator for multiple tab creation in mobile applications. Another crucial component that stores all the screens in a single stack is the stack navigator [19].

Native-Base: It is a different user interface-forming dependency that can produce user interfaces that contains elements like buttons, images, text, forms, form inputs, and lists.

8.5 WEB USER INTERFACE

The simplest kind of management is the Web User Interface (UI). It gives the end user a head start in managing the Juniper firewall due to its straightforward point-and-select functionality. The browser's left side is where the menu column is situated. You can select from the numerous configuration choices from here. This menu can be either Java-based, which is the default, or Dynamic Hypertext Markup Language (DHTML)-based. While the functionality is unchanged, the appearance and feel have been somewhat altered. In its default configuration, the WebUI only supports the Hypertext Transfer Protocol (HTTP). The CICSPlex® SM Web User Interface (WUI) offers a user-friendly interface that can be used to carry out all administrative and operational tasks necessary to manage and watch over CICS® resources [20]. Any location that can open a Web browser can link to the Web User Interface. All of your system management duties are made easier by a set of linked menus and views that come with the WUI.

To adapt views' text and look to particular business requirements, use language that is appropriate for your industry, for annotations and personalized assistance, choose the native language, use filters to restrict the data that is displayed so that users only see the information necessary for their task, provide guiding information for the user, such as contact names and phone numbers. Create a list of favorites and assign views to them for quick access. This makes it possible to access frequently used views quickly. Additionally, administrators have the authority to change user favorites. Display the information users require to do an activity. Create user profiles for user groups. These profiles provide data on the menu and result set warning count, as well as default context, scope, and CMAS context [21]. Administrators can set up the WUI in various ways to suit various user groups in order to present a more individualized interface in this way. Limit what can be altered, where it can be changed, and how it can be changed. Only show the user the information you want them to see. For example, to guarantee that the user affirms that an activity is required or that data needs to be changed. In addition to limiting entry fields to display-only or predetermined values and increasing security by presenting a confirmation panel that asks the user to confirm an action is to be conducted, the WUI can notify users before opening views that will create a large number of records. Performance is enhanced by cutting down on pointless waiting. To help people finish activities, create menus. Each menu and view has the option of linking to already-existing web-based procedures manual on an external server or hosting the help files on the CICS system, which has been designated as the Web User Interface server. Protect against unauthorized entry to the user editor, special menus, views, and help panels.

8.6 OPEN WEATHER MAP API

Weather data can be accessed via a variety of web-based services and mobile applications, including forecasts and current analysis data, using an online meteorological service known as OpenWeatherMap API. In particular, it uses unprocessed data from airport weather stations, radar stations, and other official weather stations [22]. It also makes use of meteorological broadcast services. To assess all the data and enhance the numerical weather prediction models delivered by multiple data sources, the OpenWeatherMap API in particular leverages machine learning. The OpenWeatherMap API was modeled using the Wikipedia and OpenStreetMap systems, which make information easily accessible to each person. OpenWeatherMap API frequently makes use of the OpenStreetMap to depict its spatially related weather predictions. Additionally, it offers a free, limitless JSON API with endpoints that can be called to obtain 3-hourly forecast values for up to 5 days as well as the most recent values for weather indicators, which are updated hourly.

8.7 RESULTS AND DISCUSSION

Four actions are required for smart farming. The first is observation, where we utilize sensors to feel the environment and gather data on the soil, temperature, humidity, and other factors. The next step is diagnosis, which entails sending sensor data to cloud platforms powered by the IoT for data analytics. The farmers must then decide depending on the results of the analysis and take appropriate action to produce better results. The final one is the action conducted in response to all data made available via IoT [23]. The fog, cloud, edge, and sensor device layers are the four fundamental architectural layers that are frequently in use. The many sensors, actuators, and devices correlated with the ground atmosphere are included in the sensor device layer. The many endpoint devices and sensors that transmit and receive information reside in this layer, which is home to the "things" in the IoT. These "things" can be as large as enormous factory machines or almost minuscule sensors. Their main job is to produce data and have network access to be queried and controlled.

The controller unit that various sensors and actuators interface with at the edge layer is where data is gathered and sent to the fog layer for processing. This aids in processing data at the point of collection or consumption, even before it is sent to the fog or cloud computing layer. Assembling analytics and decision-making models based on data from the edge layer, as well as sending control signals to the edge layer for actuator control, are the foremost objectives of the fog layer. This procedure, known as "fog computing," adds a layer between the cloud layer and the edge layer. Instead of relying on devices or the edge, this layer uses computers and servers that are close to the edge layer [24]. Due to the connections between these machines and servers and the cloud computing layer, processing and intelligent information can move quickly. The system becomes more efficient overall because of the fog computing architecture's reduction in the amount of data that must be uploaded to the cloud.

A centralized method called cloud computing makes it easier to distribute data and documentations to information hubs through the Internet. Accessing a variety of data and apps is made simple via a centralized cloud layer. The cloud layer, which provides more flexibility, scalability, and connectivity, can be used to store, manage, and process data. Additionally, the cloud can provide built-in machine learning, business intelligence, and SQL engines to carry out sophisticated operations necessary in the present IoT environment. Finally, in terms of this study, a user interface (UI) dashboard is utilized to exhibits actuator and sensor data on the cloud layer [25].

The ability of the suggested solution to assist farmers by providing greenhouse management using an IoT-based precision farming framework sets it apart from competing solutions. Applications for the IoT in agriculture concentrate on traditional farming practices in order to meet increasing demand and minimizing the output losses. Farmers will receive the status of soil moisture, CO2, light, and temperature, depending on soil moisture levels from a distance automatically. Figure 8.2 shows the

FIGURE 8.2 Application dashboard for smart farming.

FIGURE 8.3 Application for smart farming.

application that has been developed. Figures 8.3 and 8.4 show the application's home page and additional choices screen.

The utilization of IoT generates big data on a wide scale that offers useful information. In order to offer climate, fertilization, irrigation, and pest management, the system used wireless sensor networks to collect and analyze plant-related sensor data. Data from sensor networks were also used in another approach to data mine for knowledge. Another study examined the relationship between crop, weather, environment, and the leaf spot disease utilizing the sensors without wire and field level surveillance. To forecast the sickness, a classifier was trained, a new, practical, cloud-based online monitoring system for IoT devices. Once an agricultural IoT system had generated enough data, modeling of the pertinent functional necessities was shown to support the use of huge data investigation in farming. To extract meaningful knowledge and information, data mining has only been used in a small number of research [26]. The functionality

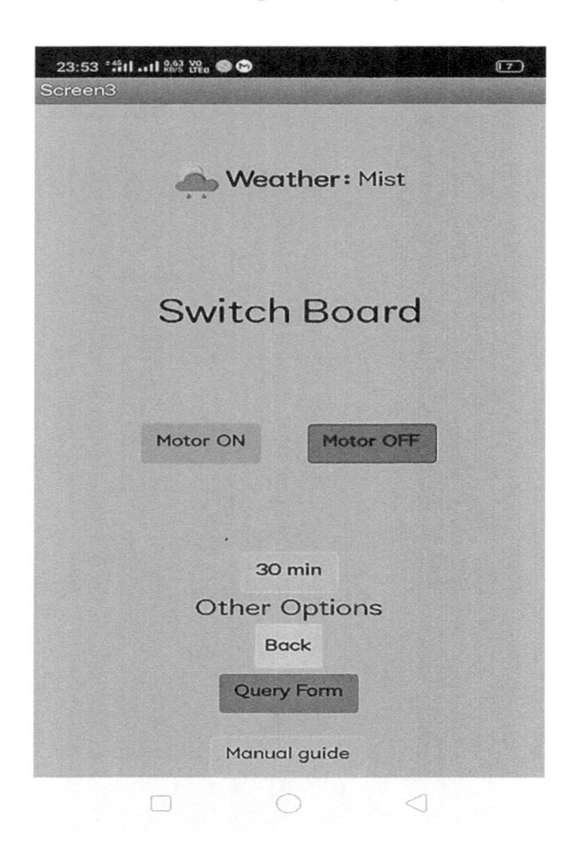

FIGURE 8.4 Application window for other options.

of IoT has been improved by a number of studies. Based on IEEE-802.15.4, the creation of a WSN for application in temperature-based frost characterization in precision agriculture was demonstrated. For scheduling irrigation in olive fields, an online decision-support tool that communicates with a WSN has been suggested. The authors accomplished this using sensors that recorded temperature, rain, humidity, sun radiation, and other variables. Temperature and soil moisture were controlled by an electronic device (Arduino), and an Android-based smart phone application was employed for flexibility and utility. In comparison with pricey components like top-of-the-line personal computers, they discovered that flexibility and low cost are advantages for agricultural control [27]. An artificial structure was established that united drought monitoring and forecasts with irrigation amount predictions were using IoT, hybrid programming, and parallel computing. WSN irrigation systems were considered with the objective of collecting environmental data and permitting smartphone-based irrigation system control. For the automation of agriculture to collect data from vegetable crops or for environmental assessments, a long-term supportable resolution was presented. For that, a transportable measurement method with soil moisture, air temperature, air humidity, and sensors was developed by the authors.

The bulk of individuals can profit from advanced technologies. IoT has just recently originated to ensure a substantial impact on daily life, expanding our senses and enabling us to change the environment around us. IoTs are especially used in the agro-industrial and environmental industries for both control and diagnostic purposes. Additionally, it can enlighten the consumer on the origins and characteristics of the product. So, the purpose of this work is to apply IoT to agricultural computer-aided optimization. The usage of IoT in the field has increased farmers' efficacy and efficiency in such agricultural optimization. It can aid in assessing field variables such soil quality, weather, and animal or plant biomass [28]. Additionally, it can be utilized to monitor and manage conditions like temperature, humidity, vibrations, and shocks while a product is being transported. WSN can also be used to track and manage variables that affect crop output and growth. They can also be used to manage machinery, find the best farmer for a certain situation, find the best time to harvest, and more. Data storage, a method to extract knowledge from a body of data, user interactions, and system development are all necessary for creating an effective system. A web-based application that uses a database will be created and put into use. To make decisions on the automatic irrigation of crops, the saved data will be used. The analysis of the agricultural data will help to improve and alter the environment, as well as forecast future crop water requirements. The agro-industrial production chain was used the IoT devices. In order to follow the enlargement of grapes for wineries, a novel architecture built on the impression of IoT was proposed [29]. It combines wireless and distributed particular sensor devices with the modeling of climatic conditions.

Suma [11] provides an overview of forecasting investigation, IoT designs with cloud presidency, and security liabilities for multi-breeding in the farm land, while keeping in mind the previous experiences of farmers. In order to monitor paddy fields haphazardly, Sethy et al. [12] proposed a method used both the deep learning and IoT. Nitrogen rank belief and paddy leaf disease detection are being surveyed using the VGG16 constructed network. According to Kaushik et al. [22], an intelligent agriculture technique that keeps an eye on the farm can assist farmers dramatically boost productivity. Fluorescence-based optical sensors are regarded as an excellent option due to their low cost, flexibility in terms of application, ease of use, and adaptability (e.g., sample rate). These sensors can then be utilized in conjunction with a security camera for weed detection [30], where the ultrasonic sensors identify plant heights and the video estimates the amount of weed and crop coverage.

8.8 CONCLUSION

In conclusion, this application will be more beneficial to farmers by enabling them to access all types of information with a single click at any time and from any location after an examination of their present understanding of modern farming practices and the growth of new techniques. All of this leads to the conclusion that every square foot of farmland is essential to maximizing crop yield. However, using supportable IoT-based devices with sensors and its communication technology is not an option; it is required if we are to deal with every inch in the right way.

REFERENCES

1. IoT Solutions World Confress. IoT transforming the future of agriculture. https://www.iotsworldcongress.com/iot-transforming-the-future-of-agriculture/.
2. Chaudhari S, Mhatre V, Patil P, Chavan S. "Smart farm application: A modern farming technique using android application," *International Research Journal of Engineering and Technology*, 5(2) (2018), 317–320.
3. Rezk ND, Hemdan EED, Attia AF, El-Sayed A, and El-Rashidy MA. "An efficient IoT based smart farming system using machine learning algorithms," *Multimedia Tools and Applications*, 80(1) (2021), 773–797.
4. Tzounis A, Katsoulas N, Bartzanas T, Kittas C. "Internet of Things in agriculture, recent advances and future challenges," *Biosystems Engineering*, 164 (2017), 31–48.
5. Perales Gómez AL, López-de-Teruel PE, Ruiz A, García-Mateos G, BernabéGarcía G, and García Clemente FJ. "FARMIT: Continuous assessment of crop quality using machine learning and deep learning techniques for IoT based smart farming," *Cluster Computing*, 25(3) (2022), 2163–2178.
6. Thombare R, Bhosale S, Dhemey P, and Chaudhari A. "Crop yield prediction using big data analytics," *International Journal of Computer and Mathematical Sciences*, 6(11) (2017), 53–61.
7. Victoria JH, Simon OK, and Jim A, "Hadoop neural network for parallel and distributed feature selection," *Neural Networks*, 78 (2016), 24–35.
8. Lin X. "MR-Apriori: Association rules algorithm based on MapReduce," in *Proceedings of the IEEE International Journal of Food Quality Conference on Software Engineering and Service Science*, Beijing, China (2014), pp. 141–144.
9. Priya R, Dharavath R, and Ekaansh K. "Crop prediction on the region belts of India: A Nave Bayes MapReduce precision agricultural model," in *Proceedings of the IEEE International Conference on Advances in Computing, Communications and Informatics (ICACCI)*, Bangalore, India (2018), pp. 99–104.
10. Suryanarayana B, Sathish S, Ranganayakulu A, and Ganesan P. "Novel weather data analysis using hadoop and MapReduceA case study," in *Proceedings of the IEEE International Conference on Advanced Computing and Communication Systems (ICACCS)*, Coimbatore, India (2019), pp. 204–207.
11. Suma V. "Internet of Things (IoT) based intelligent agriculture in India: An overview," *Journal of ISMAC*, 3(1) (2021), 1–15.
12. Sethy PK, Behera SK, Kannan N, Narayanand S, and Pandey C. " Smart paddy field monitoring system using deep learning and IoT," *Concurrent Engineering Research and Applications*, 29(1) (2021), 16–24.
13. Zhang L, Dabipi IK, and Brown WL. "Internet of Things applications for agriculture," in Hassan QF (Ed.), *Internet of Things A to Z: Technologies and Applications*. Wiley: Hoboken, NJ (2018), pp. 507–564.
14. Navulur S, Sastry A.CS, and GiriPrasad N. "Agricultural management through wireless sensors and Internet of Things," *International Journal of Electrical and Computer Engineering*, 7(6) (2017), 3492–3499.
15. Fei X, Xiao W, Yong X. "Development of energy saving and rapid temperature control technology for intelligent greenhouses," *IEEE Access*, 9 (2021), 29677–29685.
16. Dunlu P, Lidong C, Wenjie X. "Using JSON for data exchanging in web service applications," *Journal of Computational Information Systems*, 7 (2011), 5883–5890.
17. Hua L, Wei J, Niu C and Zheng H. "High performance SOAP processing based on dynamic template-driven mechanism," *Chinese Journal of Computer*, 7(29) (2006), 1145–1156.
18. Cui C, and Ni H. "Optimized simulation on XML with JSON," *Communication Technology*, 42(8) (2009), 108–111.

19. Danielsson W, Froberg A, and Berglund E. "React native application development: A comparison between native Android and React Native," Thesis, Linköping University (2016).
20. Jirapond M et al. "IoT and agriculture data analysis for smart farm," *Computers and Electronics in Agriculture*, 156 (2019), 467–474.
21. Beyshir A. "Cross-platform development with React Native," (2016).
22. Kaushik N, Narad S, Mohature A, and Sakpal P. "Predictive analysis of IoT based digital agriculture system using machine learning," *International Journal of Engineering Science and Computing*, 9 (2019), 20959–20960.
23. Cameron R, Cantrell C, Hemni A, and Lorenzin L. *Configuring Juniper Networks NetScreen & SSG Firewalls*. Syngress: Rockland, MA (2007).
24. Hsu H-H, Chang C-Y, and Hsu C-H. *Big Data Analytics for Sensor-Network Collected Intelligence*. Morgan Kaufmann: Burlington, MA (2017).
25. Musah, A, Livia MD, Aisha A, Ella B, Tercio A, Iuri VB, Merve T, et al. "An Evaluation of the OpenWeatherMap API versus Inmet Using Weather Data from Two Brazilian Cities: Recife and Campina Grande," *Data*, 7(8) (2022), 106.
26. Turgut D, and Boloni L. "Value of information and cost of privacy in the Internet of Things," *IEEE Communications Magazine*, 55 (2017), 62–66.
27. Bai X, Wang Z, Sheng L, Wang Z. "Reliable data fusion of hierarchical wireless sensor networks with asynchronous measurement for greenhouse monitoring," *IEEE Transactions on Control Systems Technology*, 27 (2019), 1036–1046.
28. Beza NG, Hussain AA. "Automatic control of agricultural pumps based on soil moisture sensing," *IEEE AFRICON Conference,* Addis Ababa, Ethiopia (2015).
29. Ruan J, Shi Y. "Monitoring and assessing fruit freshness in IoT-based E-commerce delivery using scenario analysis and interval number approaches," *Information Sciences* 373 (2016), 557–570.
30. Capello F, Toja M, Trapani N. "A real-time monitoring service based on industrial internet of things to manage Agrifood logistics," in *6th International Conference on Information Systems, Logistics and Supply Chain*, Bordeaux, France (2016), pp. 1–8.

9 Cybersecurity in Autonomous Vehicles

Balvinder Singh, Md Ahateshaam, Anil Kumar Sagar, and Abhisweta Lahiri

CONTENTS

DOI: 10.1201/9781003438588-9

9.1 INTRODUCTION

An autonomous vehicle (AV) is a vehicle that can sense the environment and operate without humans [1]. AVs can make decision by using machine learning and artificial intelligence. AVs rely on wireless data sharing to make navigation decisions. There are many types of driverless vehicles based on the levels of autonomy. AVs with low level of automation gives more control to the driver.

Vehicles with high level of automation have full control over all functions [1]. These types of AVs do not need any humans to drive them, and even there is no need for a steering wheel [1]. AVs gather information from environment by using sensors.

However, AVs can communicate with each other and share data about the surroundings. Communication is not limited to vehicle-to-vehicle (V2V) or between cars and the infrastructure (vehicle-to-infrastructure [V2I]) [2]. A hacker can attack the control technology tools that are used in AVs, for example, electrical window controls, which are now controlled by engine control units as embedded systems [3]. The engine control unit is an important part of a vehicle [4]. AVs rely on the millions of line code, and a hacker can attack the coding part of AVs to degrade hardware performance, or they can create a virus that can modify the data that are being shared between different AVs [5]. Hacker can modify the location of GPS where a person wants to go.

Some hackers use the techniques like computer viruses, Trojan horse, and worms. There are also denial-of-service attacks, phishing, and man-in-the-middle attacks. To make AVs, we have biometric authentication so that it is not easy to hack the AVs because biometric authentication is advanced-level security.

So cybersecurity is an extremely big problem for AVs. In this paper, we will show the problem of cybersecurity and types of attacks that are its main problems and roots [3]. And we will illustrate some solutions to the cybersecurity problem (Figure 9.1).

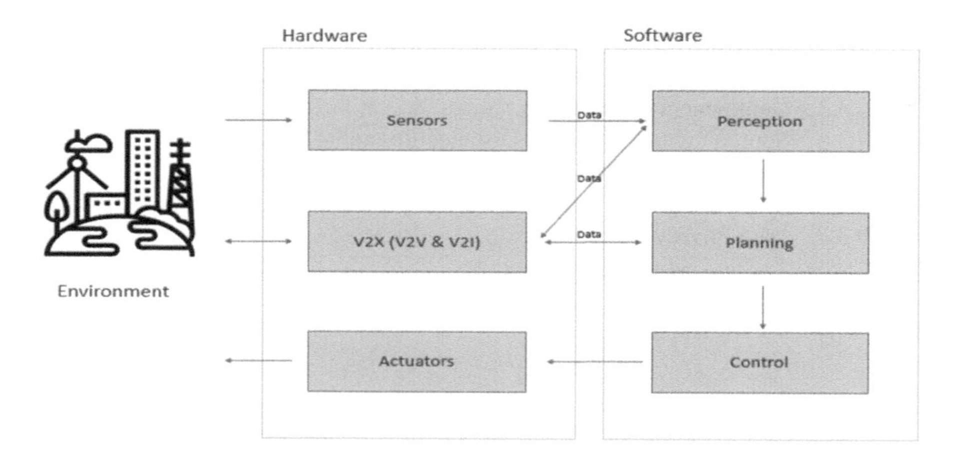

FIGURE 9.1 How automotive vehicles works cybersecurity threats in autonomous vehicles.

9.2 CYBER THREATS

Security threats are the most prominent, noticeable, and marked interests of automated vehicles. There are numerous ways to begin an automated vehicle cyberattack. This security breach will affect and target the software of the AV that is responsible for managing visual data or lane infrastructure or any kind of physical threat on the AV's hardware [4]. Due to the dependency of the automated vehicle on the sensors for sensing the surrounding environment, communicating among different systems and AI (artificial intelligence) and machine learning for their development and learning, they become more prone to the security attacks. There are two types of attacks that can be made:

i. **Internal Threats:** Attacks that are done on the in-vehicle systems and the communication channels.
ii. **External Threats:** Attacks through system external devices and other technological devices that are connected to the AV, for example, diagnostic devices or third-party apps.

Due to these attacks, the hacker may access the total control of our vehicle, and they might steal our personal data from our vehicle systems and sensitive information based on our activity in our vehicles [6]. We have explored some areas where these security attacks have influenced its way around the data, which is based on the literature survey. We will discuss those areas below.

9.2.1 Communicating Channel

AVs are under the influence of security breaches and exposed to the hackers then the non-automated vehicles. The reason for their defenselessness is due to their in-vehicle communication between systems within the vehicle and external communication between the external devices and the automated vehicles. AVs have VANET installed, which allows the vehicles to share information and communicate with one another. For example, if one automated vehicle is contaminated with malware or a virus or is already hacked by a hacker, it will make the whole connected system a target, expose them to security breaches, and make them vulnerable to further attacks.

9.2.2 LiDAR Attack

Light detection and ranging is a type of sensor that acts as the vision for AVs; it provides a 360° glimpse of the surroundings [7]. It helps in identifying the obstacles, detection of signals and lanes and used in many other different scenarios. The technology of LiDAR is based on the concept of emission of infrared rays for object detection by transmitting and receiving these rays. Now, an attacker may take help from a receiver and receive the signal generated by the vehicle and may send another malware-containing signal to the vehicle, and due to its inability to conform to its authenticity, it will welcome the coming signal and do whatever is instructed in it.

9.2.3 Packet Sniffing and Fuzzing Attacks

Packet sniffing means obtaining the data packets over a system network, and packet fuzzing means using an external software for assessment/testing of a vehicle's system liability, accountability, susceptibility, and its vulnerabilities for its exploitation [8]. The attacker will use the packet sniffing technique to get the information from the data packets and then will further transmit a malicious file under the category of a test file using a packet fuzzing technique.

9.2.4 Signal Jamming and Spoofing Attacks

In the signal spoofing attack, the attacker will send a different signal in place of the original to take control of the automated vehicle. For example, suppose you are using an AV's GPS. The attacker will continuously transmit out fake signals about the fake location and then the original one, and these signals will gradually overpower the original signals. In this way, the hacker will take over your automated vehicle.

9.2.5 DoS (Denial-of-Service) Attacks

In a DoS attack, the attackers will block all the connection between the AV and the intended server. It will overflow the server with endless traffic which will further repress the vehicle to connect with the server and restrict it to directly communicate with it [2]. Then, the hacker may apply a spoofing attack and access control of your automated vehicle.

9.2.6 Credential Acquiring Attack

Nowadays, approximately every application requires the user to sign up or sign in before they can access the application. We generally use the same id for every application for sign-up purposes, which makes it easier for the attacker to gain all our information from acquiring the login credentials and using them to login to our system.

9.2.7 Attacks via Update

Just like other electronic devices that use wireless means to update its system or software, AVs also use wireless transmission to update its system software. The hacker will hack their way into the transmission medium and transmit their own updates full of malicious content. So, after the system update, the hacker will gain access to our AV.

9.2.8 Remote Access Attacks

In today's world, most of the self-driven cars or vehicles are now available for remote access, so people can now start and open their cars with just one click. And in some countries that experience severe climates like extreme winters, they need to warm up their vehicles before use, which can all be done using your mobile phone as a remote to access functionalities of your automated vehicle. This acts as a great way for attackers to get into our systems, and there are some other ways from which an attacker might come into the system, for example, using Bluetooth for finding out some other information about the user.

9.2.9 Location- and Timing-Based Attacks

Different types of security attacks include location- and timing-based attacks. Through getting into our system, the attacker is already violating the privacy of the user, and after that, they may steal our personal information and use it for their purpose. An example of such an attack is a location-based attack in which the hacker monitors our daily activities, tracks our location at any time, knows where we frequently visit, and may sell that kind of information [4]. In the timing-based attacks, the hacker will create a lag or suspension in getting the original message, and due to this delay in getting the information about vehicle in front or an object in front, accident chances will increase.

9.2.10 Misguiding Attacks

The main purpose of these kinds of attacks is that the attacker will mislead your vehicle and trick it into rerouting their path and sending you to another destination, which might be in the attacker's favor by misleading you to another location.

9.2.11 Visual and Audio Attacks

In some automated vehicles, they are embedded with cameras and microphone (for voice assist purposes). We might need cameras along with sensors that will be used in avoiding severe catastrophic crashes and accidents, and users may also use voice assist to enjoy some of its functionalities. A hacker will target this and use this as one of the tools for finding its way around the automated vehicle.

9.2.12 Third-Party Download Attacks

Another type of attack is via downloading application from third-party sources that may contain a virus-containing file. Once the file is in the automated vehicle's system, the attacker will automatically gain access to your AV [5]. This will open a whole lot of options for the attacker to manipulate the system, for example, by doing a spoofing and jamming attack, location attacks, and other attacks that are mentioned above.

9.2.13 Threats via Wi-Fi Hotspots

Another great opportunity for the hacker to attack our automated vehicle's system is through a Wi-Fi hotspot places. The attacker will manipulate the network channel, and when the users' AV connects to such a network channel or Wi-Fi hotspot, the

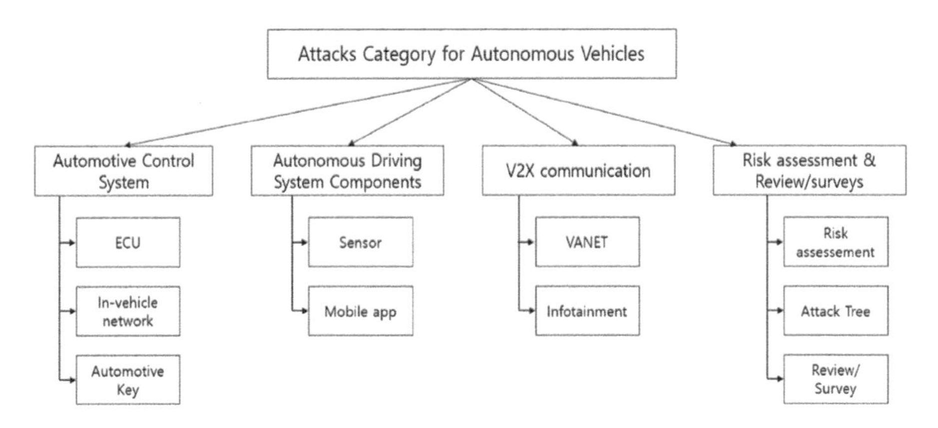

FIGURE 9.2 Category of attacks.

attacker will earn a direct access to the system [2]. And in addition to such attacks, not only one AV can be affected, but other vehicles or systems might also get infected due to functionalities like V2V and V2IOT connecting facilities (Figure 9.2).

9.3 METHODS TO ENHANCE CYBERSECURITY

Various types of cybersecurity solutions are being developed and offered by numerous companies for autonomous and connected vehicles. The mentioned solutions could address a particular side of security or provide total substructure for a vehicle to follow.

For example, Trustonic pronto proffers a particular answer for reliable digital mobile keys. At present, it offers a definite answer, which authorizes vehicle owners to replace traditional car keys and fobs with a secure digital key on their respective phones [8]. They provide a triumphant case study using the answers with famous car manufacturer Hyundai. The company, Trustonic, also provides varied remedies, such as their trusted execution environment concentrated around securely implementing code in vehicles.

9.3.1 In-Vehicle Device and Secure Communication

A proposed defense strategy of an in-vehicle gadget is suggested, just as one of the problems is through the communication channels. A vehicle ad hoc network (VANET) is built into AVs. AVs can talk to one another thanks to VANET. This is one way an attacker can join the VANET as a third party. Propose the use of trusted certificates that can be issued during the authentication process of communication between AVs in order to secure the communication channel. There are various ways to authenticate during the authentication process, some of which involve the use of passwords and occasionally certificates [9]. According to the method used to authenticate a user, in this example an AV, called certificate-based authentication, a digital certificate and public-key cryptography are both used.

9.3.2 Application for User Authentication

Having a self-driving car is sufficient, but having a self-unlocking car is an entirely different matter. Relay assaults are one of the concerns with cybersecurity. As previously mentioned, the inclusion of PKES in antivirus software creates a vulnerability that an attacker can employ a relay attack to exploit [4]. For keyless entry with AVs, PKES uses LF RFID tags and ultrahigh frequency (UHF). The hypothesis of relay attacks was put to the test by the authors, who labeled it a "two-thief model." A good user authentication is required to prevent this relay attack. In addition to using PKES, it is necessary to deploy other security measures, such as biometric user identification.

9.3.3 Deployment of Firewall

Over-the-air updates attack is one of the problems with AVs. A component for over-the-air (OTA) updates is installed in the system of AVs. OTA is a wireless method of updating a system, so they claim. Moreover, there's the SQL injection problem. The execution is SQL injection. Network traffic is watched both entering into and leaving out through a firewall [10]. Policies must be put in place for the system to accurately filter incoming and outgoing data packets in order for the firewall to work correctly. A firewall that is in place will help detect malicious material that is injected into data transmission via OTA, and it will also help reduce the problems of V2I, V2V, and V2IoT by filtering packets from trusted and untrusted sources during communication between AVs.

9.3.4 Source Signal Block and Distance Bounding

When it comes to keyless entry in cars, PKES is an excellent mechanism. However, the relay attack predominates in this mechanism. Utilizing a box to shield the key is another method of preventing relay attacks. The PKES cannot transmit or receive signals while the key is shielded in a box; therefore, if a relay attack is attempted, there will be no sender or receiver [11]. Similar to the PKES, the key fob's power source is incorporated into the device. When the key fob isn't in use, removing the battery will be a simpler solution. In a relay attack attempt, this will make the key powerless and useless to the attacker. Additionally, the impact of a relay attack can be reduced by using a technique known as distance bounding. Before granting entry access to the AV, this algorithm verifies the accuracy of the distance between the PKES and the vehicle.

9.3.5 Deployment and Installation of Gateway for CAN

A gateway is a network node that acts as an entrance point for data entering and leaving the system. On a network, a gateway acts as a communication channel. To provide communication, AVs use CAN. As previously mentioned, CAN in AVs is one of the weak spots for an attacker in the online environment. The authors in this paper choose to employ a gateway as a solution to this cybersecurity problem. The vehicle electronic control unit (ECU) is a part of AVs that acts as the brain of the device, gathering information from all other components, including cameras and sensors, to create a map of the area the AV is in Ref. [12]. The adoption of NXP enables a higher level of security when a gateway is used. A gateway device created by NXP offers additional protection for all connected AV components. To illustrate how secure, it is using the gateway offered by NXP, a picture of an NXP is presented with the components of AV linked.

9.3.6 Automated DDN Tests and Procedures

This is when AVs, using some of their components such sensors and cameras, might incorrectly label items in their environment dependent on the type of categorization the system was trained with. Due to its low resilience and inability to automate test processes throughout the training and testing phases, misclassification is a problem. This is when AVs, using some of their components such sensors and cameras, might incorrectly label items in their environment dependent on the type of categorization the system was trained with. Due to its low resilience and inability to automate test processes throughout the training and testing phases, misclassification is a problem [9]. When automated deep learning models are used in AVs, the classification model is automatically retrained to improve performance and boost decision- and prediction-making accuracy.

9.3.7 Data Privacy Prevention

Leakage of information is one of the challenges that AVs encounter. This can take place, among other things, when the AV system is updated. Future updates and training of the AV security system are in doubt given that the ultimate goal is to have a completely automated car. One of the suggested methods for protecting data privacy in AVs is to carefully select the noise you add to each gradient update. When using machine learning, where the system is updated and educated using new data, this tactic is used. This is as a result of AVs' ability to memorize models [13]. As a result, only the updates to the model parameters ever leave the system for the server. As a result, attackers are prevented from using the new updates to train the system for their own benefits.

9.4 CRYPTOGRAPHIC LIGHTWEIGHT TECHNIQUES

Future automobile security will be improved by the following lightweight protocols (Figure 9.3):

ARAN: The foundation of ARAN (authenticated routing for ad hoc network) is the Ad hoc On-demand Distance Vector (AODV) routing protocol, which sees the CA supplying a signed certificate. The CA must receive a certificate request from each new node that enters the network [6]. All approved

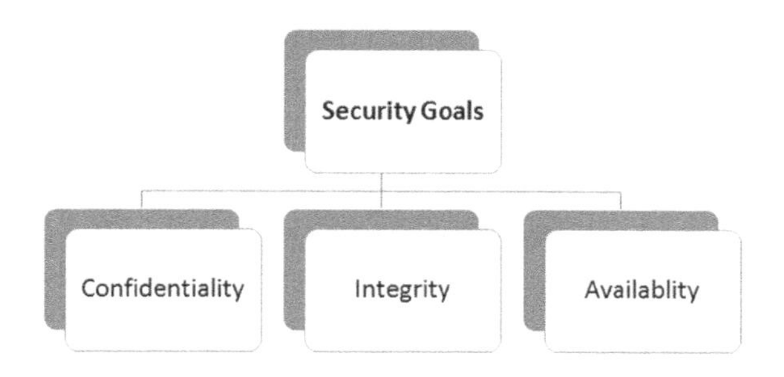

FIGURE 9.3 Security goals of cryptography.

TABLE 9.1
Table of Variables and Notations

K_{A+}	Public key of node *A*
K_{A-}	Private key of node *A*
K_{AB}	Symmetric key shared by nodes *A* and *B*
$\{d\}K_{A+}$	Encryption of data d with key K_{A+}
$[d]K_{A-}$	Data d digitally signed by node *A*
cert *A*	Certificate belonging to node *A*
e	Certificate expiration time
N_A	Nonce issued by node *A*
IP_A	IP address of node *A*
RDP	Route discovery packet identifier
REP	REPly packet identifier
t	Timestamp

nodes have access to the CA's public key. ARAN uses timestamps for safe route discovery authentication together with asymmetric cryptographic techniques (Table 9.1).

Certificate (*T*) generation formula:-

$$T \rightarrow A:\ \text{cert}\,A = \left[\,\text{IPA}, K_{A+}, t, e\,\right] K_{T-}$$

End-to-end route authentication:-

$$A \rightarrow \text{broadcast}: \left[\text{RDP}, \text{IPX}, \text{NA}\right] K_{A-}, \text{cert}\,A$$

SEAD: Dynamic destination-sequenced distance vector (DSDV) routing is the foundation of the Secure and Efficient Ad hoc Distance (SEAD) vector protocol). For the purpose of authentication, it uses a one-way hash algorithm. This protocol guards against improper routing. To prevent taking a long-lived route and guarantee route freshness, the destination-sequence number is employed. To ensure that each route is legitimate, the protocol uses intermediary node hashing.

ARIADNE: The foundation of this protocol is Dynamic Source Routing (DSR) on-demand routing protocol. Symmetric cryptographic procedures are particularly effectively used by Ariadne. Secure communication between nodes is achieved via MAC authentication and one-way hashing [14]. Using a shared key, authorization is carried out. The Ariadne protocol, which employs TESLA time interval for authentication and route finding, is based on TESLA broadcast authentication technology.

SAODV: In AODV, security was added by this protocol. All communications are digitally signed to guarantee the legitimacy of routes, and hash functions are employed to safeguard hop count. Even if the intermediate node is aware of the new route, this method forbids it from sending any route replies [13]. Double Signature can be used to overcome this issue, although doing so will make the system more complicated.

A-SAODV: Secure ad hoc on-demand distance vector (SAODV) protocol extension with adaptive reply decision characteristics has been suggested. Each intermediary node can choose whether to reply to the source node based on the threshold criteria and queue length.

OTC: For session management purposes, cookies are typically allocated per session. To safeguard the system from SID theft and session hijacking, the One Time Cookie (OTC) protocol is suggested [14]. To prevent token reuse, this protocol produces a unique token for each request and uses HMAC to link it to the request.

ECDSA: Digital signature are used by Elliptical Curve Digital Signature Algorithm (ECDSA). The system is secure and authenticated using asymmetric cryptographic operations using the hash function. Both the transmitter and the receiver need to agree on the elliptical curve's parameters.

RobSAD: An effective way to identify Sybil attacks is provided by this protocol. If two or more nodes exhibit similar motion trajectories, the node is known as the Sybil node. Because each driver operates their vehicle to suit his or her unique needs and comfort, two distinct vehicles operated by different drivers cannot maintain the same motion patterns.

Holistic Protocol: RSU is responsible for performing each vehicle's authentication under this protocol. By sending a "Hello" message, vehicles can be registered with RSU. As a result, the RSU creates and sends a registration ID to the car that includes the license number and the vehicle's registration number [13]. A certificate provided by RSU is used for further authentication. Data sharing is only possible with RSU-authenticated nodes otherwise, nodes are banned (Table 9.2).

TABLE 9.2
Lightweight Cryptographic Techniques for Automotive Cybersecurity

Lightweight Protocols	Cryptographic Solutions	Security Requirements	Attack Mitigation
	ARAN	Message authentication/ integrity	Impersonation/ eavesdropping/replay
	SEAD	Authentication/availability	Routing/DoS
	Ariadme	Availability/privacy preservation	DoS/routing/replay attacks
	SAODVIA_SAODV	Authentication/availability/ privacy preservation	Impersonation/bogus/ information/routing attack
	One time cookie	Availability	Session hijacking
	ECDSA	Authentication	Bogus information/ impersonation attacks
	RobSAD	Confidentiality/ authentication/integrity	Sybil attack
	Holistic	Authentication/ confidentiality	Impersonation attacks

9.5 CONCLUSION

This paper illustrates the problem of cybersecurity in autonomous vehicles. Fully automation and a secure automation are still the biggest challenge for AV industry. Security threats are the most prominent, noticeable, and marked interests of automated vehicles. We have shown a few hacking methods that can be used against autonomous vehicles. As a result, we have listed a number of options for securing autonomous vehicles, including the deployment of firewalls, source signal block, and distance bounds, deployment and installation of CAN gateways, automated DDN testing and procedures, and data privacy prevention. Due to their rapid movement and dynamic network architecture, traditional cryptographic methods, such as public-key infrastructure, elliptic curve cryptography, HASH functions, and symmetric key cryptography, may not be suitable for use in vehicle networks. It is crucial to develop safe, straightforward, lightweight cryptography techniques for tiny embedded devices. We argue that lightweight cryptographic techniques are crucial for addressing the rising security issues in automotive technology, notably those regarding vehicle safety and traffic efficiency. The majority of the systems we've pictured can guard level-3 automated cars. We have discussed some wise actions that can be taken to strengthen autonomous vehicle defenses. Additionally, cybersecurity will always be a major issue for AVs since hackers are constantly developing new ways to attack digital devices like autonomous vehicles. Thus, there will always be room for progress in the field of cybersecurity.

REFERENCES

1. T. Vaa, M. Penttinen, and I. Spyropoulou, "Intelligent transport systems and effects on road traffic accidents: State of the art," *IET Intelligent Transport Systems*, vol. 1, no. 2, pp. 81–88, 2007.
2. J. Shen, T. Zhou, X. Chen, J. Li, and W. Susilo, "Anonymous and traceable group data sharing in cloud computing," *IEEE Transactions on Information Forensics and Security*, vol. 13, no. 4, pp. 912–925, 2018.
3. R. Stanica, E. Chaput, and A.-L. Beylot, "Properties of the MAC layer in safety vehicular ad hoc networks," *IEEE Communications Magazine*, vol. 50, no. 5, pp. 192–200, 2012.
4. H. Trivedi, P. Veeraraghavan, S. Loke, A. Desai, and J. Singh, "Routing mechanisms and cross-layer design for vehicular ad hoc networks: A survey," in *Proceedings of the IEEE Symposium on Computers and Informatics*, Kuala Lumpur, Malaysia, pp. 243–248, March 2011.
5. B. Jarupan and E. Ekici, "A survey of cross-layer design for VANETs," *Ad Hoc Networks*, vol. 9, no. 5, pp. 966–983, 2011.
6. G. Rathee et al., "A Blockchain framework for securing connected and autonomous vehicles", *Sensors*, vol. 19, no. 14, pp. 3165, 2019.
7. K. N. Qureshi and A. H. Abdullah, "A survey on intelligent transportation systems," *Middle East Journal of Scientific Research*, vol. 15, no. 5, pp. 629–642, 2013.
8. Y. Toor, P. Muhlethaler, and A. Laouiti, "Vehicle ad hoc networks: Applications and related technical issues," *IEEE Communications Surveys & Tutorials*, vol. 10, no. 3, pp. 74–88, 2008.
9. M. Torrent-Moreno, J. Mittag, P. Santi, and H. Hartenstein, "Vehicle-to-vehicle communication: Fair transmit power control for safety-critical information," *IEEE Transactions on Vehicular Technology*, vol. 58, no. 7, pp. 3684–3703, 2009.
10. T. Litman, "Autonomous vehicle implementation predictions", *Victoria Transport Policy Institute*, Victoria Canada, 2017.
11. A. D. Smith and W. T. Rupp, "Issues in cybersecurity; understanding the potential risks associated with hackers/crackers", *Information Management & Computer Security*, vol. 10, no. 4, pp. 178–183, 2002.
12. A. A. Alkheir, M. Aloqaily and H. T. Mouftah, "Connected and autonomous electric vehicles (CAEVs)", *IT Professional* 20.6, pp. 54–61, 2018.
13. E. Some, G. Gondwe, E. W. Rowe. "Cybersecurity and driverless cars: In search for a normative way of safety", *2019 Sixth International Conference on Internet of Things: Systems, Management and Security (IOTSMS)*, Granada, Spain, 2019.
14. V. C. Giruka. "Secure routing in wireless Ad-Hoc networks", in: Y. Xiao, X. Shen, D.-Z. Du (Eds.), *Wireless Network Security: Signals and Communication Technology*. Springer: Berlin/Heidelberg, pp. 137–158, 2007.

10 Use of Virtual Payment Hubs over Cryptocurrencies

Satyam Kumar, Dayima Musharaf,
Seerat Musharaf, and Anil Kumar Sagar

CONTENTS

10.1 INTRODUCTION

In almost all industries where information technology has come, there have been problems with information security. Data that are stored electronically are easy to copy, send, modify, and replace a thread. Therefore, along with convenience, transmission speed, and huge volumes of accumulated and transmitted information, modern technologies bring with them the problems to ensure safety and reliability of data. To solve these problems in any industry where verification and prototyping is required, calling certain actions, you can use blockchain technology.

Blockchain technologies play a special role in the economy, where they are closely -related to the concept of cryptocurrency. However, the digital economy is not limited to digital fears but provides safety and convenience, the creation of financial transactions, and the conclusion of a contracts.

Cryptocurrency is emerging as the new technology that has gained great popularity over the decade. It has many advantages like it provides an infrastructure for payments without the involvement of any central authority for regulating transactions. Complex agreements can be enforced using program code such as smart contracts. The underlying technology behind blockchain is the distribution of a ledger among multiple users, resulting in a replicated state. While this replication

DOI: 10.1201/9781003438588-10

ensures security, it also limits the speed of transactions. For example, Bitcoin, a cryptocurrency, takes an average of 10 minutes to confirm new transactions and can process around seven transactions per second.

Blockchain-based cryptocurrencies, such as Bitcoin, face a significant scalability issue when it comes to microtransactions. Microtransactions refer to the transfer of small amounts of money between users, which ideally should be processed instantly. However, the confirmation of transactions in cryptocurrencies takes several minutes, making it less suitable for microtransactions. Additionally, the cost of posting these small transactions on the ledger is often higher than the value of the transaction itself, making it financially unfeasible. These limitations make it unlikely that blockchain-based cryptocurrencies will be able to support microtransactions in the future.

A technology known as Payment channels can address the above challenges in much easier way. With the use of "Off-chain" transactions, we can securely allow two parties to complete transactions rapidly. Off-chain transactions also enable users to exchange the coins without the direct involvement with the ledger. First, a channel (denoting this channel B) is opened and both Alex and Brown put coins XA and XB (respectively) into it. The channel mechanics then allows the parties to change their XA+XB distribution at will, allowing payments to be made between Alex and Brown. Either party may decide at any time to close Channel B and send money to the ledger. Only opening and closing channel require interaction with the ledger. Channel updates, on the other hand, happen without interacting with the ledger, so they can be done as often as you like, very quickly, and for free.

This work aims to address the limitations of blockchain-based cryptocurrencies in terms of scalability and microtransactions. The proposed solution is to use a concept called virtual channels. The idea is to connect two parties, Alex and Brown, through an intermediary payment hub, Ingrid, via a channel built on the blockchain. These ledger channels can then be used to establish a direct, virtual connection between Alex and Brown, allowing them to conduct transactions without the involvement of the intermediary. By eliminating the need for Ingrid to be involved in all transactions, the proposed system greatly reduces latency and costs, and also improves privacy. The primary use case of this system is to enable instant, seamless streaming of small payments.

10.2 LITERATURE REVIEW

Angelo MD and Salzer G [1] published a survey of all the tools that were used to analyze the blockchain-based cryptocurrency Ethereum smart contracts. They analyzed all the pros and cons of the smart contracts in the current scenario. Atzei N et al. [2] presented their survey on the possible attacks on the smart contracts. They presented all the weakness as well as provided the solution to make the smart contracts more secure. Bartoletti M and Pompianu L [3] presented their analysis on the design patterns of the smart contracts. With the help of this analysis, they presented the advancements made in the platforms as well as the application. Ellul J and Pace GJ [4] gave an idea of virtual machine, which they further integrated with internet of things to make the smart contract use more efficient. Gao Z et al. [5] discussed the way to scale the blockchain-based cryptocurrency for the implementation of the smart contracts. Hasan H et al. [6] used the smart contract for the efficient shipment management. Liu C et al. [7] introduced Reguard, which usually finds the reentrance of the bugs in the smart contract. Liu J and Liu Z [8] suggested a method for the security verification of the blockchain-based cryptocurrency and their application of smart contract. Hamida EB et al. [9] focused on providing the detailed analysis over enterprise blockchains core structure, technologies, and applications. They also discussed various opportunities as well as challenges in using this enterprise blockchain. Mori T [10] provided overview of this new technology which constitutes its role in securities market as well as in financial transactions. Swan M [11] explains the blockchain as public ledger that has the potential to maintain worldwide transactions whether it is financial or non-financial transactions. Peters GW and Panayi E [12] presented there's views on modern banking ledgers by the use of blockchain technologies. They provide an overview over how the concept of blockchain have the power to change the banking system with smart contracts, digital assets as well as automated banking ledgers. They also discussed the key issues in implementing this technology in the context of financial services. Pilkington M [13] presented the principal and applications of the blockchain technologies with the security and privacy services as the important concern. Salah K et al. [14] discussed about merging the blockchain with artificial intelligence (AI). They discussed all the advantages of the blockchain implemented with AI. They also discussed the things need to be kept in mind during the integration of these technologies. Yang R et al. [15] investigated the works already been done to enable the integration of blockchain system with edge computing system. They also discussed the research challenges, frameworks over this integration. Zhang R et al. [16] reviewed the security and privacy techniques including representative consensus algorithm, mixing protocols, anonymous signature, hash chained storage, and much more. Zhu H and Zhou ZZ [17] provided an analysis over the current problems faced in the practice of equity crowdfunding in China. Zavolokina L et al. [18] proposed a conceptual framework for understanding fintech and its dimensions aligned to the blockchain technologies. They examined a total of five Swiss FinTech companies for this purpose. Through this, they analyzed the nature of the FinTech Innovations in blockchain technologies.

10.3 PROPOSED METHODOLOGY

This section presents a description of the proposed system. We divided this section into three subsections. Firstly, in Section 10.3.1, we describe our system functionality as well as its properties. In Section 10.3.2, we describe its security features and at last, in Section 10.3.3, we provide the main idea as well as the full description behind our proposed system.

10.3.1 System's Functionality

The most crucial aspect of our system is the implementation of Virtual Channels. Virtual Channels are designed to remove the intermediary's involvement in the transaction process by creating a direct connection between the parties involved. We achieve this by building a virtual payment channel on top of the ledger channel, and this process is recursive in nature. By using Virtual Channels, we can greatly reduce the latency and costs associated with transactions, while also improving privacy. The use of virtual channels is the key to our system's scalability and ability to support microtransactions (Figure 10.1).

Let's consider three parties, Alex, Brown, and Ingrid, and consider three ledger channels existed between them:

According to the above given configuration, Alex and Brown have to involve the intermediary, Ingrid, for the transaction.

a) **Creating a Direct virtual Channel**: Starting from the given configuration, the Alex and Brown established a new virtual channel for payment having XA as Alex initial balance, XB as Brown's initial balance. This can be done without touching the ledger by the use of channel of BA and BB. With the opening of channel M between Alex and Brown, we can temporarily remove the underlying ledger channels of BA and BB. Figure 10.2 is shown below:

Opening of channel M is only valid if all the three parties have enough coins in their ledger channels. The initial coin balances, XA and XB, remain vanished from both parties' accounts respectively until the virtual channel is open. We don't have to keep opening and closing the virtual channels after a single transaction. Once the channel is open, we can also update it multiple times, similar to as we do with the ledger channels.

b) **Closing a Virtual Channel:** Our system ensures that the closing of any virtual channels does not affect the parties' accounts on the blockchain, but it only impacts the ledger channels BA and BB. Additionally, if a virtual

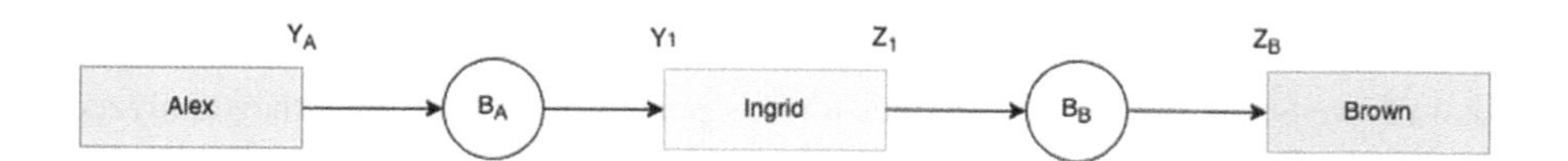

FIGURE 10.1 Payment channels between Alex, Ingrid, and Brown.

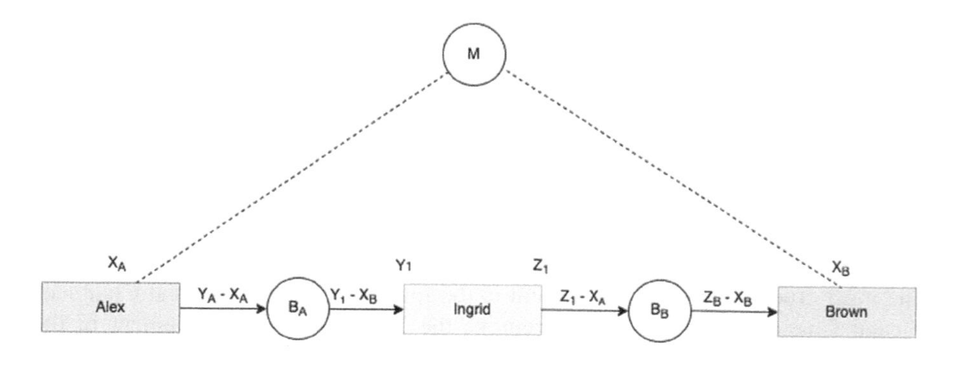

FIGURE 10.2 Creating virtual payment channel.

channel is open, our system will prevent the ledger channels BA and BB from being closed. This means that the parties responsible for opening the ledger channels must wait until all the monetary consequences from the virtual channel "M" are known before closing the channels. All parties involved in the virtual channels agree on a validity (a special real-time value) when the virtual channel is opened. Once the validity expires, the channel is closed automatically, unlike the ledger channels, which require someone to initiate the closing process. This automatic expiration and closing of virtual channels make the process more efficient and less prone to human error.

10.3.2 Security and Efficiency

Firstly, it is only possible to open any virtual/ledger channel if and only if everyone involved in the channel agrees on it. In particular, everyone has to confirm the channel's validity.

In order to use virtual or ledger channels, both parties, Alex and Brown, must confirm each update. The time required for this confirmation is constant. Suppose there is a ledger channel B. If there is no virtual channel open over ledger channel B, all parties involved have the option to request the closing of the ledger channel at any time. Once the request to close the ledger channel is made, the channel is closed in a constant time. It is guaranteed that the final balance will be distributed to all end users connected to the channel. In other words, the final balance of the channel is guaranteed to be paid out to all parties involved in the channel, and the process of closing the channel is done in a constant time, regardless of the number of parties involved or the amount of transactions that took place on the channel.

For the intermediary, Ingrid, all virtual channels are always "monetarily neutral". This means that when a virtual channel, such as Channel M, is created over two existing ledger channels, BA and BB, the intermediary's net gain or loss will be zero. For example, if Ingrid loses 'x' coins in channel BA, she will gain 'x' coins in channel BB and vice versa. This ensures that the intermediary's overall coin balance remains unchanged regardless of the outcome of the virtual channel.

10.3.3 Structure of the System

The structure of the system consists of the instructions for all the parties involved and the execution of contract code "C" by the contract instances that are deployed on the ledger. In this discussion, we consider "c" another party so that we can simplify its actions. Over this, we will make few assumptions:

Assumption 1: If any fault that is uniquely attributable to some particular party, let's say party P, is determined by the contract instance "c". Then, as a punishment, C favors the honest party, gives all the coins to that party, and terminates the process.

Assumption 2: If any party detects any fault that is uniquely attributable to some particular party, then the former party will not engage in any new protocol with the latter party.

If the party is honest, assumption 1 will never cause any harm to that party. Assumption 2 is also a genuine assumption as it makes no sense in forming new off-chain connections as it will end to a conflict sooner or later.

Now, we describe all the protocols for forming the virtual channels. In the previous section, we formed the virtual channel "m" with the help of the intermediary (Ingrid) ledger channel BA and BB. It was formed in similar way as the building of ledger channel with the help of ledger. But there is a difference between the functionalities and features that the ledger channel over the blockchain provides. While the ledger allows execution of smart contracts inside of the channel, the ledger channel only allowed to perform transactional exchanges between parties and not the execution of smart contracts. Now, the second problem is that the ledger channels we used earlier provide only a virtual ledger to two parties. We have solved these problems by implementing and extending functionalities, which these ledger payment channels used to provide.

In Figure 10.3, the relation between the contract instances and the ledger channel is presented.

If all parties involved are honest, the process of opening a virtual channel, such as channel "M", would work as follows: Firstly, Alex and Brown send their opening certificates for channel "M" to Ingrid. Ingrid will only consider the channel open once she has received both of these opening certificates and replied with her own opening certificates for channel "M". After receiving the opening certificates from Ingrid, Alex and Brown will forward them to each other. Once a party has received Ingrid's opening certificate directly or through forwarding, they will consider the channel open. Figure 10.4 of the message flow in this case is:

Let's consider the case when if some parties are behaving dishonestly. In this protocol, the opening channel "M" will not be open as the execution of the result. The main rule for our protocol is: (1) due to cheating of another party, honest party should not suffer any loss, and (2) there should be consensus between the parties for the opening or closing verification of the channel. Let's discuss the security guarantee provided by our protocol. The protocol ends up as opening certificate for channel

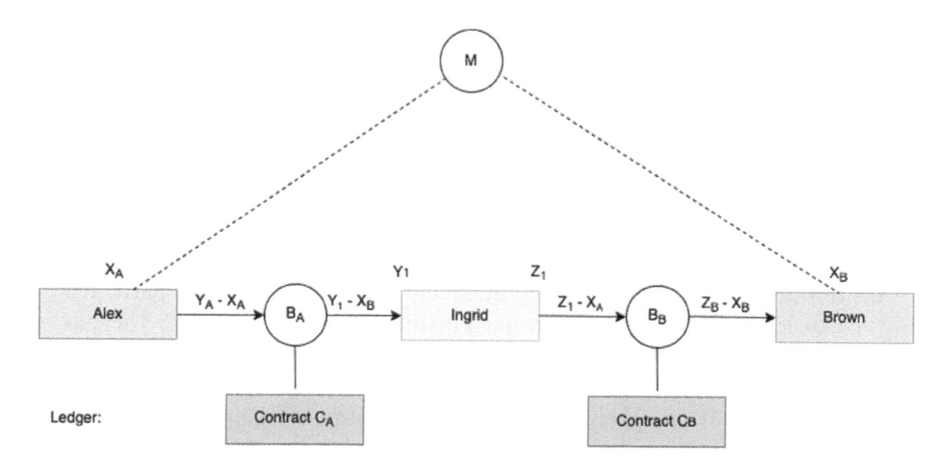

FIGURE 10.3 Introducing contracts in transactions.

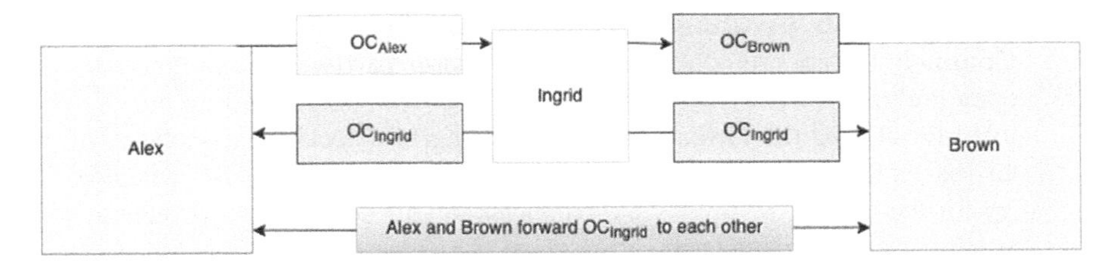

FIGURE 10.4 Flow of opening and closing certificates.

"M", which is held by every party involved in the protocol. We can use these certificates against claiming coins of party P, which belongs to Alex, Ingrid and Brown that signed it. Main security concern for any party P is that P signed a certificate through which other party can claim the coins from P while P itself doesn't receive the opening certificate for channel "M", P can never claim for the coins. It shows that this problem will never occur to Alex and Brown. This is due to their denial of considering the channel open, if they didn't receive any opening certificate for the channel "M". So, they will deny any update for the channel "M". So, even the Ingrid goes dishonest in the channel and deny the sending of opening certificate to Alex and Brown for the channel "M" then also there is neutral result of her behavior for other parties like Alex and Brown. This satisfies the property of balance neutrality for Ingrid. The problem can be larger when Ingrid sends the opening certificate to Alex and didn't receive any certificate from Brown which can lead to loss of the coin for Ingrid in the channel. This is why we let Ingrid sign the opening certificate only when she receives the certificate from both party Alex and Brown. In simple words, if Brown can be held responsible for Alex's commitment, then only Ingrid agrees to Brown's commitment in front of Alex.

Now, let's discuss for the case when Ingrid is honest. There will be consensus among all the parties, which are honest for the opening of channel "M". We allowed Ingrid to send opening certificates to both of parties or none of them. After receiving the opening certificates from Ingrid, parties in the channel will exchange that opening certificate among themselves. This guarantees that if any of them considers the channel open then other parties in the channel considers it open as well.

10.4 AN OVERVIEW OF THE TECHNICAL DETAILS

This section consists of all the technical details of our system, which consists of detailing about concurrent channel updates, simultaneously creating multiple virtual channels over the ledger channel, trigger actions on smart contracts and punishment for the dishonest parties.

a) **Concurrent Channel Updates:** Let's consider η be the virtual or ledger channel. There can be an issue when channel η. *Alex* as well as η. *Brown* initiates any updates in the same round. This can create a problem as any of them will use the value which was not updated with the new value. So, to solve this problem, we defined the update rounds of party P whenever it is possible for P to send the first message in the procedure of update.

b) **Simultaneously Creating Multiple Virtual Channels over the Ledger Channel:** In our protocol, we allowed several parties to simultaneously open multiple virtual channels over the ledger channel. Recalling our protocol for closing the virtual channel, once Ingrid detects the dishonesty of any party, then it was fine to close the related ledger channel. But when we try for the parallel settings, final balance can still be unknown if there are another virtual channel which are open. To solve this problem, we don't allow Ingrid to request the channel for closing. We let the information about x to be stored in the memory by the contract instance. There can be multiple of these contract instances, which can lead to high space requirements. So, to save the contract's storage space, we bind all of them by adding them together.

c) **Trigger Actions on Smart Contracts:** While designing the system, we assume that contract instances are self-dependent and can act by themselves. However, Ethereum contract are not self-dependent and every action of it starts with the trigger of the user. So, our protocol works for the triggering of the action. For instance, if time t comes and some party is financially interested in performing some given actions then that party sends a special message to contract instance.

d) **Punishment for the Dishonest Parties:** If the contract instances discover some parties of the channel involved in dishonest activity, then our protocol provides the rights to the contract instances for punishing all the dishonest parties. Here punishment means transferring of all the coins to the honest party and terminating the process.

10.5 CONCLUSION

In this work, we proposed a design for a system of channels for off-chain payments that utilizes virtual channels. Our system's key feature is the ability to create virtual channels over ledger channels, which eliminates the involvement of intermediaries in transactions. This approach offers several advantages over existing solutions, such as reduced latency, lower costs, and improved privacy. Additionally, we have also explored the option of slightly relaxing security requirements, which would allow intermediaries to not have to block the coins used for constructing virtual channels, further increasing the flexibility of the system. Overall, our proposed system provides a novel and effective solution for addressing the scalability and microtransactions issues faced by blockchain-based cryptocurrencies.

REFERENCES

1. Angelo MD, Salzer G (2019) A survey of tools for analyzing ethereum smart contracts. In: *IEEE International Conference on Decentralized Applications and Infrastructures, DAPPCON 2019,* Newark, CA, April 4–9, 2019, IEEE, pp. 69–78.
2. Atzei N, Bartoletti M, Cimoli T (2017) A survey of attacks on ethereum smart contracts (SOK). In: *International Conference on Principles of Security and Trust,* Springer, Uppsala, Sweden, pp. 164–186.

3. Bartoletti M, Pompianu L (2017) An empirical analysis of smart contracts: Platforms, applications, and design patterns. In: *International Conference on Financial Cryptography and Data Security*, Springer, Sliema, Malta, pp. 494–509.
4. Ellul J, Pace GJ (2018) Alkylvm: A virtual machine for smart contract blockchain connected internet of things. In: *2018 9th IFIP International Conference on New Technologies, Mobility and Security (NTMS)*, Paris, France, pp. 1–4.
5. Gao Z, Xu L, Chen L, Shah N, Lu Y, Shi W (2017) Scalable blockchain based smart contract execution. In: *2017 IEEE 23rd International Conference on Parallel and Distributed Systems (ICPADS),* Shenzhen, China, pp. 352–359.
6. Hasan H, Al Hadhrami E, AlDhaheri A, Salah K, Jayaraman R (2019) Smart contract-based approach for efficient shipment management. *Computers and Industrial Engineering* 136:149–159.
7. Liu C, Liu H, Cao Z, Chen Z, Chen B, Roscoe B (2018) Reguard: Finding reentrancy bugs in smart contracts. In: *Proceedings of the 40th International Conference on Software Engineering: Companion Proceeedings.* Association for Computing Machinery, New York, pp 65–68.
8. Liu J, Liu Z (2019) A survey on security verification of blockchain smart contracts, *IEEE Access,* 7:77894–77904.
9. Hamida EB, Brousmiche KL, Levard H, Thea, E (2017) Blockchain for enterprise: Overview, opportunities and challenges, *ICWMC 2017*, Nice, France, p. 91.
10. Mori T (2016) Financial technology: Blockchain and securities settlement, *Journal of Securities Operations & Custody*, 8(3):208–227.
11. Swan M (2015) *Blockchain: Blueprint for a New Economy.* O'Reilly Media, Inc: Sebastopol, CA.
12. Peters GW, Panayi E (2016) Understanding modern banking ledgers through blockchain technologies: Future of transaction processing and smart contracts on the internet of money, in: Tasca P, Aste T, Pelizzon L, Perony N (Eds.), *Banking beyond Banks and Money.* Springer: Berlin/Heidelberg, pp. 239–278.
13. Pilkington M (2016) 11 blockchain technology: Principles and applications, in: Olleros FX, Zhegu M (Eds.), *Research Handbook on Digital Transformations.* Edward Elgar Publishing: Northampton, MA, p. 225.
14. Salah K, Rehman MH, Nizamuddin N, Al-Fuqaha A (2019) Blockchain for AI: Review and open research challenges, *IEEE Access*, 7: 10127–10149.
15. Yang R, Yu FR, Si P, Yang Z, Zhang Y (2019) Integrated blockchain and edge computing systems: A survey, some research issues and challenges, *IEEE Communications Surveys & Tutorials*, 21(2):1508–1532.
16. Zhang R, Xue R, Liu L (2019) Security and privacy on blockchain, *ACM Computing Surveys (CSUR*), 52(3):1–34.
17. Zhu H, Zhou ZZ (2016) Analysis and outlook of applications of blockchain technology to equity Crowdfunding in China, *Financial Innovation*, 2(1), 29.
18. Zavolokina L, Dolata M, Schwabe G (2016) Fintech transformation: How it-enabled innovations shape the financial sector. In: *International Workshop on Enterprise Applications and Services in the Finance Industry.* Springer, Frankfurt, Germany, pp. 75–88.

11 Akaike's Information Criterion Algorithm for Online Cashback in Vietnam

Bùi Huy Khôi

CONTENTS

DOI: 10.1201/9781003438588-11

11.1 INTRODUCTION

Digital marketing and affiliate methods have benefited from the expansion and increasing popularity of online shopping (Christino et al. 2019). The cashback model is a shopping stimulus solution that applies to many businesses around the world. This model called cashback originated in the US and then in the UK, and is now available in several other countries, including Vietnam (Salemall 2023). In Vietnam, the cashback wave has been introduced since 2016, and putatu.com is the pioneer website in this campaign, helping Vietnamese people catch up with the trend (VTV 2017). Launched in January 2017, after 2 months of launch, over 2000 people have made thousands of transactions with a value of up to 12 billion VND via putatu.com. The amount of user savings has reached 500 million (VTV 2017). Nowadays, cashback is becoming more and more popular, especially when there is the appearance of shopping through e-wallets: MoMo, Moca, ZaloPay, and Shopback. With bill payment via e-wallet or with affiliate applications, customers can get 10%–20% cashback, even when buying data, phone cards, game cards, restaurant reservations, flight tickets, and movie tickets (Thanhnien 2020). PVcomBank is one bank that has applied this program to customers, and in the first month of opening, the bank estimated to refund customers VND 100 million (PVcomBank 2020). In December 2019, Shopback began its Beta program in Vietnam, drawing almost 800,000 users and over 150 partners. Shopback Vietnam saw a 1.5-fold increase in monthly sales and orders. To date, more than VND 4 billion has been reimbursed to Shopback users in Vietnam (CongThuong 2020). Cashback applications appeared in Vietnam as a tool to help business partners save costs in marketing and promoting products and services for the online shopping segment and attracting potential customers, potential, and increase revenue, and this is a potential form of promotion in the Vietnamese market.

In the e-commerce market, customers benefit from their transactions on linked websites or apps. Combining traditional techniques, such as word of mouth, is critical for the success of this company model because it can increase and strengthen client loyalty (Ballestar et al. 2018). By using this method, both the business and the affiliate gain from the affiliate's e-commerce transactions, resulting in a win–win situation for both the client and the business. Customers' social networks are likewise growing and becoming more active (Ballestar et al. 2016). The research of Ballestar et al. (2018) showed that the important role of customers in using cashback applications determines customer behavior and activities for e-commerce channels. Vana et al. (2018) analyzed cashback incentives that increase a consumer's likelihood of purchasing e-commerce sites and increase the size of that transaction, according to dashboard data from a large refund company. Cashback payments increase a consumer's probability of purchasing e-commerce sites and increase the size of that transaction, regardless of impact. The adoption of cashback apps is like other types of online or offline marketing in terms of effectiveness (Christino et al. 2019).

The refund application or website is already present and popular in developed countries such as the US, UK, and Singapore. In Vietnam, this type of payment method was introduced in 2016 by putatu.com, but its level of popularity has only exploded in recent years, specifically, in 2020, when many website applications are launched such as Shopback, Accumulation, Cashback, Clingme, etc. (Demaitinh 2020).

The study of Christino et al. (2019) suggests a theoretical model established on the unified theory of acceptance and use of technology 2 (UTAUT2) to identify the elements that influence customers to sign up for cashback programs. The components of ease of use (EU), personal capacity (PC), PR, and behavioral aspects were added to the original ideal.

The cashback program is a strategy that is not too new to countries around the world, and there are many studies on this strategy. However, each study gives different results on factors affecting the intention or behavior of customers. Much of the difference is in the geographic, economic, and customer segmentation that researchers choose. In Vietnam, according to the research, the author found that this is a new market, and there are still limited research topics on this issue. So from that, the author formed the idea of researching the topic of the refund program. Therefore, this study focuses on the AIC algorithm for online cashback. The following is the order in which this chapter is presented: The first section gives an outline of the study, Section 11.2 shows a review of the literature on the factor used in this study, and Section 11.3 presents the methodology. Results and analysis with some discussion and implications are offered in Section 11.4. Lastly, Section 11.5 accomplishes this chapter.

11.2 LITERATURE REVIEW

11.2.1 What Is Cashback?

A cashback program is a form of online promotion in which a consumer makes a transaction through a cashback program affiliated with a cashback company, and after a substantial period, the consumer will receive the bonus amount as the transaction from the beginning.

This cashback program increases a consumer's probability of an additional purchase and increases the size of that transaction (Vana et al. 2018). These programs, therefore, act as a tool to attract customers, similar to advertising (Christino et al. 2019). The difference here is that potential customers are identified not from search term suggestions but from visits to the cashback program (Qiu and Rao 2020). Consumers visit the websites and applications of the cashback program, not e-commerce sites.

Cashback program transactions are pre-negotiated with merchants in the e-commerce marketplace and linked and posted on websites or apps with the negotiated refund. The cashback program company will receive a commission for each successful transaction and then deposit the payments directly into the consumer's credit account (Vana et al. 2018).

Cashback is a bank's incentive when you shop and pay with a credit card that will be refunded back to the credit card based on the payment value. With today's popular shopping, we often see promotions with direct discounts on products, discount codes, gift vouchers, etc. However, another model of stimulus in shopping is derived from the US and has spread to the UK and other countries which is cashback (also known as cashback). When making a purchase, you still receive the usual discount promotion and also receive an additional refund from the system, calculated as a % of the product value. When the cashback system was born, it brought a new era of shopping for the consumers themselves. Especially for customers who are thrifty and like to buy goods with a difference from the original price, this must be an attractive program.

Cashback system has gradually ushered in a new era of online shopping. When linking with many shopping sites on one system, it will help users accumulate cashback from all stores to one account. Products purchased at a more reasonable price are still selected from reputable addresses. Some cashback systems have allowed users to link their bank accounts, so withdrawing cashback to their accounts is very easy; at will, they do not force them to use that money to make purchases next time. In Vietnam, the wave of cashback entered in 2016, and with the pioneering shot, putatu.com—a smart cashback system—has helped Vietnamese people catch up with this trend. With a system of linking over 60 e-commerce sites in many fields such as Retail & Distribution (Lazada, Tiki, FPT shop, Sendo, Zalora, etc.); Services—Tourism, Restaurants, Hotels (MyTour, Atadi, Vntrip, Wefit, etc.); Education—Online training (Edumall, Kyna.vn, VTC Academy, etc.); and Finance—Banking, Credit (Sacombank, Standard Chartered Bank, ANZ, etc.) (Salemall 2023).

11.2.2 Using Behavior of Cashback (UBC)

Commerce-related business models that connect companies with customers help them improve their purchasing decisions (Barnes-Vieyra and Claycomb 2001). Vana et al. (2018) show that cashback payments increase the likelihood of repeat purchases and that customer behavior is significantly different from those without it, as they are not only performing transactions more diversified but also important in the items that need economic investment. Customers benefit from cashback programs because they make the purchasing process easier and provide financial incentives (Ballestar et al. 2016). Consumers are easily seduced by savings, which benefits them because they have the freedom to spend or store the money in any way they see fit (Vana et al. 2018). Consumers often assume they are a customer of that business. Before making a purchase, consumers often search for the products they need to buy on e-commerce channels to see if the product is in stock. Consumers tend to remember cashback programs only when they receive a notification (reminder) from a refund program (Christino et al. 2019, Venkatesh et al. 2012). Thus, the behavior of using the cashback application is the ability to use the cashback application depending on the benefits that the cashback application brings such as helping customers save, and the convenience of using the product.

11.2.3 Ease of Use (EU)

Potential users are more likely to embrace and use innovative technology solutions that are viewed as easy to use and less difficult (Davis 1989). In the technology acceptance model, according to (Davis 1989), "perceived EU is the degree to which a person believes that using a particular system will be effortless". Perceived EU of technology will have an effect or influence on behavior and activities, that is, the higher the perceived EU of a system, the higher the users' information level (Sayekti and Wijayanti 2018). EU is implicit as the ease with which users perceive a technology to be used, which is the trust that using a specific organization will be easy (Alalwan et al. 2018). Perceived EU helps users have a positive attitude toward the service, thereby enhancing the intention to use (Widyanti and Usman 2019). Consumers often think that mobile apps, such as cashback programs, are easy to use (Arora et al. 2020). According to Bai (2015), websites, and applications with a beautiful interface will attract more customers. Davis (1989) mentioned that consumers find it difficult to remember passwords or technology-related security. Kaur and Jain (2016) argue that customers feel that online payment will be useful for online shopping. Thus, it can be seen that "EU" is an intimate factor in the behavior of using the cashback program. Because consumers in today's technology era feel it is easy and necessary to use applications and programs to serve their online shopping needs.

11.2.4 Personal Capacity (PC)

PC relates to how confident consumers are that they have the necessary skills or knowledge to effectively participate in and use a particular service area (Walker and Johnson 2006). A personal capability to estimate consumers' competence to accomplish tasks is derived from social perception (Alalwan et al. 2016). When it comes to new technology, the easier it is for a person to use a specific program, the better their cognitive–behavioral performance is (Zhang et al. 2017). Capacity is "the ability, subjective, or natural condition available to perform a certain activity" such as thinking capacity, and financial capacity. Consumer technology skills are now good. And customers easily adapt to newly launched technology applications. They believe that they can always use a new technology application. More than that, they feel comfortable when many services are connected with technology applications (Walker and Johnson 2006, Christino et al. 2019). In the era of technology and modernization, consumers always equip themselves with knowledge and updates on the launch of new technologies to serve themselves, as well as shortening the time to perform a certain job just by manipulating applications and technology on mobile devices. Thus, the "PC" of customers always goes hand in hand with user behavior. They feel that they perform well or can perform a certain technology, the higher the behavior of using that technology.

11.2.5 Perceived Risk (PR)

Compared with product risk, consumers' perception of personal information disclosure risk had a greater impact on Internet shopping decisions (Eggert 2006). Consumers may face risks throughout the purchasing process if they are aware that they have no control over their purchase or that the consequences of making a poor decision could be disastrous (Al Kailani and Kumar 2011). PR is another factor affecting personal enjoyment and motivation. The degree to which a consumer perceives the overall negativeness of an action is referred to as PR (e.g., purchasing, using, or disposing of a product/service), based on an assessment of the negative outcomes and the probability of these outcomes (Zhang 2010). Types of risks act as barriers to conducting transactions on Internet, influence online product purchase behavior, and determine consumers' choice of purchasing channels (Tandon et al. 2016). PR is defined as the possibility of a negative outcome as a result of an action (Mohseni et al. 2018). PR has a direct effect on use and intention to use (Aminu et al. 2019). Consumers' PRs will influence consumers' attitudes and interests in online shopping (Ghachem et al. 2019). Customers feel insecure when providing personal information to apply for cashback programs. They worry that when they provide personal information to refund programs, their information will be used for other purposes. Besides, they also feel worried that the refund programs will be faulty (Walker and Johnson 2006, Christino et al. 2019). According to Abrar et al. (2017), consumers feel worried when the product is not correct or unsatisfactory (damaged, wrong size, not like the picture, etc.) after completing the transactions and making payment entirely online. It can be noticed from there that when performing the activity of using consumers, they will worry about the risks when connecting to applications and technologies. These are all things that can make them hesitate or decide not to use that application or technology. Technology businesses should pay attention to this issue because it is also a way to build trust with customers.

11.2.6 Using Intention of Cashback (UIC)

According to behavioral theory, Using Intention has a direct influence on the performance of the behavior. When a customer has an intention toward a certain product or service, the ability to buy and use the product or service is very high (Lee 2010). Research on consumer intention to use has broader implications than behavior and will often have a positive impact on individual actions. Intention to use is defined as an active decision that reveals an individual's behavior depending on the product (Ross 2021). According to Ajzen (1991), the intention is energizing and indicates a person's eagerness to engage in a specific behavior. It is understood as a person's determination to act in a certain way (Ramayah et al. 2010). Using Intention is the consumer behavior of using the app in the future. They will use technology to make online shopping. And when they make good use of the technology, they will intend to continue using that technology (Venkatesh et al. 2012, Christino et al. 2019). In short, the intention to use represents what the individual wants to use in the future (Mouakket 2015).

11.3 METHODS

11.3.1 Sample Approach

The data collection method is by questionnaire, a survey designed on Google Drive, and sending surveys through social networks such as Facebook and Zalo. The answers will be updated as soon as the surveyor completes the questionnaire. The statistics of the sample characteristics are listed in Table 11.1.

Table 11.1 shows that the results for the sexes of men and women are almost equal. However, there are more men than women, with men accounting for 54.2% of the total and women accounting for 45.8%. The age group of 18–29 years accounted

TABLE 11.1
Statistics of Sample

Characteristics		Amount	Percent (%)
Sex	Male	58	54.2
	Female	49	45.8
Age	18–29	104	97.2
	30–44	3	2.8
Monthly Income (VND to USD at https://vi.coinmill.com on January 18, 2023)	Below 298.83 USD	34	31.8
	341.52–426.90 USD	35	32.7
	469.59–640.35 USD	29	27.1
	Over 640.35 USD	9	8.4

for 97.2% of the total, while the age group of 30–44 years accounted for 2.8%. The remaining age groups did not. This result for the age group of 18–29 years is the most prominent age of exposure and use of online shopping and payment. Researching results on the structure of monthly income, we see that out of 107 people surveyed, there are 34 people with income under 298.83 USD, accounting for 31.8%; people with income from 341.52 to 426.90 USD 35 people account for 32.7%; there are 29 people with incomes from 469.59 to 640.35 USD, accounting for 27.1%; and the rest accounting for 8.4% are people with incomes over 640.35 USD.

To assess the amount of consent for the linked elements, we employ a five-point Likert scale. As a result, the five-point Likert scale is used in this paper to assess the level of permission for all items, with one denoting disagreement and five denoting agreement (see Table 11.2).

TABLE 11.2
Factor and Item

Factor	*Item*
EU	Mobile tech apps, such as cashback programs, are simple to practice.
	I find apps that have a nice, easy-to-use interface and are more appealing to me.
	Passwords, technology-related security issues, and applications such as cashback programs are all difficult for me to remember.
	I feel using online payment will be useful for online shopping.
PC	My technology skills are excellent.
	I can quickly learn new technology applications, such as online shopping or online payment, and can adapt quickly.
	I believe I will use freshly released technological applications, such as online shopping and payment apps.
	When various services are linked to technology applications, such as online buying or online payment, I feel at ease.
PR	When I provide personal information to apply for a cashback program, I feel uneasy (such as bank details, phone number, and address).
	I am worried that the cashback application will fail (e.g., not receiving the money for a much longer time to receive the money than expected, etc.).
	I'm concerned that if I take part in the cashback program, my personal information will be used for unknown purposes.
	I feel worried when the product is incorrect or unsatisfactory (damaged, wrong size, not like the picture, etc.) after completing online transactions and online payments.
	I do not feel confident when providing personal data to apply for a cashback program (such as bank details, phone number, and address, etc.)
UIC	I will use the cashback program in the future.
	On my purchases, I'll try to use cashback programs.
	I continue to use regular cashback programs.
UBC	I am a cashback program customer.
	Before buying, I often search the e-commerce site to see if the product is still available or out of stock.
	I only remember the cashback program when I receive a notification (reminder) from the application or website of these programs.

11.3.1.1 Blinding

For the duration of the study, all study staff and participants were blinded. The participants in the study had no contact with anyone from the outside world.

11.4 RESULTS

11.4.1 Overview of the Cashback Program in the World

The cashback program has been successfully used on US and UK websites. According to statistics, there are between 2000 and 4000 stores that interact with about 2.5 million customers through cashback programs, having spent over 2.3 billion USD in 2013 (Ballestar et al. 2018). Ebates, the first cashback company in the United States, as soon as opened in 1998 and processed over $800 million in cashback payments to over 10,000 consumers (Vana et al., 2018). In the UK, in 2016 alone, Quidco processed over $64 million in cashback payments for 7 billion registered users and facilitated nearly $1 billion in sales to 4300 retailers (Vana et al. 2018). In Brazil, cashback models are in the development stage and gaining space, increasing visibility (Christino et al. 2019). Brazilian discount sites have traded over 1 billion BRL and about 2% of all Brazilian e-commerce transactions ride by the schemes, which have over 3500 online partners.

11.4.2 The Cashback Program in Vietnam

Realizing the popularity of cashback programs in world countries, especially developed countries, at the beginning of 2016, Putatu enterprise was the pioneer in bringing cashback to Vietnam to bring positive innovation. After 2 months of launch, over 2000 participants made thousands of transactions with a value of up to 12 billion VND (512,050,233.72 USD). The user's savings amounted to 500 million VND (21,335,426,404.83 USD), and the largest refund was 40% (VTV online newspaper, 2017). As for Shopback, an outstanding cashback application in the Asia-Pacific region, which was present in Vietnam from December 2019 to August 2020, has attracted over 800,000 participants and over 150 partners, regularly recorded revenue growth, and orders increased by 1.5 times per month. The amount of Shopback refunded on 08/2020 is 4 billion VND (170,683,411.24 USD).

Currently, this form of refund has been popular in some big cities such as Ho Chi Minh City, Hanoi, and Da Nang. However, it is still unfamiliar to many people in the countryside. Due to their low demand for online shopping, they mainly trade in the traditional way.

11.4.3 Akaike's Information Criterion (AIC) Selection

The R program used AIC to select the optimal model. The AIC has been used for model selection in the theoretic environment (Mai et al. 2021). When multicollinearity occurs, the AIC method can handle several independent variables. AIC can estimate one or more dependent variables from one or more independent variables using a regression model. The AIC is an important and useful metric for finding a complete and simple model. A model with a lower AIC is selected based on the AIC information standard. When the least AIC value is obtained, the best model will be over (Burnham and Anderson 2004, Khoi 2021).

Every step of the search for the best model is documented in R reports. The initial step is to use AIC = −142.52 for UBC = *f* (PC + PR) to analyze all 06 independent variables in Table 11.3.

The p-value of <0.05 exists for two variables (Hill et al. 2018); as a result, they are linked to UBC, which is in Table 11.4. PC and PR impact UBC. EU and UIC do not influence UBC.

11.4.4 Variance Inflation Factor (VIF)

Multicollinearity occurs when the independent variables in regression models have a high correlation. If the VIF coefficient is greater than ten, Gujarati and Porter (2009) found evidence of multicollinearity in the model. The VIF for the independent factors is less than ten according to Table 11.5 (Miles 2014), showing that the independent variables are not collinear.

TABLE 11.3
AIC Selection

Model	AIC
UBC = *f* (EU + PC + PR + UIC)	−140.17
UBC = *f* (EU + PC + PR)	−141.96
UBC = *f* (PC + PR)	−142.52

TABLE 11.4
The Coefficients

UBC	Estimate				
Intercept	**1.16025**	**Std. Error**	**T**	**p-value**	**Decision**
PC	0.29019	0.07045	4.119	0.000000	Accepted
PR	−0.42578	0.06250	6.813	0.000000	Accepted

TABLE 11.5
Model Test

	PC	PR	
VIF	**1.267508**	**1.267508**	
Heteroskedasticity	Goldfeld–Quandt test	chi^2: 0.97345	p-value: 0.5383
Autocorrelation	Durbin–Watson: 2.0046	Test for autocorrelation p-value = 0.4953	
Model evaluation	Adjusted R-squared: 0.5116	F-statistic: 56.51	p-value: 0.0000

11.4.5 Heteroskedasticity

The random error must be constant, according to one of the key concepts of the traditional linear regression model. This hypothesis is unlikely to be correct. When the variance of the random error for each observation is dissimilar, heteroskedasticity grows. Because the p-value for the Goldfeld–Quandt test is 0.97345 and >0.05 (Godfrey 1978), in Table 11.5, there is no heteroskedasticity.

11.4.6 Autocorrelation

The range of the Durbin–Watson (DW) value is 0–4. There won't be any autocorrelation if the DW value is between 1.5 and 2.5; the closer the value is to zero, the more positively correlated the error parts will be; and the closer it is to 4, the more negatively correlated the error parts will be. Because the p-value = 0.4953 is >0.05, the Durbin–Watson test (2.0046) reveals that there is no autocorrelation in the model in Table 11.5 (Durbin and Watson 1971).

11.4.7 Model Evaluation

According to findings in Table 11.5, the PC and PR impact UBC is 51.16% in Table 11.5. R^2 (R Square) has a range from 0 to 1; the closer it is to 1, the more significant the model is; conversely, the closer it is to 0, the less significant the model is. If the p-value for the F test is <0.05, the regression model's fit is tested. The dataset fits the multiple linear regression model, which can be applied. The model is statistically significant, according to the aforementioned analysis (Greene 2003).

11.4.8 Discussion

The AIC algorithm for the UBC revealed that there are two independent factors. Because their p-values are <0.05, PC has a positive impact on the UBC, and PR has a negative impact on the UBC. In descending order, compare the influence degree of these two variables on the UBC: PC (0.29019) and PR (−0.42578). So, some relationships are accepted at the 95% confidence level.

Based on the regression results, we can see how the factors affect the behavior of using the cashback program of consumers in Ho Chi Minh City, factors of which have the most influence and have the least influence on consumers' use of cashback programs in the Ho Chi Minh City area. The above results show that when other factors are not considered when the factor of "risk perception" increases or decreases by one unit, the behavior of consumers using the cashback program also increases or decreases by 0.42578 units. This level affects the strongest factor.

When the factor "PC" increases or decreases by one unit, the behavior of consumers using the cashback program also increases or decreases by 0.29019 units. This is the second or last influencing factor of the model.

Thus, after analyzing Pearson's coefficient and conducting multivariable regression analysis, the author's research model is suitable for two factors affecting the behavior of consumers using cashback programs. Users in Ho Chi Minh City are sorted according to the normalized Beta coefficient in descending order as follows: (1) "risk perception" (−0.42578) and (2) "PC" (0.29019).

The study builds on previous studies: The study by Christino et al. (2019) is based on the UTAUT2 research model, which has successfully demonstrated that customer behavior is closely related to technology and the factors such as habit, social influence, and other behavioral aspects and affects the cashback application. Ballestar et al. (2016) found that customer engagement in cashback programs has a positive impact on customer loyalty and profitability for both customers and businesses. Ballestar et al. (2018) applied the concepts of loyalty, social networks, customer growth, and interaction to find that the role of customers depends on the customer's position in the social network. This study also shows that the more customers trust, the more engaged, the more transactional, and the level of engagement with the cashback program is also related to the multi-transaction of the program. Finally, research by Vana et al. (2018) has proven that it is useful for businesses and companies to cooperate with cashback programs. This study found that the effectiveness of cashback programs was generally similar to that of previously established online and offline promotions. The studies have been successful in developed countries, but for Vietnam, the refund program is still new; with the author's research, there are two factors: "PC" and "PR", a strong influence on the "behavior of using" cashback program of customers in Ho Chi Minh City.

11.5 CONCLUSION

This chapter builds on previous studies: Christino et al. (2019) study shows the factors affecting cashback applications such as habit, social influence, and other behavioral aspects; this chapter is based on the UTAUT2 research model, which has successfully shown that customer behavior is closely related to technology. Ballestar et al. (2016) found that customer engagement in cashback programs has a positive impact on customer loyalty and profitability for both customers and businesses. Ballestar et al. (2018) apply the concepts of trustworthiness, social systems, consumer growth, and interaction to see that the role of customers depends on the customer's position in the social network. This chapter also displays that the more consumers trust, the more engaged, the more transactional, and the level of engagement with the cashback program is also related to the multi-transaction of the program. Finally, research by Vana et al. (2018) has proven that it is useful for businesses and companies to cooperate with cashback programs. This study found that the success of cashback programs was like that of previously recognized online and offline promotions. The studies have been successful in developed countries, but for Vietnam, the refund program is still new; with the author's research, the two factors "PC" and "PR" are a strong influence on the "behavior of using" cashback program of customers.

11.6 IMPLICATIONS

11.6.1 Implication for PR

This study shows that "risk perception" has a negative impact on the "using behavior" of the cashback program in Ho Chi Minh City. Specifically, the regression results show that the factor "risk perception" has the strongest impact (Beta = −0.42578). The observed variables have mean values ranging from 1.9 to 2.1. Thus, this factor is rated by consumers on average as agreeing. Since this is a variable that has a positive effect on the dependent variable, but it has a powerful impact, it is extremely necessary to give the solution implication, to improve consumers' confidence in the model.

First, with the variable PR3 "worried that if I join the refund program, my data will be used by the refund program for incorrect resolutions" has an average rating of 1.9. Currently, the problem of "trading" customer information is a concern for many customers. Often customer information will be used for different purposes but all bring damage to the customer. Therefore, to gain the trust of customers, businesses must have commitments from the business side and promises if they use customer information to create trust for customers as well as to create customer loyalty and corporate reputation.

Next, the PR2 variable "worried that the program will fail" has an average rating of 2.0. If customers when using online payment are concerned about money in their application, they will be charged unreasonable fees or their card will be locked, etc. Similarly, with the refund program, consumers are more worried. Added by a buggy program, promises of refunds that are not received or refunds will be extended, longer than intended. Therefore, businesses should perform seriously and responsibly in their promises about refund time as well as the amount to be refunded to consumers.

Last in the "risk perception" factor is the PR1 variable "feel insecure when providing personal information to sign up for a refund program" with the highest average rating of 2.19. This is considered a sensitive issue when customers join any application. Like PR3, consumers are often insecure when their personal information is exposed to the outside because it directly bothers their lives. As mentioned in PR3, businesses need to make certain commitments and promises to create a trust for customers. Because only when there is trust, the process will be long.

In short, the factor "risk perception" has a strong and negative influence on the behavior of using the cashback program. Businesses of this form need to enhance and put the reputation of the business first. Since this is a new and promising form of the future, competition is indispensable. And credibility is essential in industries related to finance as well as promotion, so show your customers your credibility.

11.6.2 Implication for Personal Capacity and Perceived Risk

Regression results show that "personal competence" has a positive impact on the behavior of consumers using cashback programs in Ho Chi Minh City (Beta = 0.29019). This is the second most influential factor towards the "user behavior". The mean is relatively high, ranging from 4.21 to 4.33.

The observed variable with the lowest mean value is PC2 (4.20) "easy to adapt to newly launched technology applications". This level shows that customers agree with this statement, which means that in the era of industrialization and modernization, consumers are always ready to equip themselves with the necessary knowledge and skills to use technology. It also means that consumers are always ready and excited about new technology features. Therefore, businesses need to invest in designing their programs to be eye-catching and convenient, creating more skills so that consumers do not feel bored when using the cashback program.

Next is the PC1 variable "good technology skills" with an average rating of 4.21. Similar to PC2, PC1 also shows confidence in equipping knowledge to use technology. This seems to indicate that the more consumers use good technology, the more eager they are for programs and applications that mention new services. Cashback program businesses should improve the quality of their websites and applications, as well as mention other related services, so that consumers can enjoy, discover, and attract customers.

For the PC3 variable "capable of using newly released technology applications", the average rating is 4.28. With a fairly high average rating, this proves that before a new technology is released, consumers will learn in advance the features, usage, as well as desire to experience. Therefore, before launching a new program or feature of the cashback program, businesses need to have a trial version for customers to experience and give their own opinions so that businesses can complete the refund program improve the program as well as the product's features.

Finally, PC4 "feels comfortable when many services are connected with technology applications" with the highest average rating of 4.33. According to the author, businesses need to create more new and more creative features for the program so that customers feel more convenient when using the business's cashback program. In order to improve competition, customers always choose for themselves utility programs and applications, to serve their own needs.

In summary, the factor of "PC" refers to the fact that consumers in the current era are always updating their knowledge and skills about technology to use new technologies. The more the service is linked with technology, the more convenient it is and it shortens the usage time of consumers. Besides, it increases discoverability. Businesses should invest as well as regularly update new features to retain customers as well as attract potential customers.

11.6.3 Limitations and Next Research Directions

First, this study was conducted in a relatively short time of only 5 months; the time limitation has more or less affected selection of the research model and building of the most suitable scale; and the time of collecting the survey collection also affects the results of the research on factors. Second, in carrying out the research, the author also encountered certain difficulties such as having difficulty in accessing leading experts, limited meeting time as well as not being able to fully comprehend all the relevant topics. Collecting secondary data also faces many difficulties because some agencies in Vietnam have not disclosed data transparently and clearly. Third, the number of samples 107 that the author studied is not large enough. This number of samples only represents a small part of the population, not the general population of the entire population of consumers living in Vietnam. And because the epidemic is quite serious, it is also limited in the direct survey process. Surveys only used on social networks are also a limitation. Therefore, the number of consumers invited to respond to the survey cannot fully assess consumer behavior in using the cashback program in Vietnam. To complete the research model and achieve better results, the next research topics on the behavior of using the consumer cashback program are as follows: research with a larger number of samples to achieve a higher level of satisfaction with better overall representation; and research on a larger scale in Vietnam, with many types of consumers to increase the diversity of survey subjects. It is possible to focus on researching each separate consumer object to devise a specific strategy for each different consumer group.

REFERENCES

Abrar, Kashif, Muhammd Naveed, and Ifra Ramay. 2017. "Impact of perceived risk on online impulse buying tendency: An empirical study in the consumer market of Pakistan." *Journal of Accounting & Marketing* 6(3):246.

Ajzen, Icek. 1991. "The theory of planned behavior." *Organizational Behavior and Human Decision Processes* 50(2):179–211.

Al Kailani, Mahmud, and Rachna Kumar. 2011. "Investigating uncertainty avoidance and perceived risk for impacting Internet buying: A study in three national cultures." *International Journal of Business and Management* 6(5):76.

Alalwan, Ali Abdallah, Yogesh K. Dwivedi, Nripendra P.P. Rana, and Michael D. Williams. 2016. "Consumer adoption of mobile banking in Jordan: Examining the role of usefulness, ease of use, perceived risk and self-efficacy." *Journal of Enterprise Information Management* 29(1):118–139.

Alalwan, Ali Abdallah, Yogesh K. Dwivedi, Nripendra P.P. Rana, and Raed Algharabat. 2018. "Examining factors influencing Jordanian customers' intentions and adoption of internet banking: Extending UTAUT2 with risk." *Journal of Retailing and Consumer Services* 40:125–138.

Aminu, Suraju Abiodun, Olusegun Paul Olawore, and Adesina Emmanuel Odesanya. 2019. "Perceived risk barriers to internet shopping." *KIU Journal of Social Sciences* 5(2):69–81.

Arora, Neerja, Garima Malik, and Deepak Chawla. 2020. "Factors affecting consumer adoption of mobile apps in NCR: A qualitative study." *Global Business Review* 21(1):176–196.

Bai, Jianyong. 2015. How the electronic commerce enterprise attract more customers. *Paper Read at 2015 International Conference on Management Science and Innovative Education*, Xi'an, China.

Ballestar, María Teresa, Jorge Sainz, and Joan Torrent-Sellens. 2016. "Social networks on cashback websites." *Psychology & Marketing* 33(12):1039–1045.

Ballestar, María Teresa, Pilar Grau-Carles, and Jorge Sainz. 2018. "Customer segmentation in e-commerce: Applications to the cashback business model." *Journal of Business Research* 88:407–414.

Barnes-Vieyra, Pamela, and Cindy Claycomb. 2001. "Business-to-business e-commerce: Models and managerial decisions." *Business Horizons* 44(3):13–13.

Burnham, Kenneth P., and David R. Anderson. 2004. "Multimodel inference: Understanding AIC and BIC in model selection." *Sociological Methods & Research* 33(2):261–304.

Christino, Juliana Maria Magalhães, Thaís Santos Silva, Erico Aurélio Abreu Cardozo, Alexandre de Pádua Carrieri, and Patricia de Paiva Nunes. 2019. "Understanding affiliation to cashback programs: An emerging technique in an emerging country." *Journal of Retailing and Consumer Services* 47:78–86.

CongThuong. Shopback cashback platform officially launched in Vietnam (Vietnamese) 2020. Available from https://congthuong.vn/nen-tang-hoan-tien-shopback-chinh-thuc-ra-mat-tai-viet-nam-141780.html.

Davis, Fred D. 1989. "Perceived usefulness, perceived ease of use, and user acceptance of information technology." *MIS Quarterly* 13(3):319–340.

Demaitinh. 2020. Six best cashback apps in Vietnam today (Vietnamese). Available from https://demaitinh.vn/6-ung-dung-hoan-tien-cashback-tot-nhat-hien-nay-tai-viet-nam.

Durbin, James, and Geoffrey S. Watson. 1971. "Testing for serial correlation in least squares regression III." *Biometrika* 58(1):1–19.

Eggert, Axel. 2006. "Intangibility and perceived risk in online environments." *Journal of Marketing Management* 22(5–6):553–572.

Ghachem, Lassaad, Costinel Dobre, Reza Etemad-Sajadi, and Anca Milovan-Ciuta. 2019. "The impact of cultural dimensions on the perceived risk of online shopping." *Studia Universitatis Babes-Bolyai, Negotia* 64(3):7–28.

Godfrey, Leslie G. 1978. "Testing for multiplicative heteroskedasticity." *Journal of Econometrics* 8(2):227–236.

Greene, William H. 2003. *Econometric Analysis*. Pearson Education India: Noida, India.

Gujarati, Damodar N., and Dawn C. Porter. 2009. *Basic Econometrics*. Mc Graw-Hill International Edition: New York.

Hill, R. Carter, William E. Griffiths, and Guay C. Lim. 2018. *Principles of Econometrics*. John Wiley & Sons: Hoboken, NJ.

Kaur, Baljeet, and Tanya Jain. 2016. "Attracting customers' to online shopping using mobile apps: A case study of Indian Market." In: Madan, Sushila (Ed.), *Securing Transactions and Payment Systems for M-Commerce*, pp. 117–140. IGI Global: Hershey, PA.

Khoi, Bui Huy. 2021. "Factors influencing on University reputation: Model selection by AIC." In: Thach, Nguyen Ngoc, Vladik Kreinovich, and Nguyen Duc Trung (Eds.), *Data Science for Financial Econometrics*, pp. 177–188. Springer: Berlin/Heidelberg, Germany.

Lee, Choongsoo. 2010. "Study on the decision factors for customer satisfaction and repurchase intention in high-speed internet access service: Focus on the company, product and customer service." *The e-Business Studies (Global E-Business Association)* 11(4):275–289.

Mai, Dam Sao, Phan Hong Hai, and Bui Huy Khoi. 2021. Optimal model choice using AIC method and naive Bayes classification. In *IOP Conference Series: Materials Science and Engineering*, Medan, Indonesia.

Miles, Jeremy. 2014. "Tolerance and variance inflation factor." In: Balakrishnan, N., Theodore Colton, Brian Everitt, Walter W. Piegorsch, Fabrizio Ruggeri, Jef L. Teugels (Eds.), *Wiley StatsRef: Statistics Reference Online*. John Wiley & Sons: Hoboken, NJ.

Mohseni, Shahriar, Sreenivasan Jayashree, Sajad Rezaei, Azilah Kasim, and Fevzi Okumus. 2018. "Attracting tourists to travel companies' websites: The structural relationship between website brand, personal value, shopping experience, perceived risk and purchase intention." *Current Issues in Tourism* 21(6):616–645.

Mouakket, Samar. 2015. "Factors influencing continuance intention to use social network sites: The Facebook case." *Computers in Human Behavior* 53:102–110.

PVcomBank. Cashback: Promotion trend with many benefits for consumers 2020. Available from https://www.pvcombank.com.vn/tin-tuc/tin-pvcombank/hoan-tien-xu-huong-khuyen-mai-nhieu-loi-ich-cho-nguoi-tieu-dung-1577.html.

Qiu, Ye, and Ram C. Rao. 2020. "Increasing retailer loyalty through the use of cashback rebate sites." *Marketing Science* 39(4):743–762.

Ramayah, T., Noor Hazlina Ahmad, and May-Chiun Lo. 2010. "The role of quality factors in intention to continue using an e-learning system in Malaysia." *Procedia-Social and Behavioral Sciences* 2(2):5422–5426.

Ross, Stanley C. 2021. *Organizational Behavior Today*. Routledge: London.

SaleMall. Cashback: Cashback system when shopping online (Vietnamese) 2023. Available from https://salemall.vn/blog/cashback-he-thong-hoan-tien-khi-mua-sam-online.

Sayekti, F., and L.E. Wijayanti. 2018. The influence of Perception of Usefulness (PoU) and Perceived Ease of Use (PEU) on the perception of information system performance. *Paper Read at Increasing Management Relevance and Competitiveness: Proceedings of the 2nd Global Conference on Business, Management and Entrepreneurship (GC-BME 2017)*, August 9, 2017, Universitas Airlangga, Surabaya, Indonesia.

Tandon, Urvashi, Ravi Kiran, and Ash N. Sah. 2016. "Understanding online shopping adoption in India: Unified theory of acceptance and use of technology 2 (UTAUT2) with perceived risk application." *Service Science* 8(4):420–437.

Thanhnien. Cashback: Promotion trend with many benefits for consumers (Vietnamese) 2020. Available from https://thanhnien.vn/tai-chinh-kinh-doanh/hoan-tien-xu-huong-khuyen-mai-nhieu-loi-ich-cho-nguoi-tieu-dung-1292140.html.

Vana, Prasad, Anja Lambrecht, and Marco Bertini. 2018. "Cashback is cash forward: Delaying a discount to entice future spending." *Journal of Marketing Research* 55(6):852–868.

Venkatesh, Viswanath, James Y.L. Thong, and Xin Xu. 2012. "Consumer acceptance and use of information technology: Extending the unified theory of acceptance and use of technology." *MIS Quarterly* 36(1):157–178.

VTV. 2017. Viet Startup brings the world's popular cashback model to Vietnam (Vietnamese) 2017. Available from https://vtv.vn/kinh-te/startup-viet-mang-mo-hinh-cashback-thinh-hanh-tren-the-gioi-ve-viet-nam-20170420080544875.htm.

Walker, Rhett H., and Lester W. Johnson. 2006. "Why consumers use and do not use technology-enabled services." *Journal of Services Marketing* 20(2):11.

Widyanti, Jovita, and Osly Usman. 2019. "Leverage of perceived usefulness, perceived ease of use, information quality, behavioral intention towards intention to use mobile banking." *Perceived Ease of Use, Information Quality, Behavioral Intention towards Intention to Use Mobile Banking* (December 27, 2019).

Zhang, Lin-lin. 2010. Measuring perceived risk of securities investment based on disposition effect. *Paper Read at 2010 International Conference on E-Product E-Service and E-Entertainment*, Henan, China.

Zhang, Peiyao, Nan Li, Dongping Fang, and Haojie Wu. 2017. "Supervisor-focused behavior-based safety method for the construction industry: Case study in Hong Kong." *Journal of Construction Engineering and Management* 143 (7):05017009.

12 Capacitated Vehicle Routing Problem Using Algebraic Harris Hawks Optimization Algorithm

Mohammad Sajid, Md Saquib Jawed, Shafiqul Abidin, Mohammad Shahid, Shakeel Ahamad, and Jagendra Singh

CONTENTS

12.1 INTRODUCTION

The vehicle routing problem (VRP), a very prominent NP-hard problem, has many applications in transportation and logistics. The effectual delivery of demands/services to customers plays a decisive role in dealing with smart logistics, UAV-assisted delivery systems, disaster management, and others [1]. Dantzig and Ramser

DOI: 10.1201/9781003438588-12

[2] proposed the VRP, which requires the efficient delivery of customers' demands at the minimum possible cost. The capacitated vehicle routing problem (CVRP), a variant of VRP, requires finding optimal routes to deliver customers' demands using a fleet of bounded capacity vehicles to optimize the total traveled distance [1–3]. The CVRP has numerous real-world applications, including e-commerce, food, dairy, beverage, solid waste, disaster management, newspaper industries, and others [4–6]. Various exact methods, heuristics, local search operators, and evolutionary algorithms are adopted to deal with the CVRP and its variants. Some well-known methods are particle swarm optimization, differential evolution (DE), ant colony optimization (ACO), firefly algorithm (FA), genetic algorithm (GA), swap, exchange, 2-Opt*, scramble operators, branch-and-bound, dynamic programming, and many others [1,7–9]. The CVRP is inherently discrete, that is, the solutions are intrinsically discrete and can be represented as customers' permutations. In the context of discrete problems, the evolutionary and swarm intelligence (ESI) algorithms can be categorized into two groups, that is, continuous-valued and discrete-valued. In discrete-valued ESI algorithms, the solutions are represented as permutations, integer-valued arrays, or bit strings, and the decoding method is not required. For continuous-valued ESI algorithms, the solutions are represented as a continuous-valued array. A decoding method is essential to transform a continuous-valued array into a discrete solution that essentially defines the solution of the problem under consideration. Various decoding methods exist like the random-key (RK) method, reverse RK method, and others to convert a continuous-valued array into an array with discrete values [10–12]. However, a single discrete-valued array can be encoded to potentially infinite numbers of the continuous-valued array due to cardinality issues, thus forcing the ESI algorithm to navigate the large plateaus in the fitness space. Numerous continuous-valued ESI algorithms are applied to resolve the discrete optimization problems [10]. However, very few algebraic ESI algorithms have been proposed for discrete optimization problems, that is, multidimensional two-way number partitioning problem (MDTWNPP), permutation flow-shop scheduling (PFSS), linear ordering problem (LOP), traveling salesman problem (TSP), quadratic assignment problem (QAP), and others [10–14]. Neither an algebraic evolutionary nor an algebraic swarm-based method exists to address the CVRP.

In this chapter, a novel algebraic Harris hawks optimization (AHHO) algorithm is proposed to resolve the CVRP to optimize the total distance traveled. Harris hawks optimization (HHO) algorithm [15] is a continuous-valued, gradient-independent, population-based, swarm intelligence (SI) algorithm which simulates the hunting behavior of Harris' hawks (Parabuteo unicinctus) to capture prey using a harmonized foraging with "seven kill" strategy (surprise pounce). It finds optimized solutions by continually refining the candidates using exploratory and exploitative phases, Harris hawk's attacking strategies, and surprise pounce. HHO consists of a mathematical model which works based on the continuous-valued vectors in the continuous-valued solution space. The proposed AHHO is the algebraic version of HHO, simulating the original movements of continuous-valued solution space in the permutation-based solution space. Since the solutions of the CVRP can be represented by the set of permutations of given customers, this set forms a symmetric group (a.k.a. permutation group) with a composition operator. As defined in Refs. [10–14], the permutation group with a composition operator allows interpreting the discrete solutions (permutations) using composition-based operators, that is, abstract addition (⊕), abstract subtraction (⊖), and abstract scalar multiplication (⊙). In the proposed AHHO algorithm, the abstract addition (⊕), abstract subtraction (⊖), and abstract scalar multiplication (⊙) operators are applied to the discrete solutions (permutations) to search for new permutation in the permutation-based space. The proposed AHHO also employs the discrete version of Lévy flights which works based on Lévy flights and local neighborhood search operators, that is, exchange, swap, relocate, 2-opt*, and inversion [9,16].

In a nutshell, the followings are the main contributions of the chapter:

- An AHHO algorithm for solving the CVRP is proposed.
- An algebraic update mechanism is developed to mimic the Harris hawks' hunting moves in permutation-based solution space.
- A discrete Lévy flight based on local search operators is proposed.
- The proposed algorithm is assessed based on 85 recognized CVRP instances.
- The performance of the proposed AHHO algorithm is superior compared to random-key HHO (RK-HHO), hybrid firefly, and cuckoo search with Lévy flights algorithms.

The documentation of the remaining chapter is mentioned later. The literature review is covered in Section 12.2, whereas the mathematical formulation of the CVRP is given in Section 12.3. The suggested AHHO algorithm, the discrete Lévy flight, and the local search operators are all explained in Section 12.4. All the experiments and the simulations are demonstrated in Section 12.5, and the concluding remarks and future research suggestions are presented in Section 12.6.

12.2 LITERATURE REVIEW

The CVRP is a commonly studied and famous NP-hard combinatorial optimization problem. There have been numerous suggested exact algorithms, heuristics, local search operators, ESI, and hybrid algorithms in the literature. The related work reports here are a few of the most recent algorithms.

Sajid et al. [3] proposed HGSA based on GA and simulated annealing (SA) to solve the CVRP. HGSA employs GA operators, that is, nearest-neighbor crossover operator and swap mutation to perform the global search in the solution space, and SA-based local search operators to local neighborhood search to improve the convergence speed. Simulation study depicts that HGSA performs better than the GA and

SA algorithms on 86 CVRP benchmark instances. HGSA suggested a novel solution for the P-n55-k15 CVRP instance that is better than the other known solutions.

İlhan [17] proposed an ISA-CO, that is, population-based improved SA algorithm with crossover operator for CVRP. The ISA-CO employs local operators, that is, scramble, swap, reversion, and insertion; 2-opt and crossover operators, that is, partially mapped crossover (PMX) and order crossover (OX) operators, to improve the convergence speed. The performance of ISA-CO was evaluated using 91 CVRP instances and many emerging algorithms.

Rabbouch et al. [18] proposed the empirical-type simulated annealing algorithm to address the CVRP with homogeneous vehicles by exploiting the fresh portion of the worst feasible solutions to update the Boltzmann acceptance rule empirically. The gamma distribution function was employed, which was adjusted incrementally as per the distribution of the rejected solutions. The parametric density function helps reduce the randomness and convergence speed of the algorithm.

Hossainabadi et al. [19] proposed a hybrid GELSGA based on gravitational emulation local search and GA to solve the CVRP. Hybrid GELSGA employs two algorithms, that is, GA and GELS, to perform the exploration and exploitation, respectively. The two algorithms' tuning helps escape the local optimum and improve the convergence speed.

İlhan [20] proposed a population-based SA algorithm for CVRP. The population-based SA algorithm developed routes based on stochastic selection and application of local search operators, that is, reversion, insertion, and exchange. On a total of 63 CVRP cases, population-based SA was found to provide optimal solutions in 23 of them.

Lin et al. [21] proposed a hybrid algorithm, that is, an order-aware hybrid genetic algorithm (OHGA) based on GA and sweep algorithm to address the CVRP to minimize the total traveled distance. The OHGA employs a sweep algorithm-based initialization strategy and best cost route crossover operator to harmonize the contradiction between diversity and convergence and to avoid local optimum.

Altabeeb et al. [22] proposed FA-based CVRP-FA for solving the CVRP by minimizing the overall traveled distance. The CVRP-FA employs the local search techniques (i.e., and 2-h-opt and improved 2-opt methods) to improve the convergence speed, PMX, and two mutation operators to improve the exploitation. CVRP-FA employs the Taguchi technique for the purpose of determining the best values for the parameters and also evaluates performance by comparing with other emerging algorithms.

Azad and Hasin [23] solved the CVRP for Bangladesh's cement company using a GA to reduce operating costs and improve the economic benefits. The proposed algorithm creates clusters of distributors/resellers and employs a weighted fitness function to search for the fittest candidates, forming routes to deliver the cement. The proposed algorithm is tested on an actual problem instance with 25 customers.

Mohammed et al. [24] employ the GA to propose GACVRP for VRP of UNITEN to schedule eight buses on two routes for transporting students to/from eight locations on each path from the early morning hours to the end of the official working hours. The goal is to find the best routes for VRP of UNITTEN to reduce time/

traveled distance which ultimately leads to a reduction in the transportation costs, including fuel and vehicle upkeep costs. The simulation results show that the daily savings are relatively low but significantly impact monthly savings.

Sajid et al. [25] have used a novel stochastic giant tour best cost crossover operator in the suggested bi-objective routing algorithm based on non-dominated sorting genetic algorithm-II (NSGA-II) to resolve the CVRP. The NSGA-II-based algorithm was tested on 88 benchmark CVRP against two different versions of NGSA-II based on edge assembly crossover and nearest-neighbor crossover operators. The bi-objective routing algorithm offered qualitative Pareto routes with optimum distance traveled and length of the longest route.

Altabeeb et al. [26] proposed a cooperative hybrid firefly algorithm (CVRP-CHFA) to address the CVRP. The CVRP-CHFA consists of multiple FA populations that communicate to exchange solutions to preserve the diversity and to avoid local optima. Each population's local search and genetic operators evolve in the search space. The CVRP-CHFA is evaluated on 108 problem instances against eight algorithms.

Dalbah et al. [27] proposed a modified coronavirus herd immunity optimizer (CHIO) based on the COVID-19 herd immunity treatment approach to address the CVRP. The population is divided into infected, susceptible, and immune cases, and a repairing procedure is also utilized to convert the infeasible solution into feasible solutions in the evolving stages. The modified CHIO algorithm is evaluated on synthetic CVRP models and 27 known benchmark instances. The authors reported the competitive performance of the modified CHIO algorithm against 13 algorithms.

Queiroga et al. [28] proposed partial optimization meta-heuristic under special intensification conditions (POPMUSICs) to address the CVRP. POPMUSIC approach utilizes the branch-cut-and-price algorithm to solve the subproblems with dimensions from 25 to 200 customers. The POPMUSIC approach is evaluated for large problem instances of 32–1000 customers. The authors reported the competitive performance of the POPMUSIC approach. The authors also utilized the POPMUSIC approach for solving the VRP with backhauls and heterogeneous fleet VRP.

Many more algorithms have also been proposed, such as the inverse optimization approach [29], ISOS [30], LNS-ACO [31], CVRP-GELS [32], GVNS [33], GLS [34], and many others [35–39].

There are algebraic algorithms to solve the MDTWNPP, PFSS, TSP, LOP, QAP, and others [10–14]. No algebraic algorithm exists to address the CVRP. This work proposes a state-of-the-art AHHO algorithm to solve the CVRP. The proposed AHHO is the algebraic version of the HHO algorithm, simulating the original movements of continuous space in the permutation-based solution space.

12.3 PROBLEM FORMULATION

CVRP consists of three elements: (1) 2-D graph $G=(N, E)$ consisting of geometric positions of depot and customers, (2) customers' demands represented by set Q, and (3) a fleet V of vehicles with their capacity limits. In the 2-D graph $G=(N, E)$, the set $N=\{i : 0 \leq i \leq n\}$ and set $E=\{(i, j) : i, \quad j \in N, i \neq j\}$ represent the set of nodes and edges between them, respectively. Node 0 represents the depot node, while the remaining nodes represent customers to be served. The 2-D geometric coordinates (x_i, y_i) characterize each node $i \in N$, and the Euclidean distance between two nodes $i, j \in N$ is represented by d_{ij}. The customers' demands are also given and characterized by $Q = \{q_i : 1 \leq i \leq N,\ q_i > 0\}$. The customers' demands are served by a fleet V of homogeneous vehicles, represented as $V = \{v_k :\ 1 \leq k \leq K\}$. It is assumed that every vehicle has an equal capacity H and starts from the depot node $0 \in N$ [1,3].

In order to solve the optimization problem, a fleet of vehicles must be used to meet the needs of physically dispersed customers in a way that minimizes the distance traveled while fulfilling constraints of service integrity, vehicle flow, and capacity limits. Let χ_{ijk} be the binary decision variable that can take a value of either zero or one. The non-zero value of χ_{ijk} indicates that vehicle v_k moves to node j after serving node i. The CVRP optimization problem seeks to minimize the total distance traveled by all vehicles, and it is given as follows:

$$\text{Min } f_1 = \sum_{v_k \int V} \sum_{i \int N} \sum_{j \int N} \chi_{ijk} \times d_{ij} \tag{12.1}$$

Having the problem constraints

$$\sum_{v_k \epsilon V} \sum_{j \epsilon N} \chi_{ijk} = 1 \ \forall i \in N \ (2) \tag{12.2}$$

$$\sum_{i \int N} \sum_{j \int N - \{n_0\}} \chi_{ijk} \times q_j \leq H \quad \forall v_k \in V \tag{12.3}$$

$$\sum_{j \int N} \chi_{0jk} = 1 \quad \forall v_k \in V \tag{12.4}$$

$$\sum_{i \int N} \chi_{ihk} - \sum_{j \int N} \chi_{hjk} = 0 \ \forall h \in N - \{0\}, \quad \forall v_k \in V \tag{12.5}$$

$$\sum_{i \int N} \chi_{i0k} = 1 \quad \forall v_k \in V \tag{12.6}$$

$$\chi_{ijk} \ \varepsilon \ \{0,1\} \forall i, \ j \in N, \quad \forall v_k \in V \tag{12.7}$$

The objective function, which is the overall distance traversed, is represented by Equation (12.1), to be minimized. The service integrity is ensured by constraint (12.2), which requires that each customer will be served once using a single vehicle. The vehicle' capacity limit is represented by constraint (12.3), that is, the total demands assigned to each vehicle must not surpass the capacity limit. Constraints (12.4)–(12.6) represent the vehicle's flow constraints. The flow constraints (12.4)–(12.6) ensure that each vehicle departs from the depot, serves the visited client, and returns to the depot, while the concluding constraint (12.7) ensures that the decision variable χ_{ijk} must be binary.

12.4 PROPOSED WORK

This section explains the group theory preliminaries and the proposed AHHO algorithm.

12.4.1 Permutation Group Preliminaries

Many permutation-based combinatorial optimization problems, such as TSP, CVRP, permutation flow-shop scheduling problem, and others, require solutions to be presented in the form of permutation. The permutation of a set is an arrangement of its members into linear order, that is, it is bijections from a set $N' = \{k : 1 \le k \le n\}$ onto N' itself. For the set N', the permutation-based solution space consists of $n!$ permutations representing CVRP solutions. It is evident that the CVRP solutions increase exponentially with the value of n. Figure 12.1 shows a permutation π_i of nine customers, graphical representation of CVRP, and the routes which can be formed based on customer's demands, vehicle capacities, and cluster-first route-second approach [25].

In mathematical theory, the set of all permutations of a given finite set of integers, that is, $N' = \{k : 1 \le k \le n\}$, is a finite symmetric group S_n with composition operator (*). For two permutations $\pi_i, \pi_j \in S_n$, the composition operator, that is, $\pi_i \times \pi_j$, produces a permutation such that $(\pi_i \times \pi_j)(i) = \pi_i(\pi_j(i))$ for all indices $i \in [n]$. The application of composition operator on permutations $\pi_i, \pi_j \in S_n$ forms the resultant solution to be a new permutation π_k which is also an element of the symmetric group S_n. The set S_n, being a group, satisfies three group axioms, that is, it consists of an identity element (π_e), inverse element (π'_e), and it fulfills the condition of associativity. The identity element π_e is the permutation such that $\pi_e(k) = k$ for all indices $k \in [n]$. Similarly, the inverse element π'_e is the permutation $\pi'_e(k) = n - k + 1$ for all indices $k \in N'$. For every group (S_n, *), there exists a generating set $\langle S_n \rangle$ of permutations, a subset of S_n such that every permutation of the group S_n can be generated

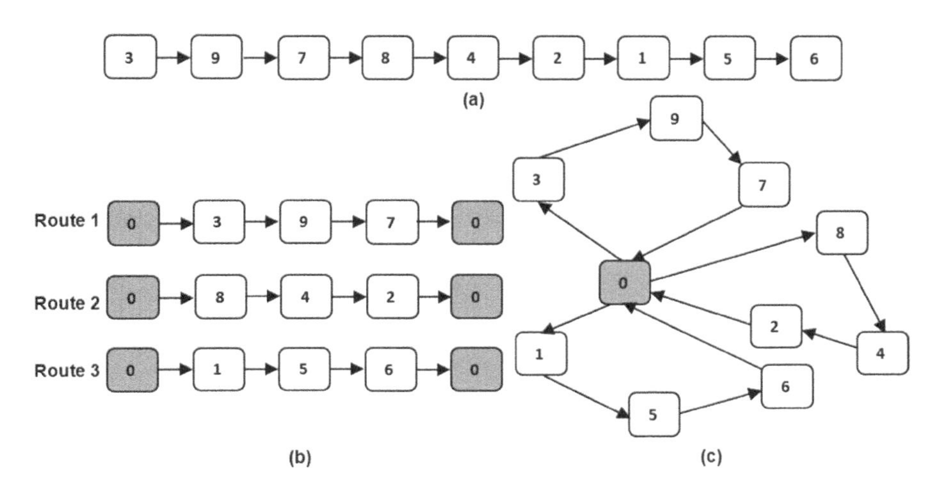

FIGURE 12.1 (a) Permutation (solution), (b) routes, (c) graphical representation of CVRP.

using elements of generating set $\langle S_n \rangle$. It has been proved that the abstract addition ($\oplus$), abstract subtraction ($\ominus$), and abstract scalar multiplication ($\odot$) in terms of composition operator (*) form the finite symmetric group S_n due to above properties. The three operators $\oplus \ominus$ and $\odot$ hold properties analogous to the numerical operations in Euclidean vector space [10–14].

As proved in Ref. [11], the algebraic abstract addition ($\oplus$) is independent of the generators and generating set. It can be defined as

$$\pi_z = \pi_i \oplus \pi_j = \pi_i \times \pi_j \quad \forall \pi_i, \pi_j \in S_n \tag{12.8}$$

Similarly, the abstract subtraction operator ($\ominus$) is also independent of the generators and generating set, that is, it can be defined as

$$\pi_z = \pi_i \ominus \pi_j = \pi_j^{-1} \times \pi_i \quad \forall \pi_i, \pi_j \in S_n \tag{12.9}$$

It is to be noted that abstract addition and subtraction operators are both non-commutative, that is, $\pi_i \oplus \pi_j \neq \pi_j \oplus \pi_i$ and $\pi_i \ominus \pi_j \neq \pi_j \ominus \pi_i$.

It is required to choose a generating set forming the symmetric group S_n to define the abstract scalar multiplication [10]. The set consisting of all permutations that swap the elements of adjacent positions is known as the set of all simple transpositions (<*ST*>). The generating set (<*ST*>) consists of $n-1$ elements, and its search space diameter is $\binom{n}{2}$. In this work, the generating set of all simple transpositions (<*ST*>) has been chosen due to its simple search space structure. There exist $n-1$ simple transpositions for n integers, and ith simple transposition (σ_i) is given as follows:

$$\sigma_i(j) = \begin{cases} j+1 & i = j \\ j-1 & i+1 = j \\ j & \text{Otherwise} \end{cases} \tag{12.10}$$

The composition of $\pi_i \in S_n$ and $\sigma_i \in \langle S \rangle$ results in the swapping of elements at indices i and $I+1$ in π_i. The successive application of this simple property helps transform permutation $\pi_i \in S_n$into the desired permutation.

The analogy for abstract scalar multiplication is similar to numerical scalar multiplication in Euclidean vector space. The multiplication of a scalar α with permutation $\pi_i \in S_n$ changes the length of permutation π_i as per the value of scalar α. For given $\alpha \geq 0$ and $\pi_i \in S_n$, the abstract scalar multiplication ($\alpha \odot \pi_i$) fulfills the following three properties to simulate the scalar numerical multiplication in Euclidean vector space [10,11]:

- P_1: $\pi_i \odot \alpha = \left[\pi_i \cdot |\alpha|\right]$
- P_2: $\pi_i \odot \alpha \leq \pi_i \quad \forall \alpha \in [0,1]$
- P_3: $\pi_i \odot \alpha \geq \pi_i \quad \forall \alpha > 1$

There may be more than one permutation in S_n due to applications of properties P_1 – P_3 and non-unique minimal decompositions of each permutation. The diameter also restricts the value of scalar α, that is, the value of scalar α must be a fraction of the diameter. Due to the above restrictions, abstract scalar multiplication can be realized in many different ways. To implement the abstract scalar operation ($\odot$) on permutations, the following three cases are considered based on properties P_1 – P_3.

$$\pi_i \odot \alpha = \pi_i \quad \text{if} \quad \alpha = 1 \tag{12.11}$$

$$\pi_i \odot \alpha \leq \pi_i \quad \forall 0 \leq \alpha < 1 \tag{12.12}$$

$$\pi_i \odot \alpha \geq \pi_i \quad \forall \alpha > 1 \tag{12.13}$$

Equations (12.11)–(12.13) follow the analogy of scalar numerical multiplication in Euclidean vector space. The abstract scalar multiplication can be computed using Equations (12.11)–(12.13) and the stochastic bubble sort algorithm [10,11].

12.4.2 Algebraic Harris Hawks Optimization Algorithm

HHO is a recently introduced (2019) continuous-valued, swarm-based, gradient-free, intelligent search algorithm which simulates Harris hawks' hunting behavior to capture prey using a harmonized foraging with seven kill strategy (surprise pounce). It finds optimized solutions by continually refining the candidates using exploratory and exploitative phases, Harris hawk's attacking approach, and surprise pounce. HHO consists of a mathematical model which works based on the continuous-valued vectors in the continuous-valued solution space [15].

The algebraic HHO is the algebraic version of HHO, simulating the original HHO's continuous space moves in the permutation-based solution space. In the proposed AHHO, the population consists of permutations as candidates, and the AHHO updates the candidate solution using algebraic operators. For the AHHO algorithm, the mathematical model of HHO is changed to an algebraic version, that is, the mathematical equations have been re-defined using abstract addition ($\oplus$), abstract subtraction ($\ominus$), and abstract scalar multiplication ($\odot$). Three operators help to move from one solution to another in the permutation-based solution space for the CVRP. Similar to HHO, the candidate solutions are considered as Harris hawks, while the best candidate is known as the rabbit. Similar to HHO, exploration and exploitation are the two main aspects of AHHO. AHHO also employs the discrete version of Lévy flights, which works based on Lévy distribution and local search operators, that is, relocate, swap, reversion, 2-Opt*, and exchange [3,9]. Figure 12.2 shows the different phases of AHHO. The pseudo-code of AHHO is similar to HHO, except all equations are algebraic versions.

12.4.2.1 Phase I: Exploration Phase

Harris hawks are patient and watchful hunters, able to wait for their prey in an arid location. The perching behavior of Harris hawks is based on the whereabouts of the rabbit and other family members, as well as at random within the group's home area.

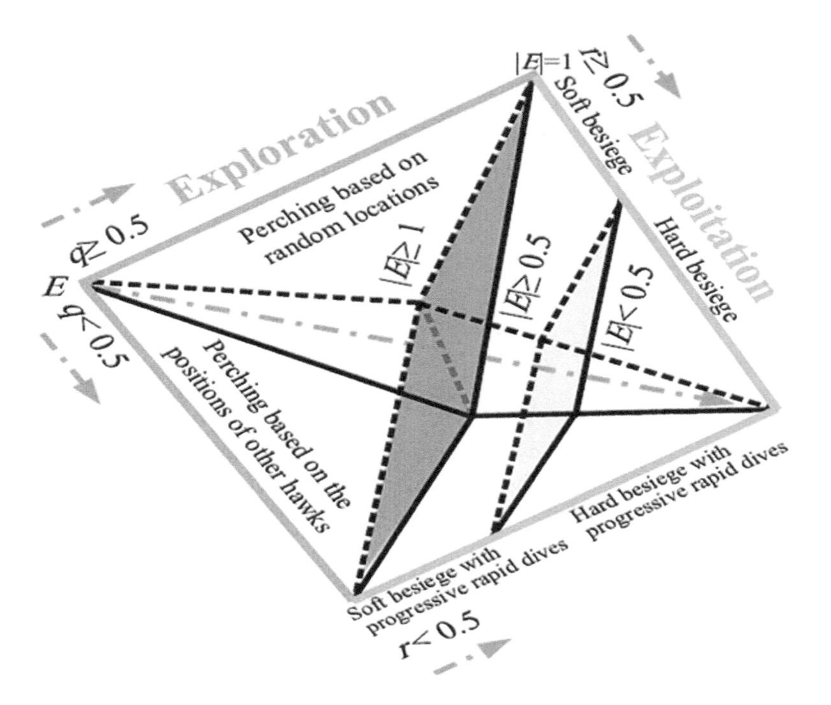

FIGURE 12.2 Different phases of HHO [15]

The former is realized using $q < 0.5$, and the latter is realized using $q \geq 0.5$, given as follows:

$$P(t+1) = \begin{cases} P_{\text{rand}}(t) \ominus r_1 \odot \left| P_{\text{rand}}(t) \ominus 2r_2 \odot P(t) \right| & q \geq 0.5 \\ \left(P_{\text{rabbit}}(t) \ominus P_m(t)\right) \ominus r_3 \odot \left(\pi_e + r_4\left(\pi'_e \ominus \pi_e\right)\right) & q < 0.5 \end{cases} \tag{12.14}$$

where $P(t)$ and $P(t+1)$ are the candidate solutions (hawks) in the current t and next iteration $t+1$, respectively. $P_{\text{rabbit}}(t)$ represents the best candidate solution (rabbit), and q, r_1, r_2, r_3, and r_4 are random numbers inside (0,1). The identity (π_e) and reverse (π'_e) elements are considered lower and upper bounds in the permutation space. $P_{\text{rand}}(t)$ is a candidate solution selected randomly from the current population, while $P_m(t)$ is the average position of the Hawks in the current population. The value of $P_m(t)$ is computed as follows:

$$P_m(t) = P_{\text{rabbit}}(t) \oplus P_{\text{max}}(t) \tag{12.15}$$

where $P_{\text{max}}(t)$ represents the candidate solution (hawk) with maximum fitness value.

The prey's escaping energy E diminishes due to their evasive behavior and can be modeled in terms of the initial state of energy E_0 and the total number of iterations.

$$E = 2E_0\left(1 - \frac{t}{T}\right) \tag{12.16}$$

Now, depending on the prey's escaping energy E, the transition between exploration and exploitation takes place. The value of E_0 belongs to the interval (–1, 1), and thus, the value of escaping energy changes in the interval (–2, 2). For escaping energy $|E| \geq 1$, AHHO performs the exploration in unexplored solution space, that is, the candidate solutions (hawks) explore different regions in search of rabbit location. For escaping energy when $|E| < 1$, AHHO performs the exploitation in their neighborhoods, that is, the candidate solutions (hawks) look for better rabbit locations in the neighborhood locations.

12.4.2.2 Phase II: Exploitation Phase

In this phase, the candidate solutions explored in the previous stage are exploited to search for a potential solution in the neighborhoods. The hawks execute the surprise attack to frighten the prey (rabbit), while the prey attempts to flee from the hawks, resulting in the prey's depleted energy level (rabbit). This depleted energy state allows hawks to easily capture their prey, and hawks will conduct soft or hard besiege to encircle their prey from multiple directions. Due to the different attacking strategies of hawks and different escaping patterns of prey (rabbit), four possible techniques are employed in AHHO as in HHO. Four methods are based on the prey's escaping chance r and escaping energy $|E|$.

12.4.2.3 Soft Besiege

The first case, $|E| \geq 0.5$ and $r \geq 0.5$, represents that the prey (rabbit) still has sufficient energy and performs random jumps to mislead the hawks. In this situation, the hawks encircle the prey softly to exhaust their energy and then perform the surprise pounce. This situation can be formulated as follows:

$$P(t+1) = \Delta P(t) \ominus E \odot \left| J \odot P_{\text{rabbit}}(t) \ominus P(t) \right| \tag{12.17}$$

$$\Delta P(t) = P_{\text{rabbit}}(t) \ominus P(t) \tag{12.18}$$

where $J = 2(1 - r_5)$ is the random jump strength of the prey (rabbit) during hunting, and r_5 is a random number in (0, 1). The $\Delta P(t)$ represents the difference between the position of prey and hawk at iteration t.

12.4.2.4 Hard Besiege

In the second case, $|E| < 0.5$ and $r \geq 0.5$, the prey consists of low escaping energy, and the hawks hardly encircle the intended prey to perform the surprise pounce. This situation is modeled as follows:

$$P(t+1) = P_{\text{rabbit}}(t) \ominus E \odot \left| \Delta P(t) \right| \tag{12.19}$$

12.4.2.5 Soft Besiege with Progressive Rapid Dives

The third case, $r < 0.5$ and $|E| \geq 0.5$, represents that the prey (rabbit) has sufficient energy to escape. In this scenario, the hawks construct a soft besiege, followed by a surprise attack. To mathematically model this situation, the Lévy flights, which represent a model of random walks characterized by their step lengths and follow a power law distribution, are employed in AHHO. The Lévy flights simulate the natural zigzag deceptive motions of prey and the abrupt, irregular, and rapid dives of raptors by simulating the evasive patterns of the prey and leapfrog movements. The hawks correct their location and direction in successive team fast dives around the prey.

In AHHO, the candidate solutions are permutations; therefore, the random jumps have been realized in permutation-based solution space [16]. It is assumed that the hawks will decide their next move as follows:

$$Y = P_{\text{rabbit}}(t) \ominus E \odot \left| J \odot P_{\text{rabbit}}(t) \ominus P(t) \right| \tag{12.20}$$

Next, the hawks evaluate their dives by comparing the results. If the move is not good, the hawks start performing deceptive motions based on Lévy flights. The Lévy flight helps to improve the quality of the search using different steps followed by occasional giant steps. The step length associated with the local search operators is generated using the Lévy flights as follows:

$$\text{LF}(s) = 0.01 \times \frac{u \times \sigma}{|v|^{\frac{1}{\beta}}}, \ \sigma = \left(\frac{\Gamma(1+\beta) \times \sin\left(\frac{\pi\beta}{2}\right)}{\Gamma\left(\frac{1+\beta}{2}\right) \times \beta \times 2^{\left(\frac{\beta-1}{2}\right)}} \right)^{\frac{1}{\beta}} \tag{12.21}$$

where u and v are random variables in (0, 1), and the value of β is taken as 1.5.

Figure 12.3a represents the three random walks generated by the Lévy flight (Mantegna distribution), while Figure 12.3b shows the value of the Lévy random variable for three different executions. It can be observed that the value of the Lévy step is very small, followed by an occasional big value step. A small jump is taken for the small value of the Lévy step, while a big jump is taken for the giant value. To convert the real Lévy random value into discrete, small and large Lévy scaling coefficients k and K are defined to take small and big jumps, respectively. The small and big jumps consist of k and $k \times K$ discrete steps, respectively. The large Lévy scaling coefficient K is much larger than the small Lévy scaling coefficient k. In the proposed algorithm, the value of k is taken as two while the value of K is considered 30. The values of k and K impact the AHHO algorithm's complexity and must be chosen carefully. In the proposed discrete Lévy flight, one step consists of applications of five local operators, that is, swap, exchange, relocate, inversion, 2-opt*, and exchange [9] as shown in Figure 12.4. The pseudo-code for Lévy flights based on the random search is given in Algorithm 1. First, the algorithm produces a random step value using Equation (12.21), and then, the iteration count (δ) is set to a small or bigger value according to the step value. In each iteration of the loop, the random neighborhood solutions of the candidate solution Y are generated using swap, exchange, relocate, 2-opt*, and inversion operators. The candidate solution Y is changed to the best neighborhood solution with a minimum fitness value. These steps are executed till the value of iteration count (δ) becomes zero.

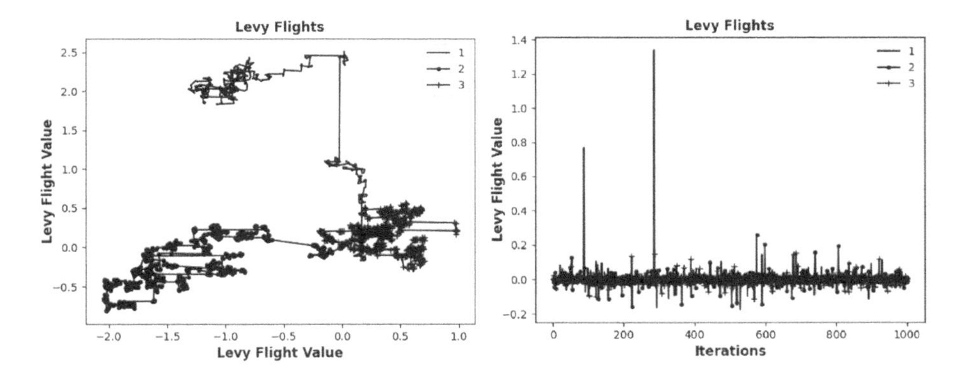

FIGURE 12.3 (a) Three random walks and (b) Lévy flight values.

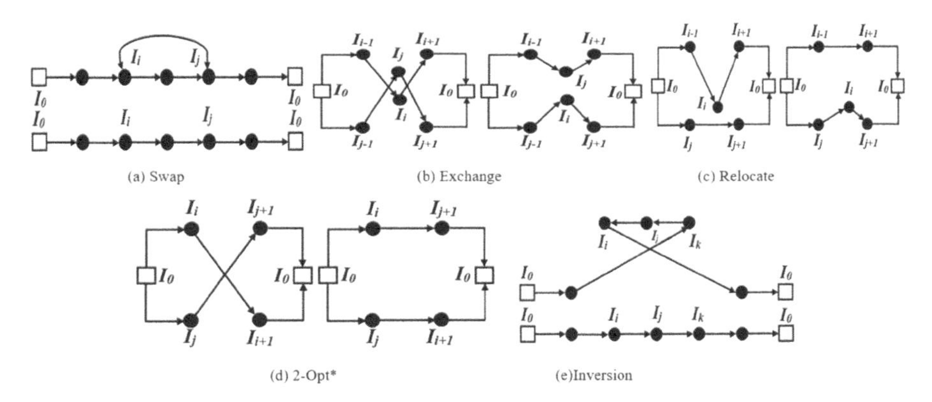

FIGURE 12.4 Local search operators: (a) swap, (b) exchange, (c) relocate, (d) 2-opt*, and (e) inversion.

Algorithm 12.1: Discrete Lévy Flights

Inputs: Candidate solution (hawk) Y
Output: A candidate solution (hawk) Z
Generate step value LF(s) using Equation (12.21)
For small LF(s), set iteration count $\delta = 2$
For large LF(s), set iteration count $\delta = 2 \times 30$
while $\delta > 0$
    Apply swap operation on Y to generate solution Y_1
    Apply relocate operation on Y to generate solution Y_2
    Apply reversion operation on Y to generate solution Y_3
    Apply exchange operation on Y to generate solution Y_4
    Apply 2-opt* operation on Y to generate solution Y_5
     Z = best of Y_1, Y_2, Y_3, Y_4, Y_5
        If fitness (Z) < fitness (Y)

            $Y = Z$

Return Y

The candidate solution, $P(t+1)$, is updated as follows:

$$P(t+1)=\begin{cases} Y \text{ if } \text{fitness}(Y)<\text{fitness}(P(t)) \\ Z \text{ if } \text{fitness}(Z)<\text{fitness}(P(t)) \end{cases} \tag{12.22}$$

12.4.2.6 Hard Besiege with Progressive Rapid Dives

The fourth and the final case, $r<0.5$ and $|E|<0.5$, indicates that the prey (rabbit) lacks adequate energy to effectively escape; therefore, the hawks conduct a hard besiege followed by a surprise attack to capture and slay the prey. The hard besiege with progressive rapid dives is comparable to the soft besiege with progressive fast dives, except that the candidate solution (hawks) attempts to decrease the average distance between their location and the prey. Thus, the first candidate solution is updated as follows:

$$W = P_{\text{rabbit}}(t) \ominus E \odot \left|J \odot P_{\text{rabbit}}(t) \ominus P_m(t)\right| \tag{12.23}$$

Now, the hawk evaluates their dives by comparing the results. If the move is not suitable, the hawks start to perform the deceptive motions based on Lévy flights and randomly generate the solution U. The next solution $P(t+1)$ is updated as follows:

$$P(t+1)=\begin{cases} W \text{ if } \text{fitness}(W)<\text{fitness}(P(t)) \\ U \text{ if } \text{fitness}(U)<\text{fitness}(P(t)) \end{cases} \tag{12.24}$$

12.5 EXPERIMENTAL STUDY

The experimental study comprises three sections. Section 12.5.1 discusses different system settings and state-of-the-art algorithms. Section 12.5.2 presents the results, while the study's observations are provided in Section 12.5.3.

12.5.1 System Settings and the State-of-the-Art Algorithms

The developed AHHO algorithm is evaluated on 85 well-known CVRP benchmark instances from four sets, that is, sets A, B, P, and E. The sets A, B, and P were proposed by Augerat et al. [40–42], while Christofides and Eilon proposed the Set E. Table 12.1 puts basic information on CVRP instances in four sets. Sets A, B, P, and E contain 27, 23, 34, and 11 CVRP instances. Each CVRP instance consists of the required number of vehicles, vehicles' capacity limit, the total number of customers, customers' geometric coordinates, and customers' demands. Each CVRP instance differs in the number of customers, customers' locations, demands, total vehicles, and capacity limits. The CVRP instance A-n44-k7 consists of 44 nodes (one depot and 43 customers) and 07 vehicles used to serve the 43 customers. The CVRP instance A-n44-k7 consists of geometric coordinates, demands of 44 nodes, and the capacity limit of vehicles.

TABLE 12.1
CVRP Instances Information

Set	Number of Instances	Smallest Demand	Extreme Demand	Nearest Customer to Central Depot	Farthest Customer to Central Depot
A	27	1	72	2.0	125.88
B	23	1	69	6.33	116.85
E	11	1	4,100	2.24	118.87
P	24	1	2,500	2.24	49.92

TABLE 12.2
System Settings

Parameter	Values
Generations	100
Population size	100
Small Lévy scaling coefficient	2
Large Lévy scaling coefficient	30
Total experiments for each instance	10
Objective value selected (table)	Mean
Vehicle route presented (graphical results)	Best

The proposed AHHO algorithm is implemented and evaluated on Desktop Intel i7 processor, 2.30 GHz, 8 GB memory, Windows 10 OS, and Python 3.10.4. Table 12.2 lists the standard parameters used to conduct the different experiments. For each problem instance, the Python NumPy module was used to generate a population of 200 random candidates. Each candidate is a random permutation of customers representing the complete solution of the CVRP instance. The performance of the AHHO algorithm is compared with the RK-HHO, hybrid firefly, and cuckoo search via Lévy flights algorithms in terms of total traveled distance. The random-key HHO (RK-HHO) is an original continuous-valued HHO algorithm with RK decoding. The RK decoding transforms the real-valued candidate into a permutation [11]. The hybrid firefly (2019) and cuckoo search via Lévy flights (2018) algorithms are recently proposed meta-heuristic algorithms to solve the CVRP [22,37].

12.5.2 Simulation Routing Results

This section discusses the results provided by AHHO, RK-HHO, hybrid firefly, and cuckoo search via Lévy flights, which are presented and discussed. Figure 12.5–12.8 shows graphical results offered by AHHO and RK-HHO algorithms corresponding to one CVRP instance in each set. Three convergence curves are used:

- BCC—The best convergence curve (BCC) represents the convergence curve leading to the best solution offered in ten different experiments.
- ACC—The average convergence curve (ACC) represents the convergence curve corresponding to the average fitness (distance) offered at each iteration

in ten experiments. The small gap between ACC and BCC means that the algorithm performs better in all experiments.
- PA—The population average (PA) curve represents the average fitness of all population candidates. This curve represents the average population fitness corresponding to the best solution offered in ten different experiments. If the PA curve and BCC gap decrease, all candidate solutions improve their fitness values.

Figure 12.5 represents the problem instance A-n32-k5, routes offered, and convergence curves by AHHO and RK-HHO algorithms. Figure 12.5a plots the customer and depot locations in 2D Euclidean space. Figure 12.5b displays the routes offered by the RK-HHO algorithm, while routes offered by AHHO are shown in Figure 12.5c. As evident from Figure 12.5b and c, minimal overlapping exists in the AHHOs' routes compared to RK-HHO algorithms' routes. The best-known solution (BKS) distance for the CVRP instance A-n32-k5 is 784, and the distances offered by AHHO and RK-HHO algorithms are 797 and 1238, respectively. Figure 12.5d represents the three convergence curves indicating the better performance of the proposed AHHO compared RK-HHO algorithm. As can also be seen from Figure 12.5d, the BCC and ACC produced by AHHO overlap, indicating the stable performance of AHHO in all ten experiments for the problem instance A-n32-k5. The BCC and ACC offered by RK-HHO have a considerable gap, showing the RK-HHO's unstable performance in all ten experiments. The BCC and PA curves offered by AHHO have a significant gap in the initial iterations while they become close to each other in the last iterations. It indicates that AHHO begins with random candidate solutions with high fitness and converges to the fitness of the best solution in the last iterations.

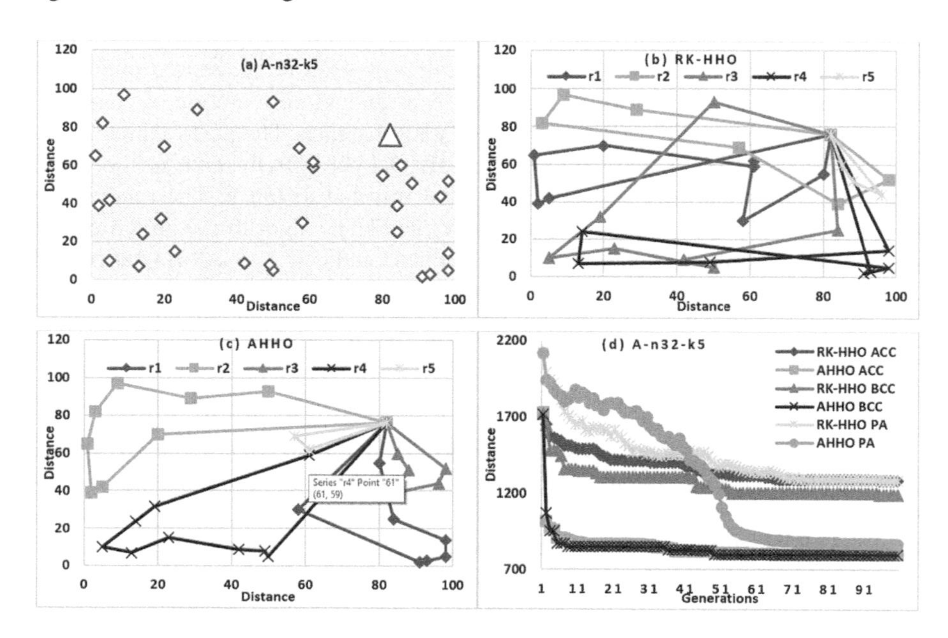

FIGURE 12.5 (a) Instance A-n32-k5, (b and c) offered routes, and (d) convergence curves.

The RK-HHO's BCC and PA curves become closer in the last iterations. It indicates that RK-HHO improves the fitness of all candidates; however, RK-HHO's PA and AHHOs' PA curves have considerable differences indicating the superior performance of the proposed AHHO algorithm.

Figures 12.6–12.8 represent the graphical results offered by AHHO and RK-HHO algorithms for the three CVRP instances, that is, B-n38-k6, P-n23-k8, and E-n33-k4, respectively. Figures 12.6a, 12.7a, and 12.8a plot the customer and depot locations on 2D Euclidean space for B-n38-k6, P-n23-k8, and E-n33-k4, respectively. Figures 12.6b, 12.7b, and 12.8b depict the routes offered by the RK-HHO algorithm for three CVRP instances. The routes offered by AHHO are displayed in Figures 12.6c, 12.7c, and 12.8c. The value for the BKS for the CVRP instance B-n38-k6 is 805, and the distances offered by AHHO and RK-HHO algorithms are 805 and 1069, respectively. For CVRP instance P-n23-k8, the distance of the BKS is 554, while the distances offered by the AHHO and RK-HHO algorithms are 531 and 563, respectively. The distance of BKS for the CVRP instance E-n33-k4 is 835, and AHHO and RK-HHO offer 837 and 1036, respectively. It is visible from Figures 12.6b and c, 12.7b and c, and 12.8b and c that the routes offered by AHHO have very few overlapping compared to RK-HHO algorithms' routes. Figures 12.6d, 12.7d, and 12.8d represent the convergence curves for the CVRP instances B-n38-k6, P-n23-k8, and E-n33-k4, respectively. The convergence curves in Figures 12.6d, 12.7d, and 12.8d depict the stable performance of AHHO in all ten different experiments because BCC and ACC produced by AHHO overlap with each other for the three problem instances. On the other hand, the BCC and ACC offered by RK-HHO indicate that the performance of RK-HHOs is not stable in all ten experiments for the three

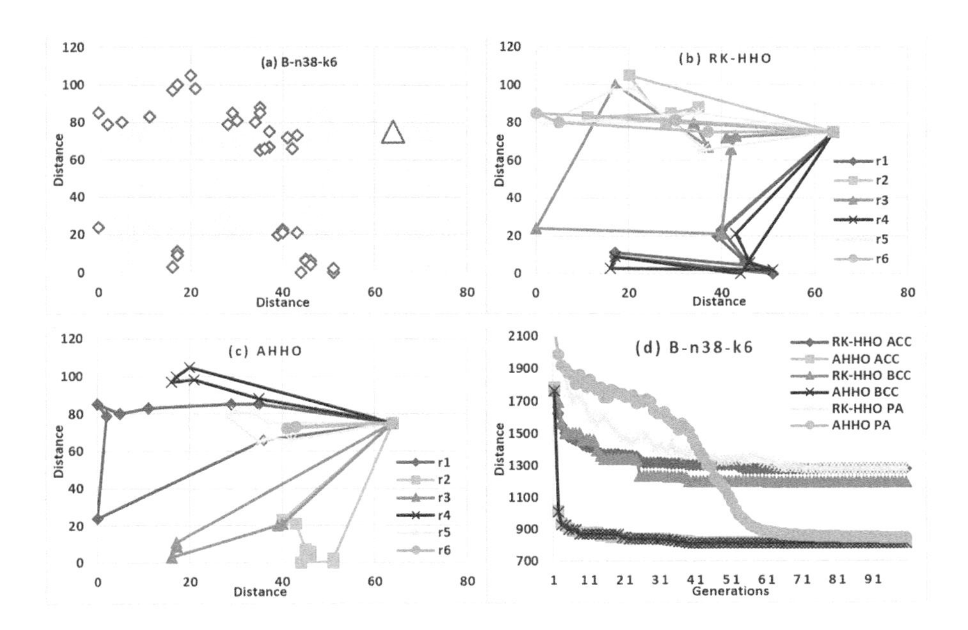

FIGURE 12.6 (a) Instance B-n38-k6, (b and c) offered routes, and (d) convergence curves.

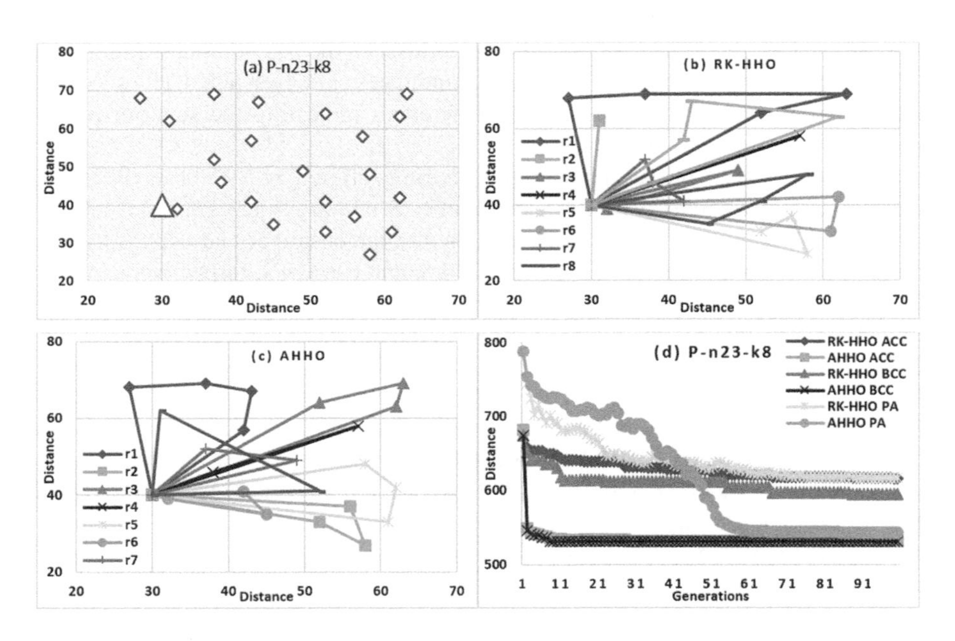

FIGURE 12.7 (a) Instance P-n23-k8, (b and c) offered routes, and (d) convergence curves.

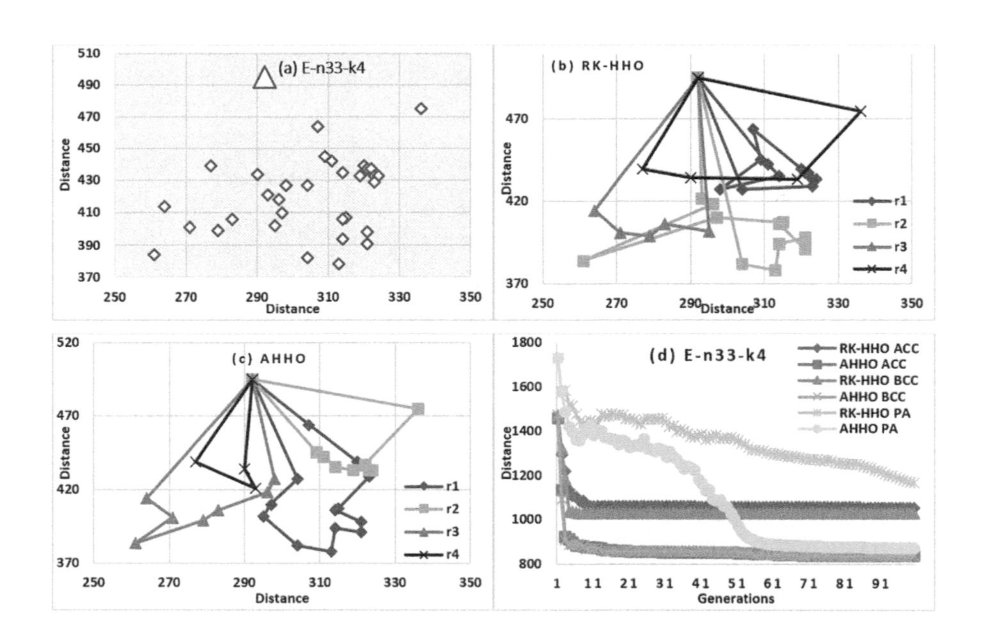

FIGURE 12.8 (a) Instance E-n33-k4, (b and c) routes offered, and (d) convergence curves.

CVRP instances. The BCC and PA curves offered by AHHO and RK-HHO have a considerable gap in the beginning and decrease as algorithms progress. It indicates that AHHO and RK-HHO start with random candidate solutions with high fitness, and the fitness of all candidates improves in the last iterations. However, there is a considerable gap between RK-HHO's PA and AHHOs' PA curves, proving the superiority of AHHO over the RK-HHO algorithm.

It is also observed for the remaining 81 CVRP instances that the proposed AHHO algorithm incessantly offers better routes and convergence curves compared to the RK-HHO algorithm. AHHO algorithm offers vehicle routes with less overlapping compared to RK-HHO algorithm. The performance of AHHO also remains more stable compared to the RK-HHO algorithm for all remaining CVRP instances. In the following part, the performance of the proposed AHHO algorithm is compared with hybrid firefly and cuckoo search via Lévy flights. For 85 CVRP instances, the results offered by AHHO, RK-HHO, hybrid firefly, and cuckoo search via Lévy flights are provided in Tables 12.3 and 12.4.

The results offered by AHHO, RK-HHO, hybrid firefly, and cuckoo search via Lévy flights are provided in Tables 12.3 and 12.4 for sets A, B, P, and E, respectively. The performance is compared by taking into account the relative best percentage gap given as follows:

$$B\_Gap = \frac{\text{Best distance offered} - \text{Besk Known Solution}(\text{BKS})}{\text{Besk Known Solution}(\text{BKS})} \times 100 \quad (12.25)$$

The B_Gap metric represents the relative performance of the algorithm. The small positive value means that the algorithm offers the solution near to the BKS. The negative value denotes that the algorithm provides better than the BKS.

Table 12.3 puts the results of 27 CVRP instances of Set A in terms of total traveled distances and B_Gaps offered by AHHO, RK-HHO, hybrid firefly, and cuckoo search via Lévy flights algorithms. The first CVRP instance consists of 31 customers, and the last CVRP has 79 customers. As it can be perceived, the AHHO algorithm provides the smallest distance compared to RK-HHO, hybrid firefly, and cuckoo search via Lévy flights algorithms for all 27 CVRP instances. The B_Gap for the AHHO algorithm is minimum compared to RK-HHO, hybrid firefly, and cuckoo search via Lévy flights algorithms. The AHHO algorithms' B_Gap is also zero for four CVRP instances, A-n33-k5, A-n33-k6, A-n34-k5, and A-n38-k5, indicating that AHHO offers the results equivalent to the BKSs. Out of 27 CVRP instances, AHHO offers 11 solutions with B_Gap less than one. However, the peer algorithms offer larger values of B_Gap for all CVRP instances. For all instances of Set A, the average B_Gap of the results provided by the AHHO, RK-HHO, hybrid firefly, and cuckoo search via Lévy flights algorithms are 1.255, 75.139, 15.318, and 65.582, respectively. It indicates that the order of performance from the best to worst for Set A is AHHO, hybrid firefly, cuckoo search via Lévy flights, and RK-HHO algorithms.

The total traveled distances and B_Gaps offered by AHHO, RK-HHO, hybrid firefly, and cuckoo search via Lévy flights algorithms for 23 CVRP instances of Set B are presented in Table 12.4. The number of customers ranges from 30 to 77 in all CVRP instances. As is visible from Table 12.4, the distances offered by the proposed AHHO

TABLE 12.3
Computational Results of RK-HHO, Hybrid Firefly, Cuckoo Search via Lévy Flights, and AHHO for Set A

		RK-HHO		Hybrid Firefly		Cuckoo Search via Lévy Flights		AHHO	
Instance	**BKS**	**Distance**	**B_Gap**	**Distance**	**B_Gap**	**Distance**	**B_Gap**	**Distance**	**B_Gap**
A-n32-k5	784	1238	57.908	831	5.994	1065	35.842	797	1.658
A-n33-k5	661	993	50.226	711	7.564	914	38.275	661	0
A-n33-k6	742	1088	46.630	783	5.525	1005	35.445	742	0
A-n34-k5	778	1073	37.917	827	6.298	1083	39.203	778	0
A-n36-k5	799	1162	45.431	870	8.886	1092	36.671	809	1.255
A-n37-k5	669	1140	70.403	669	0	998	49.178	672	0.448
A-n37-k6	949	1449	52.687	1021	7.586	1279	34.773	960	1.159
A-n38-k5	730	1244	70.410	787	7.808	1239	69.726	730	0
A-n39-k5	822	1285	56.326	898	9.245	1308	59.124	828	0.729
A-n39-k6	831	1494	79.783	868	4.452	1292	55.475	833	0.240
A-n44-k6	937	1536	63.927	1051	12.166	1417	51.227	945	0.853
A-n45-k6	944	1726	82.838	1108	17.372	1863	97.352	957	1.377
A-n45-k7	1146	1678	46.422	1281	11.780	1576	37.522	1152	0.523
A-n46-k7	914	1844	101.750	1049	14.770	1339	46.499	926	1.312
A-n48-k7	1073	1849	72.320	1218	13.513	1686	57.129	1076	0.279
A-n53-k7	1010	1807	78.9108	1200	18.811	1940	92.079	1036	2.574
A-n54-k7	1167	2036	74.464	1374	17.737	1925	64.953	1188	1.799
A-n55-k9	1073	2099	95.619	1324	23.392	1772	65.1445	1098	2.329
A-n60-k9	1408	2454	74.289	1650	17.187	2246	59.518	1417	0.639
A-n61-k9	1035	2149	107.633	1362	31.594	2200	112.561	1049	1.353
A-n62-k8	1290	2570	99.224	1566	21.395	2372	83.876	1320	2.325
A-n63-k10	1315	2434	85.095	1606	22.129	2325	76.806	1353	2.889
A-n63-k9	1634	2554	56.303	1976	20.930	2897	77.294	1656	1.346
A-n64-k9	1402	2416	72.325	1684	20.114	2468	76.034	1419	1.212
A-n65-k9	1177	2682	127.867	1522	29.311	2608	121.580	1195	1.529
A-n69-k9	1168	2565	119.606	1534	31.335	2433	108.305	1203	2.996
A-n80-k10	1764	3571	102.437	2235	26.701	3336	89.116	1818	3.061
Average	1045.259	1856.889	75.139	1222.407	15.318	1765.852	65.582	1059.926	1.255

algorithm are minimum compared to RK-HHO, hybrid firefly, and cuckoo search via Lévy flights algorithms for all 23 CVRP instances. The B_Gap offered by the proposed AHHO algorithm is comparatively minimum in the group of four. The AHHO algorithms' B_Gap is zero for four CVRP instances, that is, B-n31-k5, B-n34-k5, B-n35-k5, and B-n38-k6. The value of B_Gap is <1 for 11 CVRP instances, that is, B-n39-k5, B-n41-k6, B-n43-k6, B-n44-k7, B-n45-k5, B-n45-k6, B-n52-k7, B-n56-k7, B-n57-k7, B-n66-k9, and B-n78-k10. Moreover, the value of B_Gap is negative for two CVRP instances, that is, B-n51-k7 and B-n66-k9. Out of 23 CVRP instances of Set B, AHHO offers two solutions with B_Gap less than zero, four with B_Gap zero,

TABLE 12.4
Computational Results of RK-HHO, Hybrid Firefly, Cuckoo Search via Lévy Flights, and AHHO for Set B

		RK-HHO		Hybrid Firefly		Cuckoo Search via Lévy Flights		AHHO	
Instance	**BKS**	**Distance**	**B_Gap**	**Distance**	**B_Gap**	**Distance**	**B_Gap**	**Distance**	**B_Gap**
B-n31-k5	672	817	21.577	672	0	746	11.012	672	0
B-n34-k5	788	1052	33.502	788	0	1007	27.792	788	0
B-n35-k5	955	1391	45.654	986	3.246	1250	30.890	955	0
B-n38-k6	805	1069	32.795	834	3.602	1057	31.304	805	0
B-n39-k5	549	983	79.052	605	10.200	927	68.852	551	0.364
B-n41-k6	829	1370	65.259	919	10.856	1308	57.780	832	0.361
B-n43-k6	742	1208	62.803	793	6.873	1056	42.3180	744	0.269
B-n44-k7	909	1462	60.836	1003	10.341	1317	44.884	918	0.990
B-n45-k5	751	1364	81.624	843	12.250	1392	85.352	753	0.266
B-n45-k6	678	1010	48.967	751	10.766	1219	79.793	684	0.884
B-n50-k7	741	1637	120.917	838	13.090	1376	85.695	751	1.349
B-n50-k8	1313	1932	47.143	1414	7.692	1694	29.0175	1327	1.066
B-n51-k7	1032	1828	77.132	1230	19.186	2090	102.519	1020	−1.1627
B-n52-k7	747	1665	122.892	889	19.009	1286	72.155	752	0.669
B-n56-k7	707	1499	112.022	825	16.690	1357	91.937	713	0.848
B-n57-k7	1153	2243	94.536	1469	27.406	2373	105.811	1164	0.954
B-n57-k9	1598	2520	57.698	1802	12.765	2163	35.357	1622	1.502
B-n63-k10	1537	2611	69.876	1811	17.826	2485	61.679	1595	3.773
B-n64-k9	861	1875	117.770	1162	34.959	1977	129.617	882	2.439
B-n66-k9	1374	2558	86.171	1560	13.537	2439	77.511	1358	−1.164
B-n67-k10	1033	2211	114.036	1318	27.589	1910	84.898	1066	3.194
B-n68-k9	1304	2631	101.763	1512	15.950	2340	79.448	1325	1.610
B-n78-k10	1266	2914	130.173	1625	28.357	2621	107.030	1276	0.789
Average	971.478	1732.608	77.574	1115.173	14.008	1625.652	67.072	980.565	0.826

and 11 with B_Gap less than one. It indicates the superiority of the AHHO algorithm for the CVRP instances of Set B. For all instances of Set B, the average B_Gap of the results offered by the AHHO, RK-HHO, hybrid firefly, and cuckoo search via Lévy flights algorithms are 0.826, 77.574, 14.008, and 67.072, respectively. It establishes the performance order from best to worst for Set B: AHHO, hybrid firefly, cuckoo search via Lévy flights, and RK-HHO algorithms.

Table 12.5 shows the total traveled distances and B_Gaps offered by AHHO, RK-HHO, hybrid firefly, and cuckoo search via Lévy flights algorithms for the 24 CVRP instances of Set P. In Set P, the first CVRP instance has 15 customers, and the last CVRP contains 100 customers. As it can be observed from Table 12.5, the proposed AHHO algorithm offers minimum total traveled distances compared to RK-HHO, hybrid firefly, and cuckoo search via Lévy flights algorithms for all 24 CVRP instances. Thus, the proposed AHHO has a minimum B_Gap compared to

TABLE 12.5
Computational Results of RK-HHO, Hybrid Firefly, Cuckoo Search via Lévy Flights, and AHHO for Set P

		RK-HHO		Hybrid Firefly		Cuckoo Search via LF		AHHO	
Instance	**BKS**	**Distance**	**B_Gap**	**Distance**	**B_Gap**	**Distance**	**B_Gap**	**Distance**	**B_Gap**
P-n16-k8	450	456	1.333	450	0	451	0.222	451	0.222
P-n19-k2	212	229	8.018	212	0	234	10.377	212	0
P-n20-k2	220	247	12.272	216	−1.818	242	10	217	−1.363
P-n21-k2	211	262	24.170	211	0	229	8.531	212	0.473
P-n22-k2	216	322	49.074	216	0	235	8.796	217	0.462
P-n22-k8	603	655	8.623	603	0	632	4.809	588	−2.487
P-n23-k8	554	563	1.624	529	−4.512	542	−2.166	531	−4.152
P-n40-k5	458	783	70.960	508	10.917	691	50.873	467	1.965
P-n45-k5	510	969	90	595	16.666	855	67.647	514	0.784
P-n50-k7	554	928	67.509	689	24.3682	909	64.079	564	1.805
P-n50-k8	649	1124	73.189	828	27.580	1102	69.799	641	−1.232
P-n50-k10	696	1069	53.591	843	21.120	1098	57.758	723	3.879
P-n51-k10	745	1252	68.053	895	20.134	1304	75.033	757	1.610
P-n55-k7	524	1223	133.396	670	27.866	999	90.648	556	6.106
P-n55-k8	576	1169	102.951	741	28.645	1014	76.041	591	2.604
P-n55-k10	669	1311	95.964	853	27.503	1088	62.631	688	2.840
P-n55-k15	856	1420	65.887	967	12.967	1356	58.411	914	6.775
P-n60-k10	706	1345	90.509	940	33.144	1323	87.393	737	4.390
P-n60-k15	905	1543	70.497	1195	32.044	1473	62.762	961	6.187
P-n65-k10	792	1607	102.904	1020	28.787	1419	79.167	833	5.176
P-n70-k10	834	1802	116.067	1119	34.173	1673	100.599	876	5.036
P-n76-k4	589	1970	234.465	733	24.448	1537	160.951	627	6.451
P-n76-k5	631	1991	215.530	832	31.854	1601	153.724	661	4.754
P-n101-k4	681	2117	210.866	874	28.340	1933	183.847	719	5.580
Average	576.708	1098.208	81.977	697.458	17.676	997.500	64.247	594.042	2.411

the remaining three algorithms. Out of 24 CVRP instances, the proposed AHHO algorithms' B_Gap is zero for one CVRP instance, that is, P-n19-k2; less than zero for six CVRP instances, that is, P-n20-k2, P-n22-k8, P-n23-k8, P-n40-k5, P-n50-k7, and P-n50-k8; and less than one for four CVRP instances, that is, P-n16-k8, P-n21-k2, P-n22-k2, and P-n45-k5. On the other hand, the B_Gap offered by RK-HHO, hybrid firefly, and cuckoo search via Lévy flights algorithms is high compared to the AHHO algorithm. For Set P problem instances, the average B_Gap of the results offered by the AHHO, RK-HHO, hybrid firefly, and cuckoo search via Lévy flights algorithms are 2.411, 81.977, 17.676, and 64.247, respectively. For the Set P, the performance order from the best to worst remains the same, that is, AHHO, hybrid firefly, cuckoo search via Lévy flights, and RK-HHO algorithms.

TABLE 12.6
Computational Results of RK-HHO, Hybrid Firefly, Cuckoo Search via Lévy Flights, and AHHO for Set E

		RK-HHO		Hybrid Firefly		Cuckoo Search via Lévy Flights		AHHO	
Instance	**BKS**	**Distance**	**B_Gap**	**Distance**	**B_Gap**	**Distance**	**B_Gap**	**Distance**	**B_Gap**
E-n22-k4	375	488	30.133	375	0	375	0	375	0
E-n23-k3	569	719	26.362	569	0	580	1.933	568	−0.176
E-n30-k3	534	819	53.371	534	0	576	7.865	508	−4.869
E-n33-k4	835	1036	24.072	835	0	943	12.934	837	0.239
E-n51-k5	521	1091	109.405	626	20.154	967	85.604	542	4.031
E-n76-k7	683	1985	190.629	922	34.993	1211	77.306	730	6.881
E-n76-k8	735	2076	182.449	933	26.938	1374	86.938	786	6.938
E-n76-k10	832	1788	114.903	1186	42.548	1486	78.605	894	7.452
E-n76-k14	1032	1960	89.922	1406	36.240	1524	47.674	1087	5.329
E-n101-k8	817	2210	170.501	1619	98.164	2089	155.692	918	12.362
E-n101-k14	1077	2411	123.862	1853	72.052	2156	100.185	1159	7.614
Average	728.181	1507.545	101.419	987.091	30.099	1207.364	59.522	764	4.164

The total traveled distances and B_Gaps offered by AHHO, RK-HHO, hybrid firefly, and cuckoo search via Lévy flights algorithms for 11 CVRP instances of Set B are presented in Table 12.6. In Set E, the first CVRP instance has 21 customers, and the last CVRP contains 100 customers. As is seen from Table 12.6, the distances and B_Gap offered by the proposed AHHO algorithm are low in comparison to RK-HHO, hybrid firefly, and cuckoo search via Lévy flights algorithms for all 11 CVRP instances. The AHHO algorithms' B_Gap is less than zero for two CVRP instances, that is, E-n23-k3 and E-n30-k3, zero for E-n2-k4 CVRP instance, between 0 and 1 for E-n33-k4 CVRP instance. It again indicates the superiority of the AHHO algorithm over other algorithms for the CVRP instances of Set E. For all instances of the Set E, the average B_Gap of the results offered by the AHHO, RK-HHO, hybrid firefly, and cuckoo search via Lévy flights algorithms are 4.164, 101.419, 30.099, and 59.522, respectively. It establishes the order of performance from the best to worst for Set E is AHHO, hybrid firefly, cuckoo search via Lévy flights, and RK-HHO algorithms.

Table 12.7 presents the average B_Gap for each algorithm for Sets A, B, P, and E. The average values of B_Gap of the proposed AHHO are 1.255, 0.826, 2.411, and 4.164 for Sets A, B, P, and E, respectively. Similarly, the average values of B_Gap of the Hybrid Firefly are 15.318, 14.008, 17.676, and 30.099 for four sets. Table 12.7 also depicts the average B_Gap offered by the cuckoo search via Lévy flights and RK-HHO algorithms. Table 12.7 shows that the cumulative average B_Gap of the RK-HHO, hybrid firefly, cuckoo search via Lévy flights, and AHHO algorithms are 84.027, 19.275, 64.106, and 2.164, respectively. It established that the performance order of four algorithms for four sets A, B, P, and E is AHHO, hybrid firefly, cuckoo search via Lévy flights, and RK-HHO algorithms.

TABLE 12.7
Average Performance (B_Gap) of Four Algorithms for All Sets

Set	RK-HHO	Hybrid Firefly	Cuckoo Search via Lévy Flights	AHHO
A	75.139	15.318	65.582	1.255
B	77.574	14.008	67.072	0.826
P	81.977	17.676	64.247	2.411
E	101.419	30.099	59.522	4.164
Average	84.027	19.275	64.106	2.164

12.5.3 Observations

- As shown in Figures 12.5–12.8, the routes offered by AHHO have very few overlapping into different routes. The minimum overlap among different routes denotes that the AHHO algorithm divides all customers into a cluster of customers demanding minimal vehicles to visit the same. However, the more overlapping between routes representing the customers in one cluster requires many vehicles to serve them. The minimum overlapping occurs due to applying swap, exchange, relocate, 2-opt*, and inversion operators in the Lévy Flight.
- For the AHHO algorithm, the best convergence and ACCs over a 100 generations almost overlapped. It represents the stable performance of AHHO in all ten experiments for all CVRP instances. The best convergence and PA curves produced by AHHO have a considerable gap in the beginning and decrease as algorithms progress. It indicates that AHHO starts with random candidate solutions with high fitness values, and the fitness of all candidates improves gradually over different iterations and reaches the best solution in the last iterations. It indicates that the AHHO algorithm enhances the quality of fitness of all candidate solutions over various iterations, showing that the proposed AHHO algorithm is more effective to other algorithms. It is due to the exploration and exploitation strategy of the AHHO algorithm. The AHHO algorithm explores and exploits an equal number of solutions initially. As the algorithm progresses, it exploits more candidates than exploring new solutions.
- As can be seen from Table 12.7, the cumulative average B_Gap of the RK-HHO, hybrid firefly, cuckoo search via Lévy flights, and AHHO algorithms for 85 CVRP instances are 84.027, 19.275, 64.106, and 2.164, respectively. It established that the best-to-worst performance order is AHHO, hybrid firefly, cuckoo search via Lévy flights, and RK-HHO algorithms.
- The RK-HHO algorithm's poor performance stems from its reliance on the RK technique to convert the real-valued matrix into a transformation, which is inefficient. Due to cardinality issues, an unlimited number of continuous-valued vectors equate to a single discrete vector, so continuous-valued swarm intelligent algorithms may need to find large plateaus that result in local optimums.

- The proposed AHHO algorithm can be used to solve other permutation-based problems such as TSPs, scheduling problems, and LOPs [43–55].
- The algebraic variants of numerical algorithms like the slime mold algorithm, salp swarm algorithm, or whale optimization algorithm can also be proposed. Moreover, novel combinatorial meta-heuristics could also be developed for different permutation-based problems.

12.6 CONCLUSION

The CVRP is a well-known NP-hard combinatorial optimization problem that plays a significant role in end distribution in intelligent logistics. The CVRP solution can be represented as a permutation of given customers. The continuous-valued swarm and evolutionary algorithms require a decoding method to convert the continuous-valued vectors into permutations. However, there exist potentially infinite continuous-valued vectors corresponding to a single permutation due to cardinality reasons, thus inserting large plateaus in the fitness landscape navigated by the underlying algorithm. This work proposes AHHO algorithm to solve CVRP to minimize the total traveled distance. The proposed AHHO simulates the hunting behavior of Harris hawks to capture prey using a harmonized foraging with seven kill strategy in the permutation-based solution space. The AHHO simulates the original movements of hunting behavior of Harris hawks in permutation-based solution space using operations of abstract addition, abstract subtraction, and abstract scalar multiplication based on the composition operator. The AHHO algorithm also employs a discrete version of Lévy flights based on the swap, exchange, relocate, 2-opt*, and inversion local search operators. The performance of the proposed AHHO algorithm is evaluated on 85 well-known CVRP benchmark instances against RK-HHO, hybrid firefly, and cuckoo search with Lévy flights algorithms. The simulation results demonstrate that the proposed AHHO algorithm significantly outperforms the peer algorithms. The cumulative average performance of the AHHO, hybrid firefly, cuckoo search via Lévy flights, and RK-HHO algorithms for 85 CVRP instances are 2.164, 19.275, 64.106, and 84.027, respectively.

The proposed AHHO algorithm could be investigated for CVRPs' variants and different permutation-based problems. The AHHO algorithm can also be extended with initialization methods, local search operators, crossover operators, reinforcement learning algorithms, and clever techniques. The algebraic variants of numerical algorithms like the slime mold algorithm, salp swarm algorithm, or whale optimization algorithm can also be proposed. Moreover, novel combinatorial meta-heuristics could also be developed for different permutation-based problems.

REFERENCES

1. P. Toth, D. Vigo, *Vehicle Routing: Problems, Methods, and Applications*, Philadelphia, PA: Society for Industrial and Applied Mathematics, 2nd Ed., 2015.
2. G. Dantzig, J.H. Ramser. The truck dispatching problem, *Management Science*, vol. 6, pp. 80–91, 1959.
3. M. Sajid, A. Zafar, S. Sharma, Hybrid genetic and simulated annealing algorithm for capacitated vehicle routing problem, *2020 6th IEEE International Conference on Parallel, Distributed and Grid Computing (PDGC)*, JUIT Solan, 2020.

4. G.D. Konstantakopoulos, S.P. Gayialis, E.P. Kechagias, Vehicle routing problem and related algorithms for logistics distribution: A literature review and classification, *Operational Research*, vol. 22, pp. 2033–2062, 2020.
5. A. Mor, M.G. Speranza, Vehicle routing problems over time: A survey, *Annals of Operations Research*, vol. 314, no. 1, pp. 255–275, 2022.
6. M. Sajid, S. Pare, H. Mittal, M. Prasad, Routing and scheduling optimization for UAV assisted delivery system: A hybrid approach, *Applied Soft Computing*, vol. 12, p. 109225, 2022.
7. R. Elshaer, H. Awad, A taxonomic review of metaheuristic algorithms for solving the vehicle routing problem and its variants, *Computers & Industrial Engineering*, vol. 1, p. 106242, 2020.
8. J.-Y. Potvin, State-of-the-art review: Evolutionary algorithms for vehicle routing, *Informs Journal on Computing*, vol. 21, pp. 518–548, 2009.
9. J.F. Cordeau, M. Gendreau, G. Laporte, J.Y. Potvin, F. Semet, A guide to vehicle routing heuristics, *Journal of the Operational Research Society*, vol. 53, pp. 512–522, 2002.
10. V. Santucci, M. Baioletti, A. Milani, Algebraic differential evolution algorithm for the permutation flowshop scheduling problem with total flowtime criterion, *IEEE Transactions on Evolutionary Computation,* vol. 20 (5), pp. 682–694, 2015.
11. V. Santucci, M. Baioletti, A. Milani, An algebraic framework for swarm and evolutionary algorithms in combinatorial optimization, *Swarm and Evolutionary Computation*, vol. 55, p. 100673, 2020.
12. V. Santucci, M. Baioletti, G.D. Bari, An improved memetic algebraic differential evolution for solving the multidimensional two-way number partitioning problem, *Expert Systems with Applications*, vol. 178, p. 114938, 2021.
13. M. Baioletti, A. Milani, V. Santucci, Variable neighborhood algebraic differential evolution: An application to the linear ordering problem with cumulative costs, *Information Sciences*, vol. 507, pp. 37–52, 2020.
14. M. Baioletti, A. Milani, V. Santucci, Algebraic particle swarm optimization for the permutations search space, *2017 IEEE Congress on Evolutionary Computation (CEC),* Donostia, Spain, 2017.
15. A.A. Heidari, S. Mirjalili, H. Faris, I. Aljarah, M. Mafarja, H. Chen, Harris hawks optimization: Algorithm and applications, *Future Generation Computer Systems*, vol. 97, pp. 849–872, 2019.
16. A. Ouaarab, *Discrete Cuckoo Search for Combinatorial Optimization*, Springer Singapore Pte. Limited: Singapore, 2020.
17. İ. İlhan, An improved simulated annealing algorithm with crossover operator for capacitated vehicle routing problem, *Swarm and Evolutionary Computation*, vol. 64, p. 100911, 2021.
18. B. Rabbouch, F. Saâdaoui, R. Mraihi, Empirical-type simulated annealing for solving the capacitated vehicle routing problem, *Journal of Experimental & Theoretical Artificial Intelligence*, vol. 32, pp. 437–452, 2020.
19. A.A.R. Hossainabadi, A. Slowik, M.S. Lalimi, M. Farokhzad, M.B. Shareh, A.K. Sangaiah, An ameliorative hybrid algorithm for solving the capacitated vehicle routing problem, *IEEE Access*, vol. 7, pp. 175454–175465, 2020.
20. İ. İlhan, A population based simulated annealing algorithm for capacitated vehicle routing problem, *Turkish Journal of Electrical Engineering & Computer Sciences,* vol. 28, pp. 1217–1235, 2020.
21. N. Lin, Y. Shi, T. Zhang, X. Wang, An effective order-aware hybrid genetic algorithm for capacitated vehicle routing problems in internet of things, *IEEE Access,* vol. 7, pp. 86102–86114, 2019.
22. A.M. Altabeeb, A.M. Mohsen, A. Ghallab, An improved hybrid firefly algorithm for capacitated vehicle routing problem, *Applied Soft Computing*, vol. 84, p. 105728, 2019.

23. T. Azad, M.A.A. Hasin, Capacitated vehicle routing problem using genetic algorithm: A case of cement distribution, *International Journal of Logistics Systems and Management*, vol. 32, no. 1, pp. 132–146, 2019.
24. M.A. Mohammed, M.K.A. Ghani, R.I. Hamed, S.A. Mostafa, M.S. Ahmad, D.A. Ibrahim, Solving vehicle routing problem by using improved genetic algorithm for optimal solution, *Journal of Computational Science*, vol. 21, pp. 255–262, 2017.
25. M. Sajid, J. Singh, R.A. Haidri, M. Prasad, V. Varadarajan, K. Kotecha, D. Garg, A novel algorithm for capacitated vehicle routing problem for smart cities, *Symmetry*, vol. 13, no. 10, 1923, 2021.
26. A.M. Altabeeb, A.M. Mohsen, L. Abualigah, A. Ghallab, Solving capacitated vehicle routing problem using cooperative firefly algorithm, *Applied Soft Computing*, vol. 108, p. 107403, 2021.
27. L.M. Dalbah, M.A. Al-Betar, M.A. Awadallah, R.A. Zitar, A modified coronavirus herd immunity optimizer for capacitated vehicle routing problem, *Journal of King Saud University-Computer and Information Sciences*, vol. 34, no. 8, Part A, pp. 4782–4795, 2022.
28. E. Queiroga, R. Sadykov, E. Uchoa, A POPMUSIC metaheuristic for the capacitated vehicle routing problem, *Computers & Operations Research*, vol. 136, p. 105475, 2021.
29. L. Chen, Y. Chen, A. Langevin, An inverse optimization approach for a capacitated vehicle routing problem, *European Journal of Operational Research*, vol. 295, no. 3, pp. 1087–1098, 2021.
30. F.Y. Vincent, A.A.N. Perwira Redi, Y. Chao-Lung, R. Eki, S. Budi, Symbiotic organisms search and two solution representations for solving the capacitated vehicle routing problem, *Applied Soft Computing*, vol. 52, pp. 657–672, 2016.
31. S. Akpinar, Hybrid large neighborhood search algorithm for capacitated vehicle routing problem, *Expert Systems and Applications*, vol. 61, pp. 28–38, 2016.
32. A.A.R. Hosseinabadi, N.S.H. Rostami, M. Kardgar, S. Mirkamali, A. Abraham, A new efficient approach for solving the capacitated vehicle routing problem using the gravitational emulation local search algorithm, *Applied Mathematical Modeling*, vol. 49, pp. 663–679, 2017.
33. P. Kalatzantonakis, A. Sifaleras, N. Samaras, Cooperative versus non-cooperative parallel variable neighborhood search strategies: A case study on the capacitated vehicle routing problem, *Journal of Global Optimization*, vol. 78(2), pp 327–348, 2020.
34. F. Arnold, K. Sörensen, Knowledge-guided local search for the vehicle routing problem, *Computers & Operations Research*, vol. 105, pp. 32–46, 2019.
35. R. Baldacci, A. Mingozzi, R. Roberti, Recent exact algorithms for solving the vehicle routing problem under capacity and time window constraints, *European Journal of Operations Research,* vol. 218, no. 1, pp. 1–6, 2012.
36. R. Sadykov, E. Uchoa, A. Pessoa, A bucket graph–based labeling algorithm with application to vehicle routing, *Transportation Science*, vol. 55, no. 1, pp. 4–28, 2020.
37. J.H. Santillan, S. Tapucar, C., Manliguez, V. Calag, Cuckoo search via Lévy flights for the capacitated vehicle routing problem, *Journal of Industrial Engineering International*, vol. 14, pp. 293–304, 2018.
38. P.H.V. Penna, A. Subramanian, L.S. Ochi, T. Vidal, C. Prins, A hybrid heuristic for a broad class of vehicle routing problems with heterogeneous fleet, *Annals of Operation Research*, vol. 273, no. 1, pp. 5–74, 2019.
39. A. Pessoa, R. Sadykov, E. Uchoa, F. Vanderbeck, A generic exact solver for vehicle routing and related problems, *Mathematical Programing*, vol. 183, no. 1, pp. 483–523, 2020.
40. P. Augerat, J.M. Belenguer, E. Benavent, A. Corberán, D. Naddef, G. Rinaldi, Computational results with a branch-and-cut code for the capacitated vehicle routing problem, *Institute for System Analysis and Computer Science, CNR*, Rome, Italy, Technical report 495, 1995, pp. 1–22.

41. N. Christofides and S. Eilon, An algorithm for the vehicle-dispatching problem, *Journal of the Operational Research Society*, vol. 20, no. 3, pp. 309–318, 1969.
42. NEO Research Group. Available online: https://neo.lcc.uma.es/vrp/ (accessed on 15.06.2022).
43. E. Queiroga, Y. Frota, R. Sadykov, A. Subramanian, E. Uchoa, T. Vidal, On the exact solution of vehicle routing problems with backhauls, *European Journal of Operation Research*, vol. 287, no. 1, pp. 76–89, 2020.
44. M. Sajid, Z. Raza, Energy-efficient quantum-inspired stochastic Q-HypE algorithm for batch-of-stochastic-tasks on heterogeneous DVFS-enabled processors, *Concurrency Computation: Practice and Experience*, vol. 31, p. e5327, 2019.
45. M. Sajid, Z. Raza, Energy-aware stochastic scheduler for batch of precedence-constrained jobs on heterogeneous computing system, *Energy*, vol. 125, pp. 258–274, 2017.
46. M. Alam, M. Shahid, S. Mustajab, Security prioritized multiple workflow allocation model under precedence constraints in cloud computing environment. *Cluster Computing*, pp. 1–36, 2023.
47. S. Shiekh, M. Shahid, M. Sambare, R.A. Haidri, D.K. Yadav, A load-balanced hybrid heuristic for allocation of batch of tasks in cloud computing environment. *International Journal of Pervasive Computing and Communications,* 2022. https://doi.org/10.1108/IJPCC-06-2022-0220.
48. M. Alam, M. Shahid, S. Mustajab, Security prioritized heterogeneous earliest finish time workflow allocation algorithm for cloud computing. *In Congress on Intelligent Systems: Proceedings of CIS 2021*, vol. 1, pp. 233–246. Singapore: Springer Nature Singapore, 2022, July.
49. M. Alam, M. Shahid, S. Mustajab, Security oriented deadline aware workflow allocation strategy for infrastructure as a service clouds. *In 2022 3rd International Conference on Computation, Automation and Knowledge Management (ICCAKM)*, pp. 1–6. IEEE, Dubai, United Arab Emirates, 2022, November.
50. S. Sheikh, A. Nagaraju, M. Shahid, A fault-tolerant hybrid resource allocation model for dynamic computational grid. *Journal of Computational Science*, vol. 48, p. 101268, 2021.
51. M. Alam, M. Shahid, S. Mustajab, SAHEFT: Security aware heterogeneous earliest finish time workflow allocation strategy for IaaS cloud environment. *In 2021 IEEE Madras Section Conference (MASCON)*, pp. 1–8. IEEE, Chennai, India, 2021, August.
52. M. Shahid, Z. Ashraf, M. Alam, F. Ahmad, M. Imran, A multi-objective workflow allocation strategy in IaaS cloud environment. *In 2021 International Conference on Computing, Communication, and Intelligent Systems (ICCCIS)*, pp. 308–313. IEEE, Greater Noida, India, 2021, February.
53. S.M. Raza, M. Sajid, J. Singh, "Vehicle Routing Problem using Reinforcement Learning: Recent Advancements", *3rd International Conference on Machine Intelligence and Signal Processing*, NIT, Arunachal Pradesh, Proceedings D. Gupta, K Sambyo, M. Prasad, S. Agarwal, *Advanced Machine Intelligence and Signal Processing*, LNEE, Vol. 858, pp. 269–280, 2022.
54. M.S. Jawed, M. Sajid, XECryptoGA: A Metaheuristic algorithm-based Block Cipher to Enhance the Security Goals, Evolving Systems, 2022 [https://doi.org/10.1007/s12530-022-09462-0]
55. M. Sajid, Z. Raza, M. Shahid, "Hybrid Bio-inspired Scheduling Algorithms for Batch of Tasks (BoT) Applications on Heterogeneous Computing System", *International Journal of Bio-inspired Computation (InderScience)*, vol. 11(3), pp. 135–148, 2018. [https://doi.org/10.1504/IJBIC.2018.091698]

13 Technology for Detecting Harmful Effects on the UAV Navigation and Communication System

Elena Basan, Nikita Sushkin, Maria Lapina, and Mohammad Sajid

CONTENTS

13.1 INTRODUCTION

Today, the problems associated with the stability of the navigation signal can be associated with various factors. Firstly, there are areas where the GPS does not receive well, for example, in mountain areas. In addition, there is a situation when the area where the flight is carried out is very noisy. In addition to the natural causes of a decrease in signal quality, there are also various kinds of attacks on the GPS system. In particular, the most common ones are satellite signal spoofing and signal jamming.

DOI: 10.1201/9781003438588-13

Of course, it is quite difficult to counteract natural causes, but it is also not an easy task to counteract attacks. To begin with, it is necessary to detect the implemented attack. To do this without external analysis but autonomously and respond to it, the drone must analyze what is available to it, for example, telemetry data.

The purpose of the study is to develop a technology for detecting harmful effects on UAVs to increase the level of security and stability of the device. In addition, if you analyze a drone, you can take a response plan. After all, if you analyze the operator's module, then if you lose communication and control over the drone, you may not have time to give it the necessary instructions.

The object of research is the UAV navigation and communication system.

The subject of the study is mathematical models, methods, and algorithms for detecting harmful effects on the navigation and communication system.

13.2 UAV THREAT AND VULNERABILITY ANALYSIS

To improve the security of any information system, it is important to detect vulnerabilities before intruders do it. This is achieved through continuous analysis and risk assessment. UAV vulnerabilities can be divided into four groups:

1. Vulnerabilities of the base system (the base system is the basis of the UAV, combining its components; it is necessary to provide intercomponent communications and control the system of sensors, navigation, avionics, and communications; a certain "operating system" of the UAV);
2. Vulnerabilities of communication channels [communication is always wireless; can be divided into direct communication within the line of sight and indirect (mainly satellite communication)];
3. Vulnerabilities of sensors (cameras with different capabilities, INS (inertial navigation system), GPS, and radars);
4. Avionic vulnerabilities (avionics is responsible for the transformation of the received handling commands into commands for the driver, control surface, reactor, and spoilers) [1].

Possible consequences of an attack on UAV systems are:

- denial of service,
- loss of control over UAV,
- destruction of the system,
- disruption of the system,
- short-term shutdown of the system,
- unauthorized control of UAV,
- UAV destruction,
- wrong choice of troubleshooting mechanism,
- landing,
- self-destruction,

- automatic return,
- free flight,
- UAV landing for further research and data collection,
- UAV theft,
- disorientation,
- false estimate of the current location,
- network destabilization,
- interception of transmitted data,
- violation of the confidentiality of information sent over the radio channel (or stored in the UAV),
- violation of the availability of packets sent over the radio channel (or stored in the UAV)
- violation of the integrity of packets sent over the radio channel (or stored in the UAV)
- loss of confidential data,
- violation of the availability of network nodes,
- violation of the mission process,
- increase in mission time,
- failure of the mission,
- mission abort.

To determine the consequences, it is important to know what mission the UAV had, how the exploitation of the vulnerability can affect not only the UAV (and its components separately) but also the environment, etc., depending on the degree of the UAV's response to the situation, then, or other events [2].

13.2.1 Development of an Attack Vector for UAVs

There are extremely bad consequences for the drone and possibly the environment and humans if the satellite spoofing attack is successful [3]. Drone bugs often occur without an attack, such as firmware bugs or poorly calibrated sensors, including global navigation signal inaccuracy. Since GPS is often not accurate enough, ground stations are often used to improve accuracy, which provide additional coordination [4,5]. There are also military developments of navigation and communication systems with built-in encryption mechanisms. Then you can't fake satellites, but you can still drown them out [6]. Consequently, nondefense GPSs are very vulnerable to spoofing attacks [5]. A GPS spoofing attack brings backwash concerned with the drop (crash) of the drone or its capture, readdressing along another heading. These defects are connected with cybersecurity [7]. If analyzed from the point of view of attacks, then of course the substitution of the location by a global signal is the most dangerous problem [8]. The civilian GPS datasheet is free [9], which makes the packages more foreseeable, increasing the eventual danger of spoofing. According to what has been said, it is necessary to investigate the true causes of the substitution, to conduct an analysis, since the ease of implementation is incompatible with the likely damage. Let's assume that there are two situations for an attack on the drone's navigation system:

- The first situation is that the UAV, having a certain position, gets the steady-state axis of the object, while the attack consists of the slow shift of the drone to the opposite place.
- The second scenario is the movement along a given route and the forgery of waypoints during the flight.

13.3 ANALYSIS OF AN ANOMALY DETECTION METHOD

13.3.1 Analysis of Analogs of the Developed Technology for Detecting Harmful Effects on the UAV Navigation and Communication System

The development of attack detection systems for UAVs is carried out in different countries of the world, but the main global scientific competitors are the United States and China. This is confirmed by the fact that most of the authors who publish scientific articles work at universities in these countries.

When anomalies are detected in groups of UAVs, for example, the focus is on methods based on the evaluation of the attributes of the operation, such as sensory unit information, the state of the gear system, and the dependencies among them. At present, above all abnormality acquisition means are put on a predictive pattern, such as autoregressive model [9], state space linear dynamic model [10,11], and neural network-based regression models [12], where sensor observations are decomposed and forecasted. Predictions are generated based on values obtained over some time the system has been running and then equalized with actual determination to calculate the excess error. Deviance is inspected if the excess difference surpasses a predetermined threshold. Nevertheless, because of the presence of various noise sources in the UAV group, an accurate limit between average and anomalous sensory unit readings is usually not detected. New research has displayed that the assailant can use this uncertain limit to introduce malign sensing (named stealth attacks) that accomplish the attacker's purposes and rounded disclosure methods that are based on current solutions [13–15]. The offered other handling to the use of excess is a constant technique proved by the regulations [16,17]. Rule-based invariant techniques explore the material criteria that are prioritized for all conditions. Any observed values of material operations that disturb these procedures are divided as abnormality [18]. As a rule, these unvarying procedures are determined by system specialists, masters at the development process. This hand-feed operation is not only expensive but also imprecision outspread as the real and architect systems may not match exactly [19]. In addition, there are set masked unvarying guidelines that are greatly complicated for humans to discover, mainly those that are in multiple segments. As a result, the productivity of available abnormality detection technics based on unvarying regulations is frequently confined by either fault or incompetence of architectural procedures.

In the field of anomaly detection, several modern technics have been offered, which can be the most disintegrated into the following three classes: knowledge-based technics, model-based technics, and data-based technics. The knowledge-based technic makes it possible to receive deviance stencils by summarizing the skills of specialists in a particular field, implementing an anomaly detection system, and detecting similar deviance stencils [20]. A model-based technic is commonly built by a viewer by creating an exact physical model of the device and then comparing the observer's or filter's estimated value with the real admeasured exponent. Based on the generated residual error, anomaly detection is performed [21]. The data-driven technics automatically recognize the behavioral layout of the device based on the collected data of the system operations [22]. To implement the first two methods, it is necessary to have knowledge of the subject area or the UAV system to get a good anomaly detection result. However, it is generally difficult to create an accurate physical model for each UAV subsystem. The adaptability and anti-jamming capabilities of these two methods are limited. A comparison of methods is presented in Table 13.1.

Thus, the following can be summarized. The technics are based on the examination of the modification in the state compared to the suggestion or on the interdependence of statistics received from a few GPS sensory units. Some technics use sensor mechanisms other than GPS, such as magnetometer, accelerometer, and others, to

TABLE 13.1
Method Comparison

Method Implementation Type	Method Type	Advantages	Limitations
Software	Methods based on comparison with reference values	Performance, simple attack detection	Inability to detect advanced attacks
	Methods based on the intellectual analysis	Improving the quality of attack detection, universality for different UAV configurations	Complexity of implementation, energy consumption
Hardware and software	Methods based on the use of several GPS receivers	Harder to attack, higher detection quality	Complexity of implementation may not be supported by all UAVs. May use additional resources. Significant decision-making is required. Difficulties in implementation are possible
	Methods based on the use of other types of sensors	Increases the level and speed of attack detection	Poorly scaled since the configuration of each UAV is individual

enhance flight ability. In this case, interdependence can take place both among the units of one drone and among the units of the drone group. In addition, many authors write that their technics are outspread to false positives.

13.3.2 Implementation of Technology for Detecting Harmful Effects on the UAV Navigation and Intercommunication System

The activity of the appliance is arranged as follows. The system works on the principle of publisher–subscriber. The publisher is the sender of the telemetry data, such as a flight controller or navigation system. The subscriber is the malware detection system. The subscription module receives events, processes them, and returns the result in the form of a status code for further system response. Each event contains several parameters and their values at a certain point in time. Depending on the returned code, the system can continue to perform the flight mission or make changes if an attack is detected.

The malware detection system consists of several modules, such as a normalization module, an entropy calculation module, and an attack detection module. Each of these modules performs its own task and can be further expanded to solve more complex tasks.

The data is sent to the normalization module to bring any categories of data to a single format. After the data is normalized and the required type of probability distribution is built, they are sent to the entropy calculation module. The result of entropy calculation serves to detect an attack; this happens in the attack detection module. To detect an attack, the values of several parameters are required, that is, each of the parameters can separately signal an anomaly, but only together they can signal that an attack is unequivocally carried out to replace or jam the GPS signal. Therefore, for each parameter, its own weighting coefficient is added, depending on the importance of the parameter when an attack is detected. The application architecture is shown in Figure 13.1.

For example, the most significant parameter in a GPS signal spoofing attack is the number of fixed GPS satellites, noise level, flight altitude, and UAV coordinates. For example, in the configuration used for the experiment, a sharp change in the number of satellites used gives two conditional scores for attack detection at once, while the

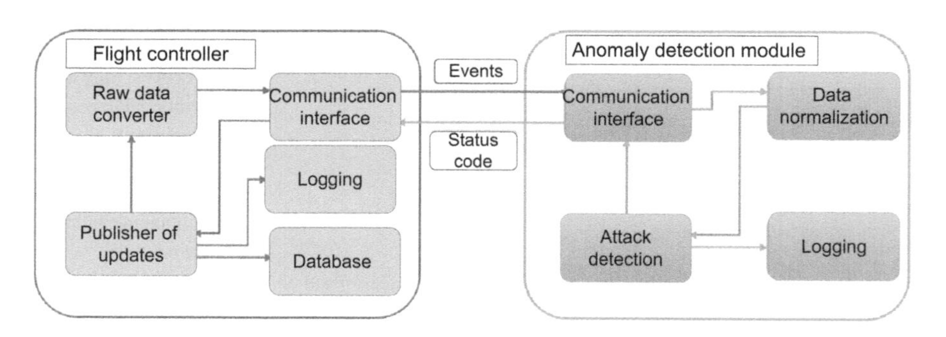

FIGURE 13.1 Technology of attack detection architecture.

remaining parameters give only one conditional score. The value of the total_score parameter will allow you to determine the presence of an attack and make a final decision about the notification. If the sum of conditional scores for all parameters is exceeded, the module gives a signal that an attack has been detected.

To present the obtained statistics in the form of a probability distribution, it is needful to construct data tables in an appropriate way. To do this, the data is collected in time series. The time series is arranged in such a way that it contains old - m and new data - i. The amount of values in the size set is set using the configuration file, namely the slice_size parameter.

The normalization module processes the raw data. Parameters such as CPU load and UAV flight altitude (ha), satellite fixation state (Gn), GPS uncertainty (Gu), and GPS noise are discrete (Gnoi), so it is impossible to use the normal distribution law for them. Let's look at the most general disposal rules and evaluate whether they are appropriate for representing cyber-physical characteristics such as CPU utilization and CPU temperature. These distributions are built for random variables obtained over a certain time interval. To construct the distribution, 10 random variables are used, and disposal is composed for them. The idea of a sliding window is then applied to actualize information about the propagation, six (/nine) va of the previous and four (/one) recent values are taken for the new propagation, and a new propagation is formed. This idea will allow you to fix a sharp increase in values. Then, to define the occurrence of an abnormality or a deviation in conditions, it is needful to fix what propagations vary from each other. It is the juxtaposition of propagations, and not unprocessed statistics, that will give a more correct effect. The unprocessed statistics can be very dissimilar, and the modifications seen in the unprocessed statistics are not indicative of an abnormality or violence. Normalizing unprocessed statistics and realizing it to a corresponding propagation type will allow you to fix a deviation in the system, and using the entropy value, that is, the difference between distributions, you can identify an anomaly.

The type of attack affects how the parameter values will change. Moreover, some attacks can affect the same fields. This means that the controlled parameter can be a marker for various kinds of attacks.

In this research, we consider two types of influence of violations on controlled parameters: this is data substitution and jamming of the data source. In the case of data substitution, most often, there is a sharp change in the values of the controlled parameter because it is far from always possible to accurately determine the current range of values of the attacked parameter, and in some cases, this is impossible. If the data source is muted, the sensor reading will be zero or undefined, depending on the source and data processing.

In some cases, different attacks change parameter values in a similar way. For example, when performing a GPS spoofing attack, the value of "number of used satellites" will be changed to the number of satellites falsely broadcast by a more powerful signal source. In the event of a GPS jamming attack, the GPS signal will be jammed, and the number of satellites will be zero. Thus, depending on how the value of the variable has changed, it can be a marker for different types of attacks. Thus, each parameter, depending on the type of malicious impact, has two sets of attacks. Each time the parameter exceeds the threshold value, attacks from these sets will be remembered. At the end of the processing cycle of all parameters of the incoming event, the attack that used the largest number of parameters, i.e., the one that occurred most often, will be determined.

In order to detect a malicious impact, it is necessary to correctly form the configuration file. The file contains the following parameters:

- Sliding window size (slize_size). The number of events that will be used to compare data and calculate entropy.
- Shear step. Number of events to be used to shift the sliding window.
- Threshold value. The sum of conditional points, the excess of which will mean the detection of an attack or the presence of anomalous behavior.
- Description of all fields that are used for anomaly detection. Each field contains:
- Names of this field.
- Entropy threshold.
- The number of points that will be summed up if the threshold is exceeded.
- Possible attacks when exposed to data substitution (DEFAULT ATTACK).
- Possible attacks on impact with data source muting (ZERO ATTACK).

Thus, the module turned out to be quite simple to use, it does not consume additional processor power. The module only requires minimal configuration when configuring thresholds. Threshold values can be determined experimentally depending on the severity of the system. The lower the threshold, the more sensitive the violation detection technics will be. That is, in fact, the threshold determines the degree to which changes are considered an attack and which are not. According to the results of the experiments, we can say that to unambiguously speak about the presence of an attack, at least two parameters must have an increased entropy value.

13.4 RESULTS AND DISCUSSION

Table 13.2 provides an example of data presentation for three parameters: noise level, signal level, and the number of satellites. Each event has its own time stamp in which the sensor readings were recorded. The module processes the latest slice_size events and calculates conditional scores for each processed parameter, if the sum of scores for all parameters exceeds the threshold value (set by the attack_score value in the configuration file). For this example, slice_size is 10 and the value is 2. The result of the points awarded is shown in Figure 13.2.

Thus, the module turned out to be quite simple to use, it does not consume additional processor power. The module only requires minimal configuration when configuring thresholds. Threshold values can be determined experimentally depending on the severity of the system. The lower the threshold, the more sensitive the violation detection technics will be. That is, in fact, the threshold determines the degree to which changes are considered an attack and which are not.

According to the results of the experiments, we can say that to unambiguously speak about the presence of an attack, at least two parameters must have an increased entropy value.

TABLE 13.2
Result of the Analyzer

Noise Level	Signal Level	Number of Satellites	Attack Detected
32	0.09236317	14	No
28	0.09236317	14	No
32	2.100334	10	No
39	2.7647352	5	Yes
30	2.7647352	5	Yes
42	2.7647352	4	Yes
32	3.0394106	11	Yes
23	0.19498521	11	Yes
33	0.19498521	12	No

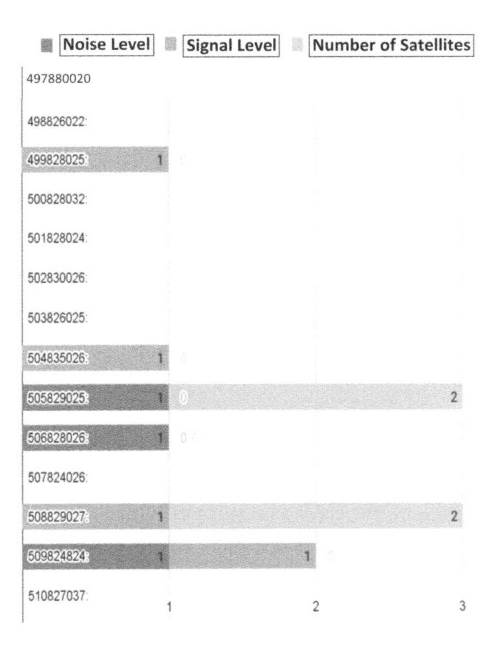

FIGURE 13.2 The result of the analyzer.

The precision of violation detection is determined by the probability of errors of the first and second kind, which should tend to be zero probability. The detection accuracy should tend to be 100%; only time delays in attack detection are possible.

To determine the accuracy of anomaly detection in the UAV navigation system, two scenarios of an experimental study are proposed.

Scenario 1: The module input receives messages in the established format, which contains telemetry data containing the above sets of UAV parameters during a normal flight. The execution of scenario 1 assumes that the attack detector receives true data from the flight controller and should eventually return a message that the state is normal [23].

TABLE 13.3
Result for Scenario 1

Files without Attack	Attack Detected	Attack Not Detected	Detection Accuracy	Type 2 Error Probability
106	1	105	99.06%	0.94%

TABLE 13.4
Result for Scenario 2

Attack Files	Attack Detected	Attack Not Detected	Detection Accuracy	Type 1 Error Probability
42	40	2	95.24%	4.76%

Scenario 2: The module receives messages in the specified format that contains telemetry data containing the above sets of UAV parameters during a flight subject to a GPS spoofing attack. Scenario 2 assumes that the attack detector is receiving data from the flight controller and should eventually report that an anomaly has been detected [24].

The result of module operation for scenarios 1 and 2 is shown in Tables 13.3 and 13.4.

13.5 CONCLUSION

The technic allows analysis of any characteristics and can work with any available data, no matter what sensors the UAV is equipped with.

With the help of the developed technic, it is potential not only to disclose an abnormality but also to define the modifications in the template of drone behavior and the modifications in its condition. If the values of the determined entropy are not too high and there is a momentary enhancement, then this may indicate a change in the flight regime. The ratio of the explored characteristics makes it possible to unambiguously identify violence and define its kind. Every violence influences a certain set of suite, so the type of attack can be described by the resulting characteristics that it affects. The statistics gathered in the form of time part can be used to train neural networks to make decisions about how to carry out an attack.

ACKNOWLEDGMENTS

This research was funded by the Russian Science Foundation grant number 21-79-00194, https://rscf.ru/project/21-79-00194/, in Southern Federal University.

REFERENCES

1. S. Low, O. Nina, A. D. Sappa, E. Blasch and N. Inkawhich, Multi-modal aerial view object classification challenge results: PBVS 2022. *2022 IEEE/CVF Conference on Computer Vision and Pattern Recognition Workshops (CVPRW)*, New Orleans, LA, pp. 349–357, 2022, doi: 10.1109/CVPRW56347.2022.00050.
2. O. Šimon and T. Götthans, A survey on the use of deep learning techniques for UAV jamming and deception. *Electronics*, vol. 11, p. 3025, 2022. doi: 10.3390/electronics11193025.
3. P. Bethi, S. Pathipati, Aparna P, Stealthy GPS Spoofing: Spoofer Systems, Spoofing Techniques and Strategies, *2020 IEEE 17th India Council International Conference (INDICON)*, 2021, New Delhi, India.
4. E. T. Lester, Military position source challenges for worldwide ads-b out compliance. In *Integrated Communications Navigation and Surveillance Conference*, Herndon, VA, USA, 2013.
5. C. A. Ericson, Software safety in a nutshell. https://www.dcs.gla.ac.uk/~johnson/teaching/safety/reports/Clif_Ericson1.htm.
6. S.-H. Seo, B.-H. Lee, S.-H. Im, and G.-I. Jee, Effect of spoofing on unmanned aerial vehicle using counterfeited GPS signal. *Journal of Positioning Navigation and Timing*, vol. 4, no. 2, pp. 57–65, 2015.
7. Global Positioning System Directorate, Systems engineering and integration interface specification IS-GPS-200G Technical Report, 2012.
8. J. S. Warner and R. G. Johnston, GPS spoofing countermeasures. *Journal of Homeland Security*, vol. 25, no. 2, pp. 19–27, 2003.
9. T. Humphreys, Statement on the vulnerability of civil unmanned aerial vehicles and other systems to civil GPS spoofing, Austin, 2012.
10. D. Hadžiosmanovíc, R. Sommer, E. Zambon, and P. H. Hartel, Through the eye of the PLC: Semantic security monitoring for industrial process. In *Proceedings of the 30th Annual Computer Security Applications Conference*, New Orleans, Louisiana, USA, pp. 126–135, 2014.
11. D. I. Urbina et al., Limiting the impact of stealthy attacks on industrial control systems. In *Proceedings of the 2016 ACM SIGSAC Conference on Computer and Communications Security*, Vienna, AT, pp. 1092–1105, 2016.
12. J. Goh, S. Adepu, M. Tan, and Z. Lee, Anomaly detection in cyber physical systems using recurrent neural networks. In *IEEE18th International Symposium on High Assurance Systems Engineering (HASE 2017)*, Singapore, pp. 140–145, 2017.
13. G. Dan and H. Sandberg, Stealth attacks and protection schemes for state estimators in power systems. In *First IEEE International Conference on Smart Grid Communications (SmartGridComm 2010)*, Gaithersburg, MD, pp. 214–219, 2010.
14. H. Yang, X. He, Z. Wang, R. C. Qiu, and Q. Ai, Blind false data injection attacks against state estimation based on matrix reconstruction. *IEEE Transactions on Smart Grid*, vol. 13, no. 4, pp. 3174–3187, 2022. doi: 10.1109/TSG.2022.3164874.
15. C. Feng, T. Li, Z. Zhu, and D. Chana, A deep learning-based framework for conducting stealthy attacks in industrial control systems. arXiv preprintarXiv:1709.06397, 2017.
16. S. Adepu and A. Mathur, Using process invariants to detect cyberattacks on a water treatment system. In *IFIP 31st International Conference on ICT Systems Security and Privacy Protection*, Ghent, Belgium, pp. 91–104, 2016.
17. S. Adepu and A. Mathur, From design to invariants: Detecting attacks on cyber physical systems. In *IEEE International Conference on Software Quality, Reliability and Security Companion (QRS-C 2017)*, Prague, Czech Republic, pp. 533–540, 2017.
18. D. Al Mohamad and A. Boumahdaf, Semiparametric two-component mixture models when one component is defined through linear constraints. *IEEE Transactions on Information Theory,* vol. 64, no. 2, pp. 795–830, 2018, doi: 10.1109/TIT.2017.2786345.

19. F. Hernández-del-Olmo, E. Gaudioso, N. Duro, and R. Dormido, Machine learning weather soft-sensor for advanced control of wastewater treatment plants. *Sensors*, vol. 19, p. 3139, 2019, doi: 10.3390/s19143139.
20. Y. M. Zhang, Fault detection and diagnosis for NASA GTMUAV with dual unscented Kalman filter. In K. P. Valavanis and G. J. Vachtsevanos (Eds.), *Handbook of Unmanned Aerial Vehicles*, pp.1157–1181. Springer: Netherlands, 2015.
21. Z. Birnbaum, A. Dolgikh, V. Skormin, E. O'Brien, D. Muller, and C. Stracquodaine, Unmanned aerial vehicle security using recursive parameter estimation. *Journal of Intelligent & Robotic Systems*, vol. 84, pp.107–120, 2016.
22. J. Pang, D. Liu, H. Liao, Y. Peng, and X. Peng, Anomaly detection based on data stream monitoring and prediction with improved Gaussian process regression algorithm. *2014 International Conference on Prognostics and Health Management*, Cheney, WA, USA, pp. 1–7, 2014.
23. E. Basan, A. Basan, A. Nekrasov, C. Fidge, N. Sushkin, and O. Peskova, GPS-spoofing attack detection technology for UAVs based on Kullback–Leibler divergence. *Drones*, vol. 6, p. 8, 2022. doi: 10.3390/drones6010008.
24. E. Basan, A. Basan, A. Nekrasov, C. Fidge, E. Abramov, and A. Basyuk, A data normalization technique for detecting cyber attacks on UAVs. *Drones*, vol. 6, p. 245, 2022. doi: 10.3390/drones6090245.

14 Current and Future Trends of Intelligent Transport System Using AI in Rural Areas

B. Iswarya and B. Radha

CONTENTS

14.1 INTRODUCTION

A subclass of mobile ad hoc network (MANET) is vehicular ad hoc network (VANET) which communicates with nearby vehicles (nodes are considered as cars), vehicle to vehicle, and also with fixed infrastructure [RSU (road side

DOI: 10.1201/9781003438588-14

unit)]. It provides safety to roadside travelers and vehicles to reduce road accidents and enhanced road traffic by passing timely services about collision warning, lane change warning, road sign alarm, road conditions, fuel services, road accidents, and locations. VANET is integrated with intelligent transport system (ITS) to enhance road safety and network efficacy (Rajadurai and Jayalakshmi 2013).

VANET will communicate through two channels to exchange information; they are (Iswarya and Radha 2021b)

1. Vehicle–Vehicle communication (V2V communication): Communicated directly with other nodes (vehicles).
2. Vehicle–Infrastructure communication (V2I communication): In this type of communication, vehicles communicate with a fixed infrastructure which is located on the road as traffic signals or towers.

The message transferred in VANET is categorized into two different applications as shown in Figure 14.1:

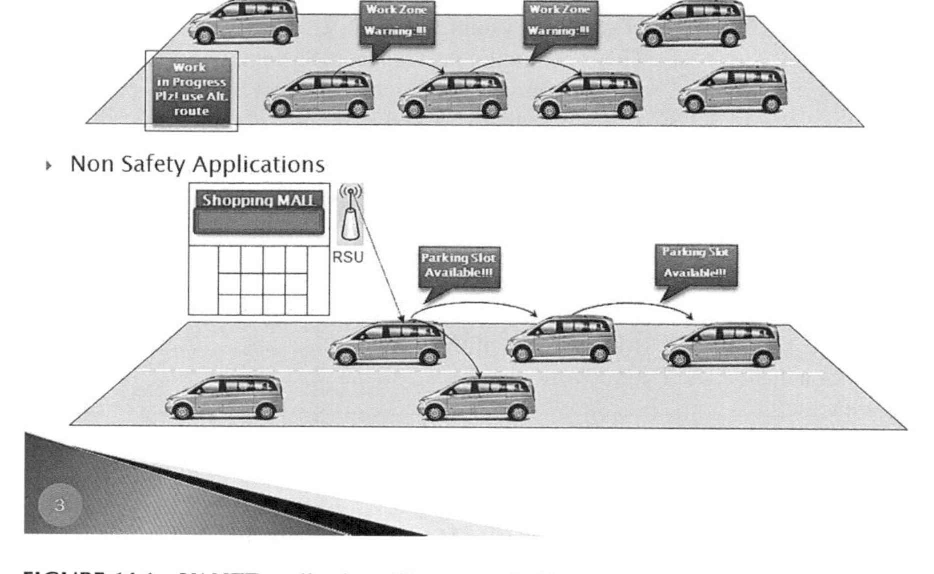

FIGURE 14.1 VANET applications (Sharma et al., 2011).

1. **Safety Applications:** The main motive of this application is to share warning messages to riders to avoid accidents. Examples are speed warning and pedestrian crosswalks.
2. **Non-Safety/Infotainment Applications:** The goal of this type of application is to provide a healthy life for the riders; it provides comfort to the travelers about hotels, restaurants, nearby hospitals, and fuel stations.

14.1.1 VANET Characteristics

An RSU or a mobility vehicle might be considered as a node in the infrastructure-less VANET network. It offers a blend of radio link techniques and the attributes of an ad-hoc network, which employs a novel architecture to interaction and platform modalities. Characteristics (Bhatt et al. 2014) of VANET are discussed below:

1. **High Mobility:** The topology of any VANET might vary often and unpredictably because of the high mobility of vehicles, which can travel at speed up to 150 km/h. As a result, a communication link between two vehicles only lasts for a relatively little period of time, especially when they are moving in opposing directions.
2. **Predictable Topology:** However, a VANET's topology can be advantageous because cars aren't supposed to leave the paved road, making car running directions somewhat predictable.
3. **Latency:** When a vehicle abruptly stops, it should convey a broadcast message information to nearby motorists of the potential risk. Taking into account the driver must react within 0.70 to 0.75 seconds at the very least, therefore the warning message should arrive with almost negligible latency.
4. **No Power Consumption:** In VANETs, reducing power usage is not a concern. The vehicle battery and dynamo are two examples of reliable power sources for nodes in VANETs, and the necessary transmission power is minimal compared to the power used by onboard facilities.
5. **Privacy and Security:** They have a significant impact on whether people will use this technology. Every node in VANETs represents a specific individual, and its position provides information about that person's location. Any invasion of privacy makes it easier for someone to watch over someone's daily activities. On the other hand, higher authorities should have access to identifying information to ensure that criminal actions are punished, as there is a concern that this function can be abused. In some circumstances, message manipulation could increase false alarms and accidents, undermining the entire purpose of this technology.

14.1.2 VANET Routing Protocols

High dynamic topology characteristics make it more difficult to build efficient VANET routing protocols represented in Figure 14.2. The VANET routing protocol can be divided (Yasser et al. 2017) into:

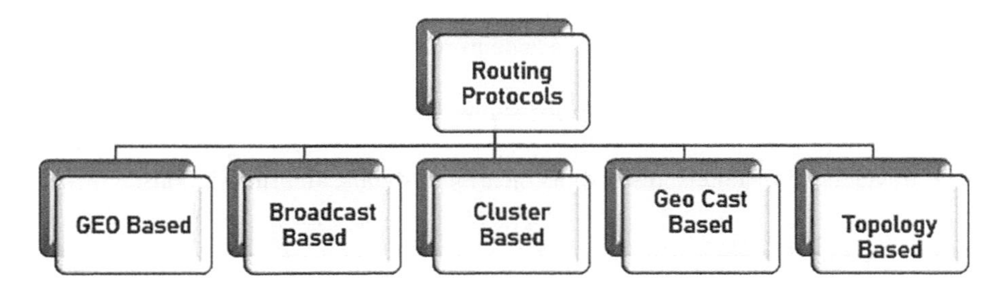

FIGURE 14.2 VANET routing protocols.

1. **GEO-Based Routing Protocols:** These protocols allow communication between sources and destinations utilizing both network addresses and geographic coordinates. The load balancing routing protocol uses node location information to calculate and customize the route. Consequently, there is no need to create routing tables. There are three parts in the protocol: beaconing, location, and forwarding services.
2. **Broadcast-Based Routing Protocols:** Broadcast-based applications rely on positioning mechanism algorithms (for example, GPS). These applications offer the data necessary for path selection. Additionally, these protocols do not maintain any tables that include routing information or any data regarding the joined state of surrounding nodes (Ghori et al. 2017).
3. **Cluster-Based Routing Protocols:** In this procedure, a cluster of vehicles with similar characteristics—such as speed, direction, etc.,—are put together. Additionally, since it is regarded as a local communication, if a vehicle node has to interact with a node within the cluster, the data will travel directly there. Additionally, if a vehicle node has to interact with another node that is outside of the cluster, it needs the assistance of its cluster head (CH) to get there. It is a smart option for network designers because of its scalability characteristics, but one of its downsides is traffic jams (Ghori et al. 2017).
4. **Geo-cast-Based Routing Protocols:** Zone of relevance (ZOR) and zone of forwarding are the two key zones that make up the geo-cast protocol (ZOF). ZOR is the space set aside for the region's nodes. This protocol's major objective is to enable communication between the vehicles existing in ZOR. A vehicle will enter ZOF if the source vehicle wants to connect with one that is not already in ZOR, and any vehicle that enters ZOF is required to broadcast the data to other ZORs. A connection disconnection can occur frequently as a result of frequent zone changes, which places the aforementioned point in the category of disadvantages (Ghori et al. 2017).
5. **Topology-Based Routing Protocols:** Topology-based routing techniques use connectivity information to route data packets between nodes throughout VANET. The proactive approach, which relies on table-driven routing techniques, and the reactive approach, which relies on on-demand routing techniques, are two subdivisions of this mechanism.

14.1.3 Classification of Ad-Hoc Routing Protocol

VANET routing protocols are classified into three categories (1) proactive, (2) reactive, and (3) hybrid as in Figure 14.3.

1. **Proactive:** In proactive routing, the background maintenance of routing information, such as the next forwarding hop, occurs regardless of requests for communication. The proactive routing protocol has the benefit of avoiding route discovery because the target route is already stored in the background, but it also has the drawback of having low latency for real-time applications. Within a node, a table is built and maintained. Each entry in the table denotes the subsequent hop node leading to a specific destination.
2. **Reactive:** Whenever a node needs to connect with another, only reactive routing does open the route. It minimizes the load on the network by maintaining the routes that are currently in use. When a route is found, the route discovery phase of reactive routing comes to an end. During this phase, a flood of query packets is sent out into the network to hunt for a path.
3. **Hybrid:** Comparing hybrid routing to conventional reactive and proactive routing protocols, which integrate elements of reactive routing with location-based geographic routing, may result in a significant reduction in the routing overhead.

In VANET, where numerous nodes combine to form a cluster depending on shared characteristics, clustering is a key idea. However, machine learning and fuzzy logic algorithms are also the foundation of many VANET clustering methods. Mobility-based clustering strategies are the most popular in VANET clustering. Some VANET clustering algorithms combine fuzzy logic and machine learning techniques to increase the cluster's efficiency and stability. VANET clustering also makes advantage of multi-hop-based methods and network mobility (Mukhtaruzzaman et al. 2020).

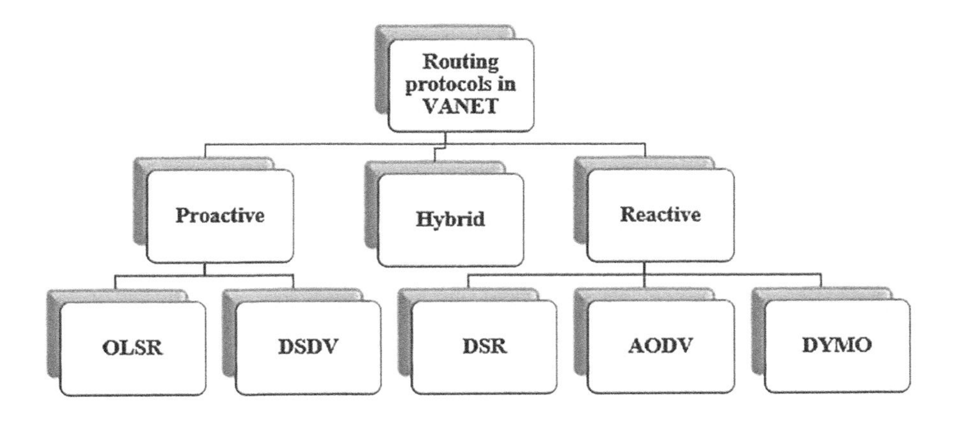

FIGURE 14.3 Classification of routing protocols.

In the current scenario, more numbers of riders are using their own vehicles to ride from one location to another. This is the cause of increased traffic flow and death rates in the road transport system. Traffic flow and road accidents can be reduced by communicating suitable information regarding the condition of the road and its surroundings to the riders in a trusted way. To provide a solution to this kind of problems, nodes (vehicles) must be integrated with sensors to transfer the message to the riders (Azees et al. 2019).

AI applications have given a new initiative for ITS using VANET (Tong et al. 2019). Sensors mounted within vehicles are working smarter within the stipulated time to give new exposure to the riders. This new development will lead to driverless-based vehicles with new innovation by utilizing AI-based techniques to provide safety to travelers (Liang et al. 2017). ITS plays a vital role in smart city utilization for school-going children, college students, office workers, and other employers. Smart city concept can also be implemented in rural areas to save the time of travelers and reduce road accidents.

ITS can provide prior information about any traffic problems such as ongoing road constructions, road accidents, and bus timing to the road travelers. There are various applications of ITS used in various countries; this can also be implemented in India to control traffic and provide safety to road travelers. Sensors like GPS and onboard unit can collect information about vehicles such as speed of the vehicle, location, weight of the vehicle, and time of travel with the help of hardware devices that are connected to the servers at the data collection center.

ITS is utilized in many places. It not only alerts the travelers but also possible to provide information like vehicle timings, seat availability, vehicle number, location of the vehicle, time at which the vehicle will reach the destination, the number of passengers available inside the vehicle, and so on. Nowadays, like other countries, our government has also taken necessary steps to provide safe and comfortable travel to roadside travelers, but the implementation is a slow process; it should be implemented in all places.

The cameras inside vehicles with onboard sensors scan the interstate highway to reduce the accidents. Sensors can be used to track the safety of drivers or road travelers; when they meet any dangers, the sensors will pass communication to nearby patrol police and emergency vehicles. This system warns the drivers with a sound about collision warning, lane changing, road accidents, road construction, and if a vehicle collides with pedestrians, two-wheeler riders, or any animals. If the rider changes the lane without giving any alert to other riders, this system will give an alert sound to the travelers.

When comparing the number of road accidents and death rates in India to other countries in 2021, nowadays, the death rates are reduced due to the implementation of safety measures in vehicles. Table 14.1 and Figure 14.4 show the number of death rates.

In the present world, trending vehicles are equipped with multiple systems and numerous modern technologies, including AI and automotive communication processes (Seuwou et al. 2013). VANET brings more observation due to a greater number of applications relevant to road safety and traffic flow. In smart cities, there are

TABLE 14.1
Road Accidents and Death Rates

S. No.	Country	No. of Accidents	No. of Death Rates
1	India	480,652	150,785
2	China	212,846	63,093
3	Indonesia	106,129	26,185
4	Russia	173,694	20,308
5	Brazil	60,228	6,398
6	France	57,522	3,477
7	Iran	293,305	15,998
8	Morocco	80,680	3,785
9	Turkey	185,128	7,300
10	Korea	220,917	4,292

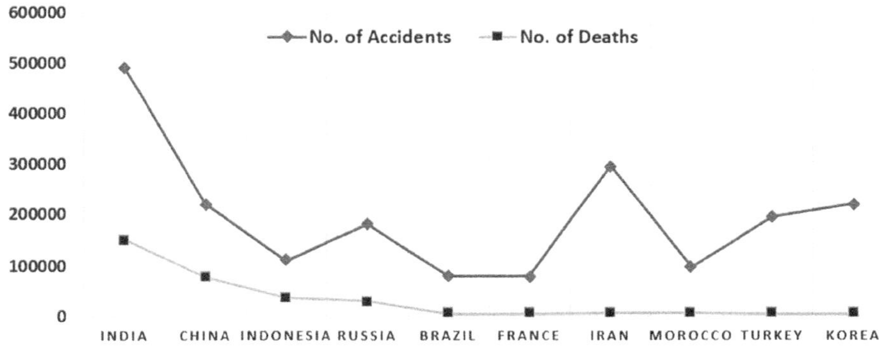

FIGURE 14.4 Comparison chart of road accidents and death rates.

various problems because of the large number of vehicles increased in roadside; there may be a chance of malicious nodes (vehicles) within the web and it can damage and lose the data transferred by the vehicles. ITS will overcome this type of issue in the network (Satyanarayana Raju and Selvakumar 2022).

A major threat for automated vehicles is to detect road damage or accidental situations of the road in both urban and rural zones. To overcome this type of issue and to exchange the emergency situation, ITS is to establish VANET to provide a safe and secure traveling situation to riders (Bibi et al. 2021). A fuzzy logic approach is used to detect the malicious node attack in the cluster via VANET (Alkhalidy et al. 2022).

ITS with VANET plays a main role in various domains all over the country. ITS is a digitalized system to equate the transport system to deliver massive services to the public. Before beginning their journey, drivers must be aware of the road's conditions; ITS can help with this. Roadside users can benefit from ITS that was created to reduce traffic and road accidents (Garg 2022). This is possible by passing significant instructions to the nodes (vehicles). Simulators offer a safe and cost-effective environment for investigating VANETs (Weber et al. 2021). In this proposed system, SUMO tool and network simulator 3.5 are used.

Section 14.2 briefs the literature review of various authors, Section 14.3 will take a short review of AI and ITS in VANETs, Section 14.4 describes the background study of this proposed system, and at last Section 14.5 concludes this proposed system.

14.2 LITERATURE REVIEW

Gayathri and Gomathy (2022) developed a concept of AI-based trust authentication Sugeno fuzzy inference system used to compute the weights of a node for selecting the trusted CH and CM (cluster member) which will prevent the nodes from various attacks. Simulation results will reduce the E-E delay and enhance the packet delivery ratio (PDR) and network throughput.

A study by Seth et al. (2022) provides insight into machine learning techniques that are frequently employed in the automotive industry. The automobile industry's use of machine learning techniques has advanced the development of a driverless car. It is fairly possible to apply machine learning algorithms and give the user its benefits with the aid of sensors and cameras. It is beneficial to give the car the ability to carry out particular duties that could eventually replace the driver. The vehicles' built-in AI processors let them traverse the highways.

Satyanarayana Raju and Selvakumar (2022) proposed an approach to integrate K-nearest neighbors and random forest with proximate analysis. The primary goal of this proposed system is to analyze the falsification of vehicles within the network and enhance the mobility of the nodes by producing malicious communication from source to destination. Malicious nodes (vehicles) will attack the grid and provide wrong communication to the driver. This system is used to assist the traffic department to track the road conditions and support the travelers while they are in danger.

Alkhalidy et al. (2022) revealed the new innovative approach to exchange roadside information to vehicles. Intruders may send false messages to the cluster to prevent the vehicle from getting the right road situations. This system will detect the intruders with fuzzy logic to evaluate the node's trustworthiness. Simulation tools are utilized to check and detect the false nodes from the cluster.

Garg et al. (2022) made a comparative study with the previous study about the importance, advantages, and disadvantages of ITS. This study relates various problems with ITS relevant to security, design, and their challenges. They made a study with more than 70 research articles and this paper will help young research minds to do research.

Iswarya and Radha (2021a) researched about clustering techniques. This article proposed a clustering-based optimization technique called energy-efficient clustering technique with the ad hoc on-demand distance vector (AODV) protocol's K-medoids clustering algorithm in order to cluster vehicle nodes and find nodes that are convincing to interact in a defined secured and reliable path, which were detailed in previous works. Aiming for energy-efficient communication, efficient nodes from each cluster are chosen in order to maximize the parameter as low energy consumption in VANET.

Weber et al. (2021) studied the current scenario of simulators used with VANET to provide safety measures to pedestrians and roadside travelers. They identified that there are numerous hurdles to be labeled to enrich the quality of VANET simulations. The development of driverless vehicles in the modern day is made possible by cutting-edge technologies like edge computing and 5G, which will open up VANET research.

Bibi et al. (2021) proposed an automated detection system using automated vehicles to exchange the emergency message with nearby vehicles using edge AI and VANET. Concepts of ResNet-18 and VGG-11 are integrated for the automatic detection and classification of road conditions. Conditions of road images are detected using the cameras mounted inside the vehicles. This will reduce road accidents and provide a safe travel experience to the riders.

Niestadt et al. (2019) explained about EU action and AI-powered autonomous mobility, for example, it could help reduce human error, which is the main cause of many traffic accidents. These opportunities can bring them unanticipated consequences and misuse, such as cyberattacks and biased transportation choices. Additionally, there are implications for employment and moral concerns surrounding responsibility for judgments made by AI instead of humans.

Raja Kumar et al. (2019) proposed a system and tested it with various traffic conditions in Chennai like Kathipara, T. Nagar, Highway, and outback areas in Tamil Nadu. They analyzed and reviewed each scenario with dense and moderate traffic conditions with many numbers of vehicles. RSU is involved to analyze the delay and PDR in each scenario for land and marine. The results are generated using the SUMO tool and network Simulator 2.

Srivastava et al. (2019) explained how ITS can be used to reduce traffic congestion in the present and prevent accidents on the road. Congestion-related issues are what gave rise to the concept of ITSs. India's transportation system is under tremendous pressure as a result of the country's rapid vehicular growth, population growth, rural-to-urban migration, and economic growth. Traffic congestion causes efficiency to decline, travel times to lengthen, air pollution to occur, and increased fuel consumption. Expansion of the transportation system also contributes to a rise in the number of traffic accidents across India.

Tong et al. (2019) explained how to solve numerous research issues in V2X systems. This paper gives a thorough assessment of the research efforts that have used AI. These research works' contributions have been summed up and divided into groups based on the application domains. Finally, we outline unresolved issues and research hurdles that must be overcome if AI is to fully progress V2X systems.

Ghori et al. (2017) analyzed and discussed a number of papers in this paper that are related to routing protocol in order to determine which routing protocol is suitable for video applications in VANET. Additionally, after looking at several systems developed by researchers, we critically examined them and identified their benefits and drawbacks for upcoming projects. Additionally, simulation is used to compare delays and throughput between different routing systems. Additionally, they demonstrated through research that in a VANET context, AODV performs better than other ad hoc protocols.

Liang et al. (2017) first went through how resources are allocated for automotive communications before talking about whether millimeter wave bands could be used for vehicular communications. The primary underlying properties that distinguish vehicle communications from other kinds of wireless systems are initially introduced, along with modeling of vehicular channel characteristics. They then proposed methods for estimating time-varying vehicular channels as well as several modulation strategies for high-mobility channels and finally listed the difficulties and potential benefits of vehicle communications.

Azees et al. (2016) explained the VANET system model, its characteristics, and security and service-related problems in VANET. In addition, they summarized various attacks and countermeasures in an easy way. A binary confirmation and key exchange strategy for secured data transmission in VANET is proposed in this system.

Due to the huge scale of networks and the significant node mobility, VANET differs from MANET. When developing a VANET, security and traffic are the two key concerns. Even though there are many ideas that have been put forth to enhance VANET security, security is still a difficult study topic. In this article, rural and urban networks are created for the VANET scenario for NS-2, and different routing protocols are then created using various performance matrices, including residual energy and throughput, along with the two routing protocols—AODV and dynamic source routing.

Seuwou et al. (2013) gave a view of the emerging technologies in the field of VANET and investigated the issues relevant to security and also discussed the productive and legal suggestions of VANET in today's era. Security issues which remain unsolved due to the system are not fully automotive, and within a few years, there will be an improvement in road safety by implementing VANET.

Rajadurai and Jayalakshmi (2013) reviewed the characteristics, structure, and heterogeneous attacks faced by VANET. They discussed protocols that will enhance the scalability and security with the usage of vehicular public-key infrastructure, signature, and Regional Trusted Authority (RTA).

14.3 ARTIFICIAL INTELLIGENCE AND INTELLIGENT TRANSPORT SYSTEM

14.3.1 Artificial Intelligence and VANET

AI will have numerous applications for gaming, speech recognition, and automated vehicles. AI changes the transportation system because automated vehicles swap the lifestyle of the universe and also help in traffic management. Nowadays, transportation is rapid, capable, authentic, and safer with the advancement of using AI.

14.3.1.1 AI and Driverless Vehicles

The transportation sector is one of the industries where AI has been most effectively applied, opening up entirely new levels of collaboration among various road users. Automakers, IT firms, and research institutions throughout the world are looking into AI technology to develop driverless cars for use in both commercial and personal transportation. These vehicles are supported by a variety of sensors, such as GPS, cameras, and radar, as well as actuators (equipment that transforms an input signal into motion), control systems, and software. Others aim to totally replace the human driver, while others just automate specific driving tasks (like parking). In the EU, AI technologies that reinstate some driving tasks are already widely available and test fully automated vehicles (including transport packages) in a specific few driving circumstances and locales. In fact, automated testing is more difficult. Vehicles in metropolitan locations must anticipate considerably more (often unpredictable) indicators of movement because there are many diverse actors, sophisticated road networks, and infrastructure (intersections, traffic signs, etc.) (Niestadt et al. 2019).

Different automakers, like Ford, Tesla, and BMW, throughout the world are developing and testing autonomous vehicles utilizing AI with sensors and cameras. A recent estimate indicates that production of driverless automobiles is anticipated to reach 8,000 units globally by 2023–2030 although there are concerns regarding technological improvements and their potential to protect travelers from danger and assist in bringing down accident rates to a minimum.

Every single member of the public deals with traffic congestion on a daily basis as they move from one location to another. AI can be used to avoid traffic flow and reroute travelers in order to overcome these difficulties. Road conditions can be detected by built-in cameras and sensors put on city streets, saved in the cloud, and sent to travelers via AI applications to prevent traffic jams, accidents, and delays. To ensure safe and secure travel, sensors positioned inside the vehicles must be able to track and send messages to the passengers. AI-enabled systems make it possible to update real-time information on the vehicles.

14.3.1.2 Operations and Difficulties of AI in Transport

Although AI benefits road traffic, there are potentially substantial disadvantages, especially in mixed-use environments. AI has the ability to reduce traffic congestion, free up drivers' time, make parking simpler, and promote ride and car sharing. AI can minimize fuel consumption caused by vehicles idling when stationary, improve air quality, and help with urban planning since it aids in maintaining traffic flow. However, decreased transportation expenses and relieving the driver of driving responsibilities would also encourage more individuals to use a car (rather than a public transportation system), which would consequently increase traffic and air pollution (Niestadt et al. 2019).

Challenges faced in the transportation industry using AI are a value of adoption, reliability of structure, security issues in the cloud, and manpower consumption.

14.3.1.3 Benefits of AI in Road Transport

AI use in the transportation sector has the potential to improve universal transportation, gather traffic data, and reduce congestion. Traffic light algorithms will be able to function based on the volume of traffic, thanks to this approach. At a time, if there is heavy traffic, the "red light" may be on for extended periods of time, and if there are few vehicles on the road, the "green light" may be on. AI can be used in public transportation to schedule and route the fleet properly.

1. **Traffic Management:** Cameras and sensors can be set up at various locations in the future. These gadgets gather data and instantly send it to the cloud. Data inputs and AI-powered system can be used to do analysis. Perceptive facts about traffic can then be obtained from the processed data. For example, data about the number of vehicles using a specific stretch of road at a given moment, the types of vehicles, and the frequency—if either accident can be acquired to accurately foresee gridlock, incident spots, and barricades, a pattern can be found.
2. **Automated Process:** The use of automated vehicles is becoming more popular. Before that goal is entirely reached, other aspects of transportation can be automated. Equipment that provides tickets to passengers as soon as they board the vehicle can be deployed at the entrances to public transit. As a result, hiring a conductor will not be required for the job. Aside from the cost savings, the manual fare collection system can be replaced because it is laborious.
3. **Parking System:** Cameras are the IoT devices used to collect data. In a busy city, for example, the sensor's ability to discern whether a parking space is filled helps cars locate spots faster.
4. **Licensed Number Plate Recognition Software:** It uses computer vision-based camera systems set on street poles, overpasses, and highways to record the license plate, time, date, and location of the vehicle. Police officers will find this highly developed system with AI elements to be very useful in detecting and preventing crime. For instance, investigators can establish whether a particular vehicle was present at the crime site. Future traffic pattern detection, toll management, parking management, and asset tracking for auto dealership organizations will all be made possible by the same technology. This can stop autos from being utilized for illicit purposes.
5. **Drone Taxis:** The travelers will have access to get to their depot as quickly as possible with the use of drone taxis. AI-based drone taxis are the genuine answer to every issue a city planner is now facing. Recently, nearly 17 passengers in China enjoyed smart air mobility, thanks to unmanned aircraft. This suggests that comparable AI technologies that facilitate mobility will be positively adopted in the future. Additionally, a lot of delivery companies have begun adopting drones for delivery services. With GPS navigation, obstacle identification and avoidance, delivery drop, emergency, and contingency management, AI helps drone deliveries.

14.3.2 Intelligent Transport System

The potential of new "intelligent" technologies to increase the energy efficiency and environmental effect of existing transportation systems is drawing increased research interest at the driver, vehicle, and transportation system levels. These technologies are able to provide communication between transportation infrastructure, vehicles, and people that can result in system-level efficiencies because of their high computer capacity, analytical software, connection, and other aspects. Connectivity, including vehicle-to-vehicle and vehicle-to-infrastructure communication, as well as mobile devices the majority of travelers carry in their wallets is necessary for efficiency.

In an effort to solve the increasing traffic congestion in urban areas, the idea of ITS was first suggested in the 1960s. ITS is now a vital part of any smart city, transforming urban areas into digital societies that make life easier for its residents in every way. Mobility is a major issue in any city; whether people are traveling within the city for work, school, or other reasons, they use the transportation system.

ITS can help individuals save time while also making the city smarter. By reducing traffic issues, ITS strives to improve traffic efficiency. It offers consumers access to historical data on traffic, nearby conveniences, real-time running information, seat availability, etc., which shortens commuters' trip times and improves their comfort and safety (Choudhary, 2019).

Today, ITS applications are extensively accepted and employed in numerous nations. The use includes improving road safety and making optimal use of infrastructure in addition to controlling traffic congestion and providing information. Due to its limitless potential, ITS has evolved into a multidisciplinary conjunctive field of study. As a result, numerous firms all over the world have created solutions for ITS applications to suit the need.

Example 14.1:

The city's Intelligent Transport System provides daily commuters with information on public transportation, including timetables, seat availability, the bus's present location, the time it takes to get there, the bus's next location, and the volume of passengers it is carrying.

The sensors are in the buses that the city's bus operators operate. In order to ensure that the bus arrives at the next bus stop on time and not early, the bus is momentarily and very slightly slowed down at the red light for a little bit longer than it should. Because there are so few delays, both drivers and passengers are ignorant of them, thanks to the system's clever architecture.

14.3.2.1 Goals of Intelligent Transport System

The main goals of the ITS (Parmar et al. 2018) were discussed below:

- Enhance the region's rural transportation network's safety and security.
- Improve service accessibility, personal mobility, and the system user's ease and comfort.
- Boost the system providers' productivity and operational effectiveness.
- Boost the economic productivity of people, companies, and organizations.
- Reduce your influence on the environment and energy use.
- Create and support long-lasting associations that will enable the display of ITS efforts and conventional solutions that cater to the region's rural needs.
- Make sure that your ITS programs are in line with regional and universal initiatives.
- ITS should be taken into account while planning and developing transportation programs.

14.3.2.2 Applications of Intelligent Transport System

The major role of ITS is considered traffic management. All the information is gathered and evaluated for use in future operations, real-time traffic management, or data on local transportation vehicles. Traffic management operations depend on the automated gathering of location-specific data, processing of that data to produce accurate information, and finally dissemination of that information back to travelers.

1. Data gathering
2. Data tracking
3. Data processing
4. Information on travelers

The data is gathered through the hardware devices mounted on the vehicles. The main purpose of sensors is to sense the traffic condition and the speed of the vehicles, their time, location, and delay time. The sensors are connected to various cloud data centers to store the information for further processing. Travelers will have announcements pertaining to traffic via the internet, SMS, or onboard units of vehicles. The information that was gathered is then processed further in a number of phases: they are data cleansing, data synthesis, error correction, and adaptive logical analysis.

Data is then further adapted and gathered for analysis. This updated total amount of data is further revealed to track traffic flow that is available to provide consumers with pertinent information. The system provides real-time data on things like routine time, travel speed, delay, traffic accidents, route changes, detours, and the state of work zones, among other things. Numerous different electronic devices offer this information.

14.3.2.3 Current Scenario of ITS in India

Many Indian towns, mainly the metropolises, have ideas for ITSs in the works. But most of the time, each endeavor just made use of a portion of ITS rather than fully integrating it. While some efforts achieved some level of success, many fell short of making a meaningful difference. Some of the key contributing elements to these versions of ITS failing were improper implementation, regulation, and legal compliance. AIS-140 was one of the ITS implementations that the Automotive Research Association of India developed that was successful. Additionally, one successful ITS endeavor has been making toll collection booths cashless with the aid of radiofrequency identification and detection tags. This is included in the list of ITS initiatives (Sharma 2019).

In India, there have been a few ITS projects done, mostly in metropolises and major cities like Delhi, Ahmedabad, Bengaluru, Chennai, etc. These initiatives are individual in form and concentrate on certain ITS features including parking lot management, traffic light management, public transportation, and toll collecting. In cities like Pune, Delhi, Indore, Ahmedabad, Bengaluru, Hyderabad, Chennai, Coimbatore, Jaipur, Madurai, Nagpur, Vijayawada, and Visakhapatnam, bus rapid transit projects frequently use ITS (Rawal 2015).

Example 14.2

1. Chennai has implemented a traffic management system that includes surveillance cameras at intersections for anyone breaking the law, an automatic number plate reader, and CCTV cameras placed throughout the city to monitor traffic flow without human intervention in the automated system's decision-making and adjustment of signal waiting times.
2. Mumbai has developed an area traffic control scheme that manages traffic flows at important intersections with the use of cutting-edge technology.
3. In Bengaluru and Hyderabad, real-time traffic scenarios of key intersections and their subsidiary link roads may now be contained online, thanks to a test project. The real-time photos are accessible 24 hours a day and 7 days a week, and they are refreshed every 15 seconds.

14.4 BACKGROUND STUDY

14.4.1 Problem Statement

This proposed model is developed with nodes as vehicles (cars) and fixed infrastructure. Fixed infrastructure is said to be RSU that is used to deliver the information to all nodes in the cluster. Here, RSU is considered the gateway and it maintains the centralized data packets which it receives. RSU transmits the message to vehicles directly in the cluster via VANET using the AODV protocol. It stores information related to:

1. traffic flow
2. other RSUs
3. centralized data center which stores the number of nodes in its region, emergency alerts received from nodes, and broadcasts.

Let us take a sample case; if there is an accident in the surroundings, the RSU receives the warning messages from the nodes and this RSU will transfer the warning messages to the nearby RSU to alert other vehicles far away from the accident-prone zone. This is a main advantage to other riders to save their time and it will help others to avoid road accidents and prevent traffic jams in that zone. This application will pass this information to nearby patrol police, ambulance services, and hospitals through RSUs about the accident. Using this type of model will prevent the traffic jam and time consumption can be reduced by transferring messages through RSU.

When traffic is heavy in backcountries during peak hours and the road is in a poor condition to allow a vehicle to take a detour around ongoing work, accidents do happen. The proposed model will help to overcome this type of issue while implementing this system in real time.

14.4.1.1 Challenges in Implementing ITS in India

- The poor state of the roadways increases the likelihood of accidents.
- Inadequate traffic control methods.

- The Indian traffic police system is similarly subpar.
- Coordination between vehicles and the use of new technologies.

Solutions for the above issues are discussed below (Srivastava et al. 2019):

1. The problem of traffic congestion can be alleviated with the implementation of a properly planned traffic management system, which includes the use of GPS, geographic information system, and remote sensing. This system should be included as standard equipment in Indian automobiles, much like airbags and braking systems currently.
2. Vehicles can pass through toll gates at regular traffic speeds, thanks to electronic toll collection (ETC), which also automates toll collecting and eases congestion at toll plazas.
3. The inter-vehicle call is a crisis call that can be made physically by the occupants of the cars or inevitably following an accident thanks to the activation of in-vehicle sensors.
4. Using a camera and a vehicle tracking device, a highway patrol camera system is used to detect and record vehicles that are moving over the posted speed limit or breaking another traffic law.

A connected car system would increase the safety of moving a car since it would be easier to handle emergencies because the driver would always have access to aid, regardless of the time or situation.

14.4.2 SUMO Tool

SUMO tool is used to analyze vehicles from the range of 100 meters down in rural and urban areas (village) through two models:

1. Traffic model,
2. Modification of trust signals.

In the first model, the correct and alternate routes are established and analyzed with the moving vehicles in a particular region with a real-time map. Next, the messages are communicated within vehicles and are analyzed with the nodes of performance as average end–end delay, PDR, and network throughput.

14.4.3 Simulation Results

Project simulation time in a network is referred to as simulation time. The length of the simulation is expressed in seconds.

14.4.3.1 Traffic Model

Analysis was done in two locations: one in traffic-dense areas like Coimbatore and another in rural areas. SUMO tool images are shown in Figures 14.5 and 14.6.

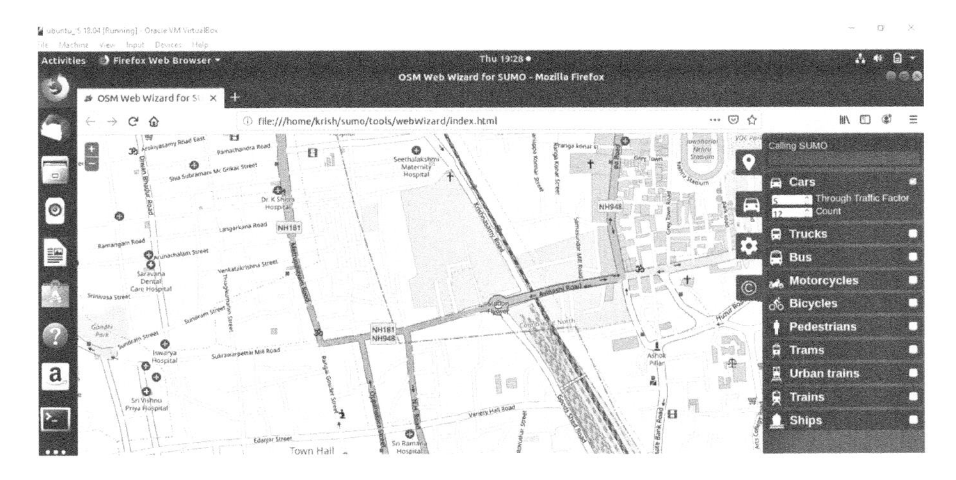

FIGURE 14.5 SUMO results with Google location.

FIGURE 14.6 Real-time scenario.

14.4.3.2 Modification of Trust Signals

NS 3.5 is used to stimulate the results of a communication model. Communication is done with vehicle to vehicle and vehicle to infrastructure to produce optimal solution. AODV routing protocol is utilized to analyze the real-time results in urban and rural areas.

14.4.3.2.1 Throughput

Throughput is the amount of data sent from one vehicle to another per unit of time. Throughput can be measured in bits per unit or packets per unit. By using a communication link, delivery and data transmission are both possible. Packet transmitting and receiving can be done with high throughput. Less throughput in a network indicates packet loss (Figure 14.7).

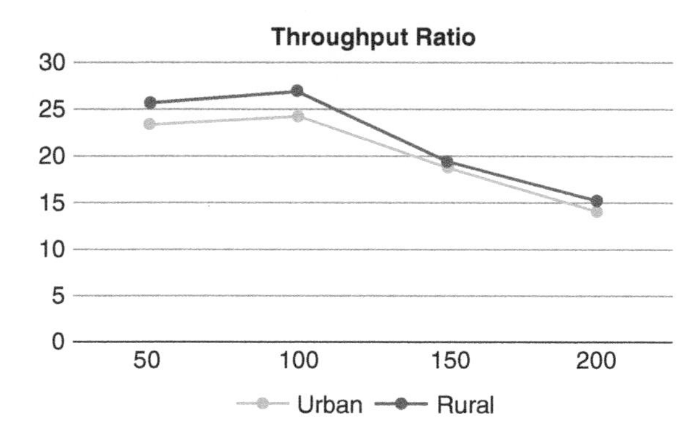

FIGURE 14.7 Simulation throughput results for urban and rural regions.

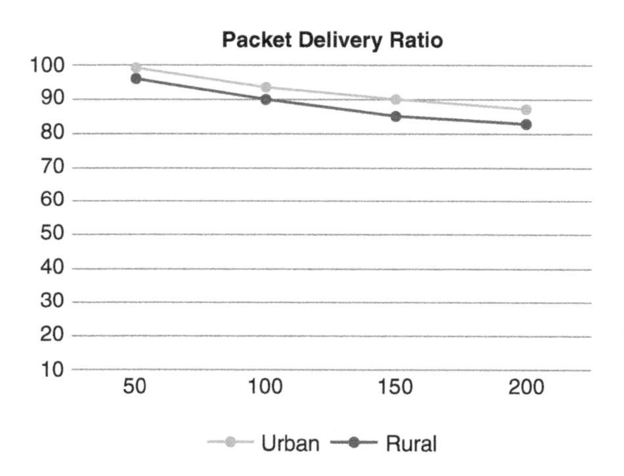

FIGURE 14.8 Simulation PDR results for urban and rural regions.

14.4.3.2.2 Packet Delivery Ratio

This is defined as a ratio of the total number of data packets created by the source to the number of data packets received at the destination (Figure 14.8).

14.4.3.2.3 End-to-End Delay

End-to-End delay is a parameter used in vehicle networks. An overall delay in packet transmission could be provided. The source and destination nodes, as well as the application agent, are all utilized. To deliver a rapid communication, a vehicular network with low end-to-end latency is needed (Figure 14.9).

14.5 CONCLUSION AND FUTURE WORK

This paper briefs the aspects of VANET research issues, routing protocols, and security challenges. The potential value of VANETs with regard to applications for safety and entertainment in a future intelligent society is uncertain. Recent years have seen

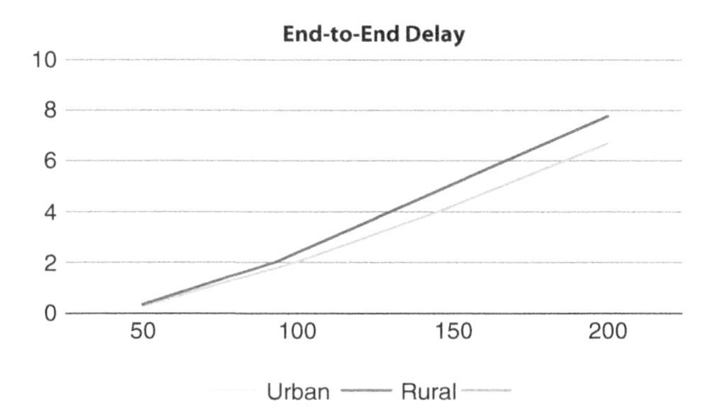

FIGURE 14.9 Simulation end-to-end relay results for urban and rural regions.

the emergence of new car applications in a variety of fields, including navigation safety, location-aware content delivery, commerce, and games.

Safety applications continue to be the main area of research interest in the mobile vehicle environment since they are necessary for maintaining constant awareness of the road ahead. The subject of "how to use model checking to automatically explore if these apps fulfill the standards" should be investigated by researchers as they continue to work on the standards and security of VANET applications.

AI is without a doubt the most incredible technology ever created by civilization, but as with every amazing creation that makes life easier for humans, it can be seen that, up to this point, AI has not been used to its full potential and that many things still need to be investigated. The use of AI in transportation that is discussed in this article just offers a small sample of options and potential that technology may bring to the field of transportation in the future. AI-powered transportation will be fascinating and will be geared toward resolving long-standing transportation issues. The amount of vehicles on the road is rising as urbanization spreads quickly. Combining the two places tremendous pressure on cities and rural areas to maintain a better traffic infrastructure so that the areas can continue to function without any difficulties. The only answer is to apply an ITS.

The introduction of an ITS in India will undoubtedly improve the quality of our rides. It will cut down on the likelihood of traffic accidents, the amount of time spent on traveling, and fuel usage, all of which will benefit the environment. ITS is helpful in ensuring the comfort and security of our journey.

REFERENCES

Alkhalidy, Muhsen, Atalla Fahed Al-Serhan, Ayoub Alsarhan, and Bashar Igried, "A new scheme for detecting malicious nodes in vehicular ad hoc networks based on monitoring node behavior", *Future Internet*, vol. 14, p. 223, 2022. doi: 10.3390/fi14080223.

Azees, Maria, Pandi Vijayakumar, and Lazarus Jegatha Deborah, "Comprehensive survey on security services in vehicular ad-hoc networks", *ITE Intelligent Transport System*, vol. 10, no. 6, 2016. doi: 10.1049/iet-its.2015.0072.

Bhatt, Anjali and Naresh Purohit, "Performance evaluation of Urban and rural areas for VANET", *International Journal of Technology Research and Management*, vol 1, no. 5, pp. 1–5, 2014.

Bibi, Rozi, Yousaf Saeed, Asim Zeb, Taher M. Ghazal, Taj Rahman, Raed A. Said, Sagheer Abbas, Munir Ahmad, and Muhammad Adnan Khan, "Edge AI-based automated detection and classification of road anomalies in VANET using deep learning", *ComputIntellNeurosci*, 2021. doi: 10.1155/2021/6262194.

Choudhary, Mahashreveta, "What is intelligent transport system and how it works?" 2019.

Garg, Tanya and Gurjinder Kaur, "A systematic review on intelligent transport systems", *Journal of Computational and Cognitive Engineering*, pp. 1–14, 2022. doi: 10.47852/bonviewJCCE2202245.

Gayathri, M. and C. Gomathy, "AI-TASFIS: An approach to secure vehicle-to-vehicle communication", *Applied Artificial Intelligence, International Journal*, vol. 36, no. 1, 2022. doi: 10.1080/08839514.2022.2145636.

Ghori, Muhammad Rizwan, Ali Safa Sadiq, and Abdul Ghani, "VANET routing protocols: Review, implementation and analysis", *International PostGraduate Conference on Applied Science & Physics*, 2017. doi :10.1088/1742-6596/1049/1/012064.

Iswarya, B. and B. Radha, "Energy efficient clustering technique for VANET", *Advances in Parallel Computing*, EbookVolume 40: Advances in Parallel Computing Technologies and Applications, pp. 105–113, 2021a. doi: 10.3233/APC210129.

Iswarya, B. and B. Radha, "Reliable path identifying protocol in VANET", *Indian Journal of Science and Technology*, vol. 14, no. 29, pp. 2410–2419, 2021b. doi: 10.17485/IJST/v14i29.438.

Liang, L., H. Peng, G. Y. Li, and X. Shen, "Vehicular communications: A physical layer perspective," *IEEE Transactions on Vehicular Technology*, vol. 66, no. 12, pp. 10647–10659, 2017.

Mukhtaruzzaman, Mohammad and Mohammed Atiquzzaman, "Clustering in VANET: Algorithms and challenges", *Computers & Electrical Engineering*, vol. 88, 2020. doi: 10.48550/arXiv.2009.01964.

Niestadt, Maria, Ariane Debyser, Damiano Scordamaglia, and Marketa Pape, "Artificial intelligence in transport Current and future developments, opportunities and challenges", *European Parliamentary Research Service*, pp. 1–12, 2019.

Parmar, Nayan, Ajay Vatukiya, Mayanksinh Zala, and Shweta Chauhan, "Advanced rural transportation system (ARTS): A review", *International Journal of Research in Engineering, Science and Management (IJRESM)*, vol. 1, no. 4, pp. 59–61, 2018.

RajaKumar, R., R. Pandian, P. Indumathi, and Sheik Mohammed Shoaib, "Detection of traffic congestion and marine border intrusion using vehicular ad-hoc networks", *International Journal of Innovative Technology and Exploring Engineering (IJITEE)*, vol. 8, no. 10, 2019. doi: 10.35940/ijitee.J9008.0881019.

Rajadurai, R. and N. Jayalakshmi, "Vehicular network: Properties, structure, challenges, attacks, solutions for improving scalability and security", *International Journal of Advance Research*, IJOAR.org, vol. 1, no. 3, pp. 41–50, 2013.

Rawal, Tejas and V. Devadas, "Intelligent transportation system in India: A review", *Journal of Development Management and Communication*, vol. 2, no. 3, pp. 299–308, 2015.

SatyanarayanaRaju, K. and K. Selvakumar, "An intelligent transport system in VANET using Proxima analysis", *International Journal of Advanced Computer Science and Applications*, vol. 13, no. 7, pp. 116–122, 2022.

Seth, Ishita, Kalpna Guleria, and Surya Narayan Panda, "Introducing intelligence in vehicular ad hoc networks using machine learning algorithms", *ECS Transactions*, vol. 107, no. 1, 2022. doi: 10.1149/10701.8395ecst.

Seuwou, Patrice, Dilip Patel, and George Ubakanma, “Vehicular ad hoc network applications and security: A study into the economic and the legal implications”, *International Journal of Electronic Security and Digital Forensics*, vol. 6, no. 2, pp. 115–129, 2013.

Sharma, Rashi, “Intelligent transportation system: The big idea in India”, August 2019.

Srivastava, Shubham and Siddharth Jain, “Intelligent transportation system in India”, *Journal of Emerging Technologies and Innovative Research (JETIR)*, vol. 6, no. 6, pp. 25–29, 2019.

Sumra, Prshad Ahmed, Halabi Hasbullah, J. Ab Manan, Iftikhar Ahmad, M. Y. Aalsalem, “Trusted computing in vehicular ad hoc network”, *AWER Procedia Information Technology & Computer Science*, vol. 1, pp. 928–933, 2011.

Tong, Wang, Azhar Hussain, Wang Xi Bo, and Sabita Maharjan, “Artificial intelligence for vehicle-to-everything: A survey”, *Special Section on Emerging Technologies on Vehicle to Everything (V2X)*, 2019. doi: 10.1109/ACCESS.2019.2891073.

Weber, Julia Silva, Miguel Neves, and Tiago Ferreto, “VANET simulators: An updated review”, *Journal of the Brazilian Computer Society*, vol. 27, p. 8, 2021. doi: 10.1186/s13173-021-00113-x.

Yasser, Ahmed, M. Zorkany, Neamat Abdel Kader, and Kun Chen, “VANET routing protocol for V2V implementation: A suitable solution for developing countries”, *Cogent Engineering*, vol. 4, no. 1, 2017. doi: 10.1080/23311916.2017.1362802.

15 Future Technology
Internet of Things (IoT) in Smart Society 5.0

Arun Kumar Singh, Mahesh Kumar Singh, Pushpa Chaoudhary, and Pushpendra Singh

CONTENTS

DOI: 10.1201/9781003438588-15

15.1 INTRODUCTION

The idea of a 'Smart Society' has been around for quite a long period of time; however, the success is witnessed in the last decade, which has been a major move forward for humanity. Smart communities, which operate on the implementation of smart appliances and smart houses, are the benefit in smart societies (Gerlitz, 2015). The expectation of a smooth flow of traffic as a consequence of traffic control is based on detailed and accurate information gathered by these vehicles and evaluated using sophisticated algorithms (e.g., based on artificial intelligence (AI)), and the ubiquity of the Internet of Things (IoT) (Singh et al., 2020a) at the smallest level is the most prominent characteristic of smart communities (Frank et al., 2019).

In the imminent years, towns will become more populated, and metropolises of populations over ten million inhabitants will emerge (Pervez et al., 2018). These metropolises will face problems such as developing safe and cost-effective ecosystems, enhancing residents' standard of health, and coping with non-static definitions that vary enormously.

Information and Communication Technology (ICT) provides a concept that promotes long-term economic growth and a high standard of living while also ensuring prudent resource management, for smart city infrastructure value of ICT must be capable of integrating smart homes into a cohesive smart city perception (Tay et al., 2018).

Components of ICT to develop smart societies are the IoT, Clouds of Things (CoT), and AI. This chapter reflects on the incorporation of a smart city, its integrated smart houses, and it provides a utility system with cohesive ICT-based approach. As a result, data from smart homes would be critical in the implementation of future smart city programs.

Various forms of home-related data will be collected by an ICT-based infrastructure in the near future to provide an intelligent society with customized healthcare services (Mohanty et al., 2016). Adaptive AI, IoT, and CoT are all study fields that collapse under the umbrella of ICT. Both areas are quickly emerging in their own right, but there is currently no research work that integrates them with smart homes and smart cities.

15.2 SMART SOCIETY

Smart society is made up of 'smart' and 'society,' and understanding of 'smart' and the composition and definition of society are therefore important. This chapter attempts to illustrate what is considered 'smartness' and identify what falls outside of civilization. This chapter details the reach and capacities currently found in a smart society, as well as the characteristics found in development and distribution directed toward smart services (Levy and Wong, 2014, Arias-Oliva et al., 2020). This community in such a position that it has 'smart automobile', 'smart house', and 'smart agriculture' ICT (or ICTs) is a key enabler for smart societal progress. Any healthy community would choose to adopt a city management that is 'wise' or 'sophisticated' (Uskov et al., 2019). Figure 15.1 describes the evaluation of smart society in pictorial format.

Features of smartness can be recognized by the following:

a. The introduction of sensor technologies for autonomous service
b. The use of an AI (or machine learning) algorithm
c. Delivering globally ubiquitous facilities every time and everywhere using mobile technologies
d. For the user-centric services, making continuous contact possible between providers and customers

An intelligent society uses the strength and ability of technology to make them more prosperous, concentrates their energy on important tasks and relationships, and eventually improves fitness, wellness, and life satisfaction (Bridgman, 1938).

The way people live, function, and play changes a series of technical advancements. Our life is completely interlinked with physical and virtual devices and it is increasingly affected by their interactions (Menon et al., 2020). All emerging fields that are of growing importance include the IoT, M2M (Machine-to-Machine), hyper networking, connected devices, smart living, and all-around computing. New modes of networking and new kinds of digital connections and anchoring possibilities offered through increased convergence of the connected technology into daily life are commonly the underlying agenda (Salgues, 2018).

In this way, an intelligent society is defined as 'one which successfully exploits the potential of digital technologies, connected devices and the use of digital networks to better the lives of citizens' (Salgues, 2018).

15.2.1 Pillars of Intelligent Society

a. Connectivity includes networks (such as mobile, wired, and others dependent on radio frequencies), a significant enabler, and a contributor to Machine-to-Machine (M2M) software and services, such as traffic management.

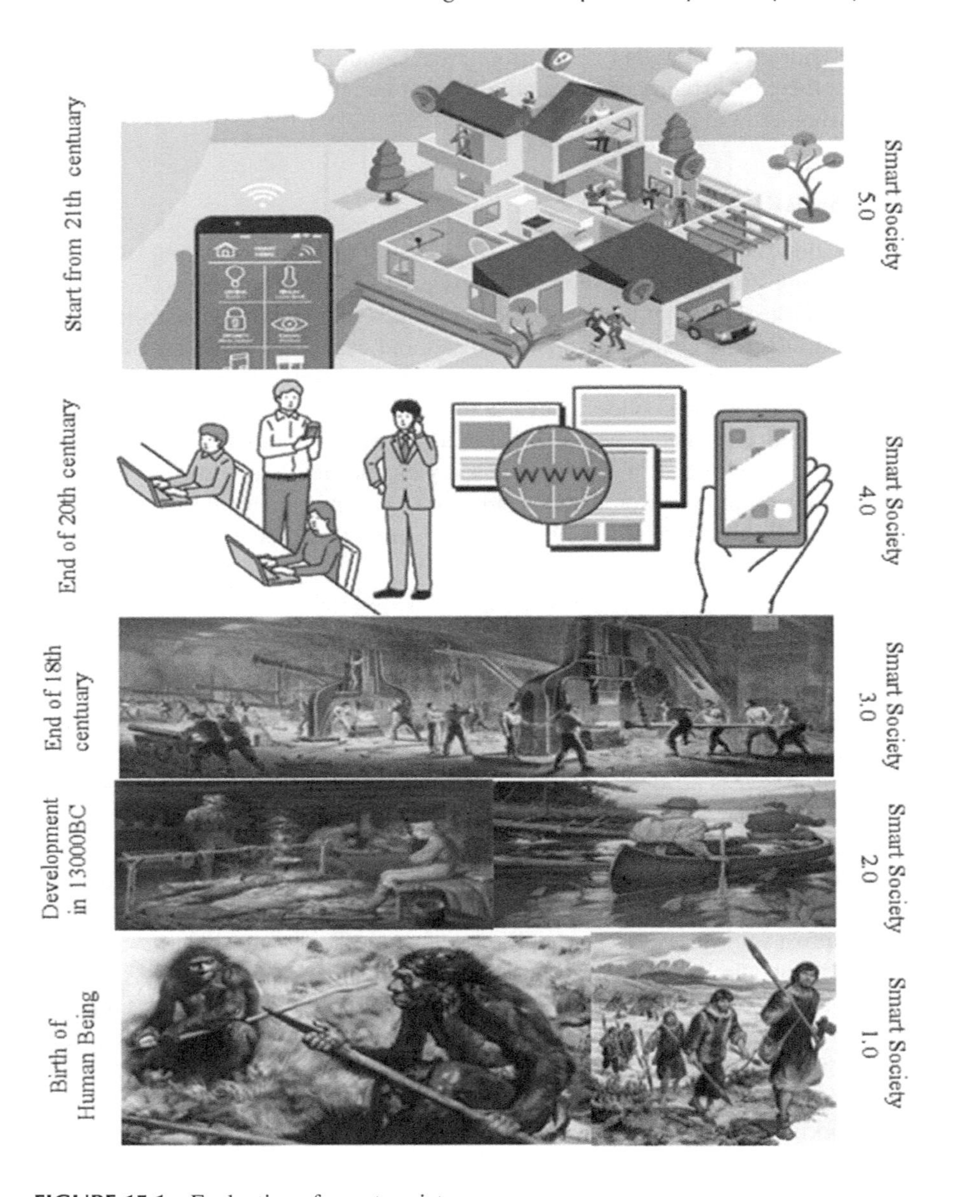

FIGURE 15.1 Evaluation of smart society.

b. Innovations help build smart communities. Cars, traffic signals, equipment, health sensors, and street lights are needed to become intelligent in order to advance sustainability and social and economic progress. This is crucial in developed countries.

c. Operating in tandem, promote emerging and novel programs. These new services are working on improving everything that range from electricity, protection, water, food, and manufacturing to government services.

Human, societal, and technical resources are all of equal importance in this paradigm of urbanization. There are the idea of a smart city and even the idea of a smart society. Society 5.0 is the next autonomous system that can help human training by enabling a modern approach to preventive maintenance (Mallapaty, 2018).

Intelligent algorithms are important in society 5.0 since they help make data processing feasible in the future. However, society 5.0 is focused on industry geostrategic change brought about by digital and web-based technology.

15.2.2 Characteristics of Smart Society

A community is composed of six main components: government, industry, education, culture, public service, and people (Aldabbas et al., 2020).

a. Suppliers that are self-propelled by sensors and intelligent machines. The manifestation of the citizenry demands and interests of humans.
b. Achieving increased inclusion of people in everyday processes including community intelligence. The implementation of student-led learning in the classroom.
c. Develop the unification method (attitude and citizens-oriented lifestyles) for varied personal and cultural preferences and beliefs to fulfill the various requirement such as ethnicity, gender, age, salary, and location.

In the context of this chapter, a smart society may be described as 'a smart society in which governance, administration, manufacturing operations, knowledge generation, community, and civilians exist and work with effective public involvement by not only from the need for advanced ICT as well as the facilitation of change in legislation' (Salgues, 2018).

15.2.3 Smart Society and Sustainable Development

The smart society described above leads to the achievement of Sustainable Development Goals (SDGs). This society-goals based ICT strategy can be realized by including policy integration, educational interventions, and financial accessibility (Trindade et al., 2017).

Smart societies have attention as a prominent response to the sustainable development which has arisen mostly as a result of urbanization. They are widely recognized for a promising future. Considering their obvious success, the research indicates a lack of contextual consistency around the word 'smart society' as a result of the numerous meanings currently in use. Definitions of smart societies are ranked as per the aspects of sustainability people regard, such as economic, social, and environmental as well as the importance they place on the principles of sustainable development, given in Figure 15.2.

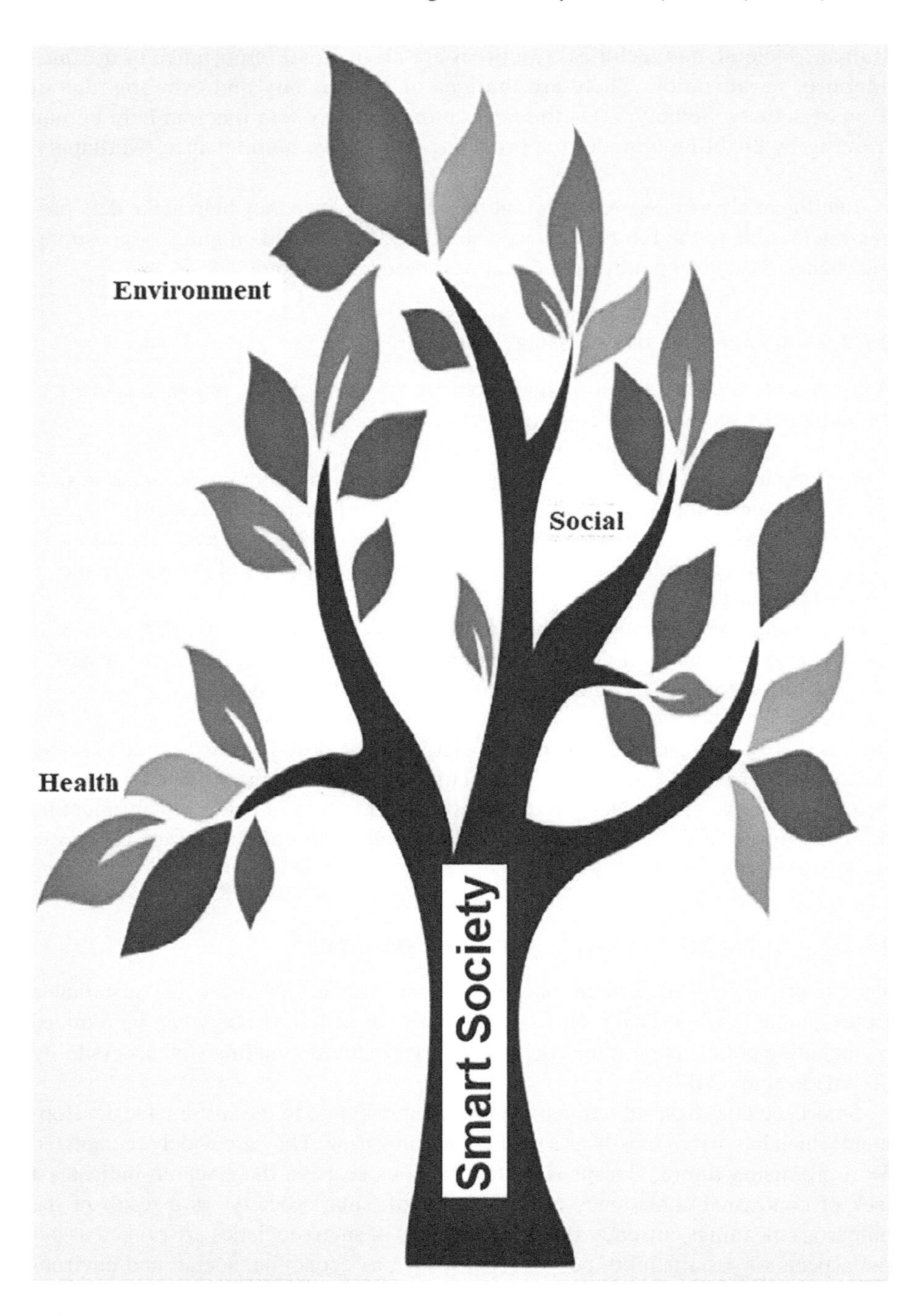

FIGURE 15.2 Characteristics of smart society.

15.3 INTERNET OF THINGS (IoT)

Most devices, items, objects (of both types), or various bits of bits that can be interconnected by a network are being referred to as the IoT.

The IoT may be provided with a wide variety of devices (e.g., smartphone, Wi-Fi, or Wired) and has to be implemented using an approach that is impartial in technology. In my opinion, industry players should be permitted to use the most suitable IoT technologies. For example, shorter and longer lengths, indoors and outdoors, as well as mobile and stationary application scenarios can involve a separate spectrum deployment strategy. On the other hand, the availability of fifth-generation networks comes with many advantages. Conversely, satellite offers convergent telecommunication services (voice, info, and audiovisuals) (Madakam et al., 2015; Li et al., 2015; Patel and Patel, 2016). Figure 15.3 shows several applications of IoT for a smart society.

Technological advancement of IoT has enabled all industries to innovate and develop. Wireless sensor network (WSN) architecture is used in a smart agriculture

FIGURE 15.3 Uses of IoT in smart society.

solution which has three layers: a sensing layer, a data transmission layer, and an application layer. Agriculture has seen major improvements in the use of IoT applications (Kumar et al., 2018).

15.3.1 Interaction between Rural and Urban Regions through ICT

In our rapidly changing society, Information & Communication Technologies (ICTs) are playing an enormous role. The number of internet users, the percentage of wired devices among inhabitants, and the use of cell phones shape the foundation for ICT developments and show a significant disparity between the developed and developing countries.

An observation is made of whether a country's economic growth is driven by new technology. In all its iterations, digital differential development is a minor part of the inequality (Dlodlo et al., 2016).

Several people in society have positive perceptions regarding their capacity to adapt through opportunities and issues in the modern generation as a result of the opportunity presented by ICTs. ICTs, in particular, have raised living standards in the society.

ICTs provided opportunities to people and customers in terms of improving their standard of living and the possible benefits for achieving maximum facilities and culture; restricted and periphery communities in terms of improving their opportunities in life in a dynamic environment and removing any linguistic distance barriers; and public administrative institutions in terms of their ability to more efficiently, transparently, and responsively to fulfill the needs of residents, corporations, and other entities; large and small and medium-sized businesses for more efficient monitoring and evaluation processes; privileged access to employee training and other resources; direct online contact with consumers and vendors; and opportunities for collaboration (European Commission, 2004).

15.3.2 Digital Gap between Rural and Urban Areas

ICT is a wonderful for helping us to change our lives, by improving our everyday routines, and informing, educating, and keeping us well. It allows us to extract the maximum benefit from the natural and cultural capitals. To make this possible, we must have the required infrastructures (Dlodlo et al., 2016).

A significant inequity exists between rural and urban areas, with respect to the adequacy of and accessibility of infrastructure. The void appears in the following.

15.4 LITERATURE REVIEW: PAST CHALLENGES IN THE SMART SOCIETY IN DEVELOPING COUNTRIES

In current scientific knowledge research, smart cities are heavily debated. The following are the main characteristics of smart cities: looking to find workable policies to control investors, unleashing economic and social development, retaining benefits for the city's population, and allowing development in data analysis investment, which are all essential challenges for policymakers.

a. On the one side, people are concerned with personal data's safety. The addition of IoT-enabled cameras and sensors to intelligent street lights will give the impression that they are being monitored continuously by local councils.
b. Aside from these privacy issues, the most challenging obstacles for smart cities are attracting finance to launch a project and having adequate resources to support the program over time. One of the most common investment forms used to address these financial troubles is governmental partnerships.
c. Smart projects, but at the other side, rely heavily on data generated by stakeholder, which can contribute to a competitive revenue model that manages investment costs and also provides revenue to support other businesses.

There are lots of challenges in the developing country for smart society, and by using the latest technology, security, interoperability, and funding are such issues that can be overcome by making a policy and implementing it. In this context, few of the challenges are as follows.

15.4.1 Policies and Regulations of Information & Communication Technology

Next-generation networks are essential requirements for creativity in smart societies through ICT sectors, powering the implementation of M-services and new innovative technologies. Therefore, governments can support policies and frameworks such as fostering advanced network deployment and developing robust technologies for the growth of IoT and other smart services, such as those required for smart society in the future. Interoperability is made possible through cooperation among all public agencies, that is, national, federal, state, and local levels to provide the domestic and cross-border data flows. Policymakers and regulators must be aware of the role that spectrum and assignment play in the smart society.

Additionally, we call on business to ensure that its facilities are accessible, even at an affordable prices for servers, computers, and applications, to realize SDGs goals.

15.4.2 Financial Ambitions

Different smart society programs do not have an allocated budget for their operations. Instead, planners ought to find all funds that were earmarked for application areas. This helps the implementing company to deal with the problems that society has already encountered. One of the most important engines of the smart city is the government's financial capability. Emerging countries are more likely to be revenue constrained, necessitating a variety of traditional and creative funding instruments to boost infrastructure investments in smart developing countries of the world. Land-based funding instruments (such as property tax, vacant land tax, growth principle of responsibility, and mutual benefit charges), congestion charging mechanisms (such as welfare payments, passenger vehicles taxes, and expressway tolls), and financing have been suggested as the perfect combination for the government in financing smart societies in India (Mishra, 2019).

15.4.3 Standardization

One of the most critical ways to prepare for intelligent culture is to standardize our activities and behaviors. The ITU-T Study Group 20 is developing IoT technologies to meet functional standards, with an emphasis on smart communities (SC&C). The ISO standards about sustainable societies were created by the committee "Sustainable cities and communities — Indicators for city services and quality of life, ISO 37120:2018," which focuses on the quality of life and the efficiency of public services. It was being published in 2014, and in July 2018, an updated version was published with the inclusion of 28 new measures, the elimination of 24 old ones, and minor changes to ten indicators. ISO 37122, which offers metrics for smart cities, was first made public in June 2018 as an internationally accepted draft. The third accurate measure standard for robust cities, ISO 37123, is currently being developed. These principles were created with sustainability as a governing principle in mind, and can thus be used in tandem to provide an economic sustainable community. The World Council on City Data (WCCD) becomes interested in ISO indicator growth and certifies cities based on the number of ISO 37120 indicators measured and published by WCCD (WCCD, 2018).

15.4.4 Human Capital

Human capital is a key element in the development of a smart society. While the City of Portland had a high labor market in human resources, there were basic skills required to operate IoT that are too advanced to be easily accessible. The approach of Portland's planners to this issue included forming a relationship with the Portland State University, Portland, to manage the theoretical dimensions of their IoT applications. Another advantage of this collaboration was the ability it provided for planners to harness the high R&D capability of universities.

The shortage of human resources to allow IoT to be deployed is also a concern in developed countries. Insufficient abilities imply that the advantages of such IoT resources, such as data processing and research, cannot be maximized, thereby minimizing the benefits to society and preventing the creation of beneficial externalities that can be realized in a world of big data.

15.4.5 Sustainable Development via ICT

Sustainable Development is an additional obstacle facing several developed nations to achieve a smart society via ICT programs. The careless reflection of sustainability in the designing & planning level could contribute to the breakdown of the task or project. Consequently, decision-makers must consider environmental issues in the implementation of the ICT program. In order to ensure the sustainability of ICT programs, attention should be given to the growth of local human resources, the appropriateness of ICT technologies implemented, the effect of the financial strain on people, the varied interests of social communities, and metrics for calculating these four sustainability apparatuses. These metrics should be used as benchmarks to ensure the feasibility of ICT programs.

15.4.6 The Role of Artificial Intelligence in a Smart Society

The main function of machine learning in the holistic deployment of the smart society is inspired by the basic belief that human beings are unable to interpret all of the data in a conventional manner. The term 'machine learning' often points to how much the challenges with many of these techniques exist (Rochman et al., 2020; Holroyd, 2020). The computer has always been learning, which means it always adopts logical reasoning, which further provides us with such an exciting challenge of finding out how AI and techniques affect everyone and the smart society framework.

15.4.6.1 How AI-Based Smart Home Systems Work

The main objective of smart society is to minimize the need for manual controls; it will substantially reduce the anxiousness for interconnected householders through using AI and machine learning, which take a closer look into some of the AI technologies in smart homes. Figure 15.4 describes the role of AI in smart society.

15.4.6.2 Smart Devices with a Location Function

Smartphone applications can monitor smart societies' systems remotely, and settings can be controlled automatically. When we enter our home, and within a certain distance, for example, the machine will set up your room with your desired lighting,

FIGURE 15.4 Smart home system.

curtains, and air-conditioning temperature. Convenience and relaxation are improved by these functions through innovative technologies.

15.4.6.3 Voice-Enabled Devices

An intelligent voice assistant device like, Alexa Or Siri, can provide speech recognition through automation systems, that may be used to monitor lighting, formulate voice or video calls, examine the place orders, reports, and participate in music, among other things. These functions might be used to alert friends and family members in the case of an emergency.

15.4.6.4 Intelligent Security System

Security services shall provide gesture tracking, biometric access control, and speech and face recognition. The IoT can be thought of as a network of networks that connect and integrate heterogeneous forms of electronic communication (Singh et al., 2011), such as smartphone networks, social platforms, and industrial networks. To deal with the newest issues in this form of complicated society, innovative successful techniques are designed. Understanding malware dissemination features in IoT-based services and infrastructure, modeling knowledge spread trends in wireless communication, and designing successful preventive strategies, for example, are all important (Xu et al., 2016).

15.4.6.5 Face Detection

Smart home solutions provide Wi-Fi-facilitated videos, which provide for surveillance devices. These can be accessed remotely by some features on the mobile from any location. ML-based smart devices integrate facial identification technologies into linked video cameras. AI-inbuilt neural networks will recognize the eyebrows, cheekbones, nose, and chin of people from the appropriate camera, correlate them with the faces of the occupants of the house, and deliver reminders of unusual behavior to the user's smartphone.

15.4.6.6 Detecting Motion

Motion sensors can track random movements by video monitoring, such as detecting intruders and alerting home owners. End of the day, the whole smart city must be continuously tracked and studied, but data processing and uncovering illegal activities are exceedingly difficult. The infrastructure for smart surveillance systems is provided by traditional closed-circuit television (CCTV) systems. They are, however, attached to a recording device and lack intelligent computing capabilities. Furthermore, human operators can overlook a scene and cause an error. Through smart surveillance, this is possible to track people's movements in order to spot any violent acts and even identify perpetrators. Smart surveillance systems may sound an alarm if a potentially dangerous incident occurs. It can be used as a guide for potential pedestrian facility design or modification by tracking people's attitudes and identifying pedestrians' crossing trends. A population system model for traffic population surveillance and emergency management system is required at a concert or in a public place such as airports where there are large crowds. Infrared cameras are being used to detect and monitor people at night because they operate based on temperature (Wang et al., 2012).

Another feature of this device is the ability to identify the kinds of objects people can carry in order to detect any unauthorized or prohibited items. To do this, video sequence structures are used, which operate by taking into account any inconsistencies in person's silhouettes. This is validated by considering a prototype of an average person moving in quite the same way to see if there is any protrusion or deviation that might be used to hold objects. Other camera monitoring systems can detect unusual circumstances such as pedestrians crossing the street without using crosswalks or vehicles traveling in the wrong direction by using motion detection method to identify and format video data.

15.4.6.7 Regulation of Biometric Access

With Amazon's and Google's smart home solutions, secure door lock which integrate biometrics, such as Smart Lock, can indeed be conveniently installed. They usually use biometric scanners to take images of the user's fingertips and use them for identification, unless they use two-factor authentication for passwords and fingerprints.

15.4.6.8 Recognition of Voice

Natural Language Processing (NLP) is used in AI-based speech recognition, and this technology segregates the users' speech from surroundings noise, which changes it to an electronic medium, convey it to the cloud for NLP investigation, and prompt a suitable response. It is possible to train a code that recognizes a series of different words or messages using machine learning.

The expression 'speech identification' or 'speaker recognition' consigns to a definition of the speaker rather than the voice's meaning or comprehension. Which may be skilled on real-life speech or used to authenticate or double-check the speaker's identity as part of a security protocol?

Aside from tracking and notifying warning signals, smart home automation products often optimize energy efficiency. Smart systems have patterns of energy expenditure to facilitate users minimize the use of machines and equipment while they are not in use, thanks to the integration of AI and machine learning.

Climate change and global energy challenges have made fuel savings and sustainability hot topics. Multiple home automation technologies, such as smart thermostats, smart plugs, and automatic lighting controls, can be monitored by AI-powered connected devices to reduce carbon footprint.

15.5 SMART SOCIETY CHALLENGES

You need to merge the smart systems of various vendors. This raises questions of interoperability. However, several vendors are working to address the gaps and interoperability problems. For example, a radio frequency-based communication associated with a protocol named Z wave helps manufacturers from home to combine their hardware with various intelligent home solutions.

Nonetheless, there are a slew of goods to facilitate that are incompatible through one another in any way. That is, many intelligent devices, like Google Assistant, pair a coffee maker with WeMo, and Wi-Fi enabled gadget, can't just make a cup of coffee for you.

The AI-powered automation system for home gives energy savings, home security, convenience, and peace of mind. Yet, many customers are reluctant to invest in emerging technologies because of the higher initial investment costs.

Another critical consideration is safety. To deter hacking, smart systems need a strong firewall as well as an extremely secure technique to safeguard sensitive data and ensure privacy. However, bear in mind that no protection device is absolutely foolproof. Facial recognition, for example, will operate now but not in 20 years when your face will no longer be recognizable. Furthermore, this isn't entirely true.

15.5.1 Challenges Resolved by AI and IoT

The challenges that can be solved by AI and IoT are explained in the following.

15.5.1.1 The AI in a Smart Town

Water security, energy shortage, traffic management, health care, and industry are only a few of the issues that AI and machine learning collective through IoT have the ability to solve. The main goal of a smart city is to build a 'smart society' that can function efficiently and effectively by using city infrastructures with AI. It also features advanced technology and information analysis to optimize major characteristics and boost economic development all through improving peoples' standard of living.

To capture and analyze data, Smart Cities employ AI, cloud-based services, and IoT technologies like embedded devices, lights, and meters. The data are then used by the cities to improve the infrastructure, utility services, and facilities, among other things.

With the aid of IoT technologies, any city can be turned into some kind of smart city. Many sectors, including production, telecommunications, infrastructure, healthcare, and electronic products, have realistic examples.

The number of people who have access to the internet is increasing all over the world. Furthermore, connection costs are declining, and new devices have built-in Wi-Fi technologies with quality and diversity. Phones are getting smarter, with reduced form factors, lesser energy usage, and internet access. IoT is ripe for creativity under these conditions.

15.5.1.2 Smart Management of Water

Water, fuel, housing, and illumination are some of the most critical necessities when we transition from rural to urban areas. Water and food are two of the life's most important necessities. Technology assists us in obtaining basic necessities and lowering the cost of life in urban areas. The city's use of water and electricity can be simplified with AI. Smart grid technology can be used to properly control the use of electricity and water sensors, allowing leaks to be detected and unnecessary water consumption to be avoided.

15.5.1.3 Smart Lighting System

Everybody likes their neighborhoods to be well-lit. Although street lamps are required, they consume a considerable amount of electricity. Smart lighting can help save energy by allowing only certain places to be lit when they are required.

Depending on the location of pedestrians, cyclists, or cars, the brightness of the lamps may be changed. A large mesh network could be used to light up nearby lights to provide a secure sphere of lights from an approximately humanoid participant.

15.5.1.4 Smart Traffic Control

Automobiles abound in today's towns. Smart traffic systems based on AI can be used to show that users can travel from one location to another securely and effectively. At global level, several overcrowded cities have adopted smart traffic solutions to control traffic flow. Sensors are installed on the roads and also on IPTV cameras to deliver real-time traffic flow information to the framework for unified traffic control. Camera-fed information are investigated and users are informed of any obstruction on the roads or any malfunctioning of traffic signals.

15.5.1.5 Smart Parking Space

Not only managing traffic congestion problems, but finding a parking spot, especially during holidays, can also be difficult. To assess whether parking spaces are open or filled, smart parking technologies can be used. Smart parking uses road surface sensors mounted in the ground on the parking area to create a real-time parking map. This will cut down the time it takes to find a vacant parking spot, which helps to minimize traffic and energy.

15.5.1.6 Smart Management of Waste

It is impossible to eliminate waste, but it can be better handled. The city's waste collection operation is vital. Smart waste management approaches are needed for the ever urban population. Not only does an AI technique help for smart waste management, but it also allows for long-term waste management. Smart sensors mounted in waste bins, for example, transmit alerts to officials when the bins are about to be loaded, enabling waste disposal trucks to be dispatched. Paper, plastic, bottles, and food waste can all be held in separate containers. These waste products can be recycled and reused in other ways.

15.5.1.7 Smart Police Force

Everyone always likes to live in a place with negligible violence. Evidence-based and data-driven techniques are used by smart law enforcement officials to efficiently and economically track law enforcement. A system of sensors, for example, may be placed in every corner to help detect people who commit crimes, smoke in restricted areas, or appear suspicious. Cameras may be used to track, among other things, crowd density, vehicle movement, and hygiene in a specific location.

15.5.1.8 Smart Governance

Smart governance refers to the effective use of information and communication technologies in government- and public-sector organizations to eliminate corruption. The key objective is to enhance decision-making by fostering recovered association among various stakeholders, together with the government, and residents. E-governance improves decision-making and citizen enforcement by using data, facts, and other tools.

15.5.1.9 Smart Society Reflect to Smart Nation

In 2014, the Singapore government introduced the Smart Nation concept. Its key goal is to create technology-enabled solutions by using information and communication systems, networks, and big data technology. In the creation of a modern country, AI is important.

Singapore has so far recognized five nationalized AI interventions: health care, smart cities, education, security, and transportation and logistics. Such types of businesses have all the potential to make significant changes by developing innovative solutions to almost all of the challenges facing the world today.

In several ways, digital technology and AI (Singh et al., 2021) will significantly improve the modern world, along with emergence of smart communities and smart cities. In today's world, a smart society is described as a place where people in a specific geographic area have access to all of the benefits and solutions provided by technology, such as easy transportation, safety, security, modern essentials, and social well-being. Smart communities and smart cities, in theory, make up a smart country.

15.6 THE CASE STUDY OF SOCIETY 5.0 IN THE REAL WORLD

'Data analytics,' 'AI,' and 'IoT' are only examples of the innovations we use every day in our modern world. In the age of information technology, we grow both our personal and technical concepts as well as business enterprises that need constant engagement with cloud-based data and communication tools. We can hardly fathom how much our lives have changed over the past decade due to the recent appearance of the smartphone or other innovations. The contribution of these technologies at the social level is described below in an elaborate way.

15.6.1 Society 5.0 Enables a Commitment to Sustainability

The exponential development of Information and Communication Technologies (ICTs) is carrying about significant changes in culture and business. Manufacturing processes and policies have consistently shifted toward higher competitiveness and production, even at the expense of social and environmental aspects, as shown by successive industrial revolutions. Technology, on the other hand, is constantly being created in order to change and promote human life.

Industry 5.0 and Industry 4.0 have effect on any country's overall economic in short-term and long-term demand growth, as well as in fiscal, social, and environmental terms. Social responsibility through sustainability is becoming profoundly relevant in corporate and organizational discourses. As a result, a number of organizations around the world are focusing on their influence, in addition to the internal governance processes. Organizations place a premium on activities relevant to the philosophy and social responsibility.

Enterprises must also incorporate sustainability values into their organizational structures and strategic agendas, since sustainability impacts an organization's environmental and long-term viability. As a result, there is an increasing importance of promoting ICT as an innovative and productive way of addressing societal problems.

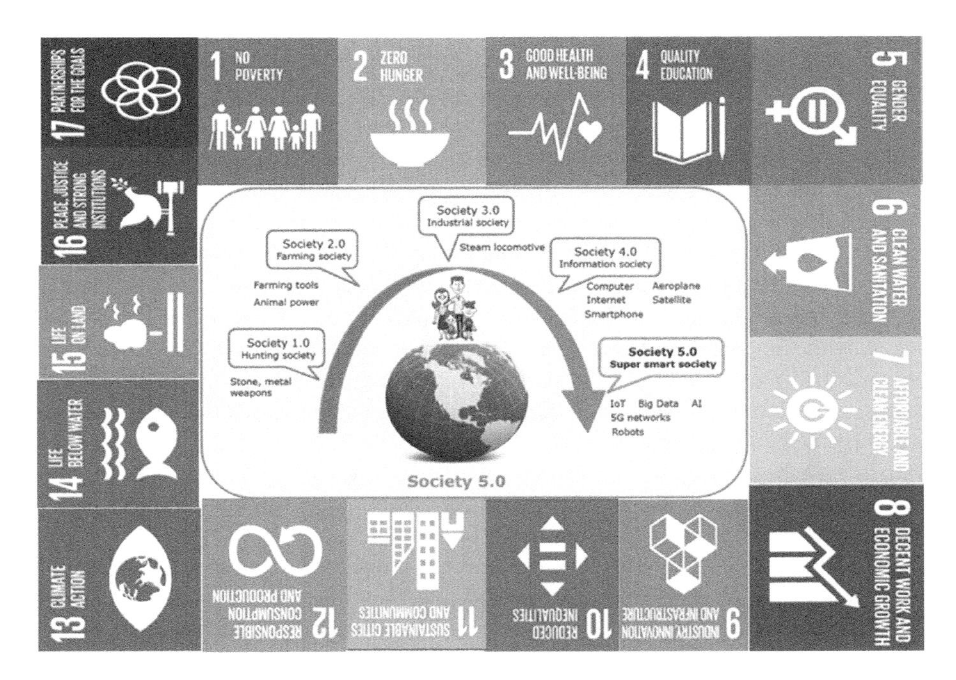

FIGURE 15.5 Sustainable development in Society 5.0 through SDGs.

The aim of this essay is to better understand the meaning of Society 5.0 in industry and its capacity for proactive behavior inside these SDGs shown in Figure 15.5.

Due to the numerous issues society faces relating to limited resources, exponential urbanization, industrial development, and global capitalism, sustainable development has now become progressively important to global fiscal policies. Throughout this context, such issues lead to the rise of new paradigms centered on industrial organization with an inherent focus on resource performance and better social responsibility, along with Industry 4.0 or 5.0. This is one of the core elements to promote Sustainable Development Goals around the globe and it is one of the core features to achieve Sustainability Globally.

The Japanese government coined the word 'Society 5.0' in January 2016 and described it as a development plan for the nation. Throughout human experience, we will identify distinct levels in society: 'Society 1.0,' 'Society 2.0,' and 'Society 3.0.' The terms 'Society 4.0' and 'Society 5.0' are interchangeable. Both of these technological revolutions culminated in economic prosperity, higher production, and improvements in a group of countries that reaped the majority of their benefits, including high-quality goods and services.

The fundamental tenet of Society 5.0 is equilibrium or the alignment of economic growth and social issue resolution. This relates to the SDGs' 17 development goals, which seem to be currently the community's global targets for resolving global problems. As a result, there are anticipated possible services in emerging markets that the city will need, leveraging the capacity for technical growth.

15.6.2 Case Study: Hitachi-UTokyo AI-Based Modern Society

Data mining, AI, and the IoT are only a few of the R&D items that have been ubiquitous in our everyday lives. Our academic and interpersonal lives are filled of online information and data technologies, which enables us to create and exchange innovations, which in turn generates new businesses. Consider how our lifestyles have changed dramatically over the last decade, with the advent of the internet, modern modes of shopping, and new ways of functioning.

Japan has developed a plan for planning such a society focused on a segmented approach to each business market. Their principles are based on specific steps influenced by the Sustainable Development Targets (United Nations, 2018). Each of these incremental steps is anchored in reality by financial assistance, established milestones, and human preparedness. This strategic strategy ensures that their dreams are implemented successfully (Shiroishi et al., 2018). The Sustainable Development Targets have been formally established and can be listed on the United Nations' official website. Society 5.0 is a prospect populated by humans, and it is self-evident that the amount of variables controlled cannot be limited.

On January 22, 2016, the Japanese government unveiled the country's fifth scientific and technological reasonable model. The proposal introduces the concept of 'Society 5.0,' a hypothetical society led by science and technical advancements. The following describes the rationale behind this concept: Through an effort that unites physical space (the contemporary age) and cyberspace by maximising the use of ICT, we propose an optimal form for our future community: a 'super-smart society' that generates prosperity for its people. The sequence of measures aimed at achieving this perfect world is now being deepened and aggressively marketed under the banner of 'Society 5.0.' According to an annotation, the word Society 5.0 was coined to refer to the modern society developed as a result of science and technical innovation, following predator culture, agricultural society, industrialized civilization, and knowledge society.

The government launched the 'Comprehensive Science, Technology, and Innovation Strategy for 2016' in 2016, and it published the 2017 version of its own coherent, wherein it expanded on Society 5.0 as follows: 'Society 5.0, the vision of the future society whereby the fifth basic Plan recommends they strive, would be a human-centered society capable of achieving a high degree of integration between cyberspace and physical space.' In other terms, Society 5.0 serves as a vehicle for the government to express its vision for the development of society to business and the wider community. This model was developed as a result of multiple discussions between experts from a variety of fields. Additionally, it was founded on historical studies into technology and social progress. However, the government research quoted above offers only a cursory overview of this academic discourse. Without an awareness of the underlying concepts, it is impossible to obtain a complete image of Society 5.0. For instance, to know about cyberspace, concept of physical space is needed for an AI-enabled society.

What does it imply for these two spaces to be merged? What does it imply to strike a compromise between economic growth and social justice? A human-centered society—doesn't it seem self-evident? Readers are justified in posing those concerns. To find the solutions, we must first comprehend the fundamental thought and narratives of Society 5.0. Thus, this book serves as a guide on Society 5.0 by delving into the words' hidden interpretations and the context in which they originated.

15.7 CONCLUSION

The smart sensor is in the center of the smart community. Distributed in tens of thousands through smart city test neighborhoods or even whole boroughs, IoT sensors track anything from traffic flows and footprints to utility use and pollution. The actions currently taken by states, businesses, and telecommunication providers can form human existence over the next century. If rural neighborhoods, as well as underserved regions of urban areas, can be linked using 5G and IoT, and if smart city projects can be a vehicle for leveling the playing field rather than increasing economic disparity, then today's businesses and policymakers can be seen to be moving for a future of smart cities and smart societies. While more societies move to smart city capabilities, the susceptibility of smart city networks to data theft, unauthorized data access, device breaches, virus-based attacks, and other risks to organizational integrity is likely to grow. As advancements such as IoT-based networks and greater use of AI become fundamental to system architectures, the task for smart societies' designers and developers is to keep humans in the loop in order to engender the requisite levels of confidence in protection and privacy from users.

15.8 LIMITATION

This research becomes constrained by its emphasis on security and privacy (Singh et al., 2020a), which could leave out such a range of human-centered considerations that may affect the implementation of smart societies in the future. More analysis is needed to help explain the 'lived-in' reality of smart cities from the viewpoint of citizens, in order to measure the multiple interactions and day-to-day organizational challenges and engender greater public confidence. Furthermore, only technical solutions to enhance stability, privacy, and organizational risks were included in this report. However, legal and administrative aspects of city infrastructure are also present. The use of blockchain in smart cities needs further study. It is critical that new government and sector legislation be implemented with the assistance of experts in order to prevent conflicts between the parties involved in the transaction. Furthermore, the cost of installing and running a blockchain-based infrastructure in smart cities must be considered. As a result, it's critical to conduct a tailored pilot to assess the possible cost of a blockchain-based smart city infrastructure, as well as to examine comprehensive models of societal value co-creation that take into account a rigorous cost–benefit analysis to help governments make informed decisions.

REFERENCES

Aldabbas, M., Xie, X., Teufel, B. and Teufel, S., 2020, October. Future security challenges for smart societies: Overview from technical and societal perspectives. In *2020 International Conference on Smart Grid and Clean Energy Technologies (ICSGCE)*, Kuching, Malaysia, pp. 103–111, IEEE.

Arias-Oliva, M., Pelegrín-Borondo, J., Murata, K. and Palma, A.M.L., 2020. Societal challenges in the smart society.

Bridgman, P.W., 1938. The intelligent individual and society.

Dlodlo, N., Gcaba, O. and Smith, A., 2016, May. Internet of things technologies in smart cities. In *2016 IST-Africa Week Conference,* Windhoek, Namibia, pp. 1–7, IEEE.

European Commission, 2004, Growth, competitiveness and employment, White Paper, Brussels, Luxembourg, ISBN 92-826-8547-ECSC-EC-EAEC.

Frank, A., Al Aamri, Y.S.K. and Zayegh, A., 2019, January. IoT based smart traffic density control using image processing. In *2019 4th MEC International Conference on Big Data and Smart City (ICBDSC)*, Muscat, Oman, pp. 1–4, IEEE.

Gerlitz, L., 2015. Design for product and service innovation in Industry 4.0 and emerging smart society. *Journal of Security & Sustainability Issues*, *5*(2), 181–198.

Holroyd, C., 2020. Technological innovation and building a 'super smart' society: Japan's vision of society 5.0. *Journal of Asian Public Policy*, *15*(6), 1–14.

Kumar, A., Zhao, M., Wong, K.J., Guan, Y.L. and Chong, P.H.J., 2018. A comprehensive study of IoT and WSN mac protocols: Research issues, challenges and opportunities. *IEEE Access*, *6*, 76228–76262.

Levy, C. and Wong, D., 2014. Towards a smart society. The Big Innovation Centre (The Work Foundation and Lancaster University), London. Available from internet: https://www.researchgate.net/publication/263083596_Towards_a_smart_society.

Li, S., Da Xu, L. and Zhao, S., 2015. The internet of things: A survey. *Information Systems Frontiers*, *17*(2), 243–259.

Madakam, S., Lake, V., Lake, V. and Lake, V., 2015. Internet of Things (IoT): A literature review. *Journal of Computer and Communications*, *3*(05), 164.

Mallapaty, S., 2018. Pillars of a smart society. *Nature*, *555*(7697), S62–S63.

Menon, V.G., Jacob, S., Joseph, S., Sehdev, P., Khosravi, M.R. and Al-Turjman, F., 2020. An IoT-enabled intelligent automobile system for smart cities. *Internet of Things*, *18*, 100213.

Mishra, A.K., 2019. Henry George and Mohring–Harwitz Theorems: Lessons for financing smart cities in developing countries. *Environment and Urbanization ASIA*, *10*, 13–30.

Mohanty, S.P., Choppali, U. and Kougianos, E., 2016. Everything you wanted to know about smart cities: The internet of things is the backbone. *IEEE Consumer Electronics Magazine*, *5*(3), 60–70.

Patel, K.K. and Patel, S.M., 2016. Internet of things-IOT: Definition, characteristics, architecture, enabling technologies, application & future challenges. *International Journal of Engineering Science and Computing*, *6*(5), 6122–6131.

Pervez, S., Abosaq, N., Alandjani, G. and Akram, A., 2018, January. Internet of Things (IoT) as beginning for jail-less community in smart society. In *IEEE International Conference on Electrical, Electronics, Computers, Communication, Mechanical and Computing (EECCMC)*, Vellore, Tamil Nadu, pp. 28–29.

Rochman, G.P., Chofyan, I. and Sakti, F., 2020, February. Understanding the smart society in rural development. In *IOP Conference Series: Earth and Environmental Science*, Surakarta, Indonesia, vol. 447, no. 1, p. 012016, IOP Publishing.

Salgues, B., 2018. *Society 5.0: Industry of the Future, Technologies, Methods and Tools*. John Wiley & Sons: Hoboken, NJ.

Shiroishi, Y., Uchiyama, K. and Suzuki, N., 2018. Society 5.0: For human security and well-being. *Computer*, *51*(7), 91–95.

Singh, A.K., Firoz, N., Tripathi, A., Singh, K.K., Choudhary, P. and Vashist, P.C., 2020a. Internet of things: From hype to reality. In Balas, V.E., Solanki, V.K. and R. Kumar (Eds.), *An Industrial IoT Approach for Pharmaceutical Industry Growth*, pp. 191–230. Academic Press: Cambridge, MA.

Singh, A.K., Tewari, P., Samaddar, S.G. and Misra, A.K., 2011, February. Communication based vulnerabilities and script based solvabilities. In *Proceedings of the 2011 International Conference on Communication, Computing & Security*, Rourkela, Odisha, pp. 477–482.

Singh, A.K., Tripathi, A., Choudhary, P. and Vashist, P.C., 2020b, January. Research and challenges of security & privacy in Internet of Things (IoT). In *2020 International Conference on Computation, Automation and Knowledge Management (ICCAKM)*, Dubai, United Arab Emirates, pp. 487–492, IEEE.

Singh, A.K., Tripathi, A., Singh, K.K., Choudhary, P. and Vashist, P.C., 2021. Artificial itelligence in medicine. In Singh, K.K., Elhoseny, M., Singh, A. and Elngar, A.A. (Eds.), *Machine Learning and the Internet of Medical Things in Healthcare*, pp. 67–87. Academic Press: Cambridge, MA.

Tay, K.C., Supangkat, S.H., Cornelius, G. and Arman, A.A., 2018, October. The SMART initiative and the Garuda smart city framework for the development of smart cities. In *2018 International Conference on ICT for Smart Society (ICISS)*, Semarang, Indonesia, pp. 1–10, IEEE.

Trindade, E.P., Hinnig, M.P.F., Moreira da Costa, E., Marques, J.S., Bastos, R.C. and Yigitcanlar, T., 2017. Sustainable development of smart cities: A systematic review of the literature. *Journal of Open Innovation: Technology, Market, and Complexity*, *3*(3), 11.

United Nations, 2018. The Sustainable Development Goals Report. https://www.un.org/en/desa/sustainable-development-goals-report-2018

Uskov, V.L., Bakken, J.P., Gayke, K., Jose, D., Uskova, M.F. and Devaguptapu, S.S., 2019. Smart university: A validation of "smartness features—main components" matrix by real-world examples and best practices from universities worldwide. In Uskov, V.L., Howlett, R.J. and L.C. Jain (Eds.), *Smart Education and e-Learning*, pp. 3–17. Springer: Singapore.

Wang, J., Chen, D., Chen, H., and Yang, J., 2012. On pedestrian detection and tracking in infrared videos. *Pattern Recognition Letters*, *33*, 775–785.

World Council on City Data (WCCD), 2018. http://www.dataforcities.org/.

Xu, K., Qu, Y. and Yang, K. 2016. A tutorial on the Internet of Things: From a heterogeneous network integration perspective. *IEEE Network*, *30*(2), 102–108.

16 IoT, Cloud Computing, and Sensing Technology for Smart Cities

Kazi Nahian Haider Amlan, Mohammad Shamsu Uddin, Tazwar Mahmud, and Nahiyan Bin Riyan

CONTENTS

DOI: 10.1201/9781003438588-16

16.1 INTRODUCTION

A smart city can be defined as a high-tech advanced metropolitan region where a variety of electrical devices and sensors are interconnected and specific data are gathered together. Internet of Things (IoT), cloud computing, and sensor technology all play significant roles in this. The phrase "Internet of Things" has a wide definition. It describes a group of physical items or items with sensors, software, and other technologies integrated into them that can connect to and exchange data with other hardware and software over the internet. The "Internet of Things" is essentially any device that is connected to the internet [1]. The next stage in the development of internet-based computing is cloud computing. Instead of using the hard disk in your computer, cloud computing involves storage and accessing data and software through the internet. The interconnection between the devices in smart cities and the sharing of information between them are made more effective by cloud computing, which makes servers, storage spaces, databases, and a variety of application services freely accessible through the internet. Besides, the use of information technology as a service is made possible by cloud computing [2].

Smart cities use various types of sensors to determine the physical characteristics of any objects or situations. The major types of sensors are biosensors, electrical sensors, chemical sensors, smart grid sensors, etc. With the development and organization of several technological phases during the past few years, the smart city concept has become increasingly important in education and business. A new approach to reducing the issues brought on by urban population expansion and fast urbanization is "smartening" a city. Massive amounts of real-time data may be used to gather actionable information and generate new insights with IoT, sensing, and cloud technology. The usage of IoT and cloud computing technology in smart cities also reduces the human work required to manage and monitor the system [1]. The creation of a smart city is built on six primary pillars that make up the fundamentals of sustainable urban development. These are given below:

Smart Governance: Enhancing planning and decision-making via the use of technology is the aim of smart governance. Its primary goals are to strengthen democratic processes and modernize public services delivered through e-government and a mobile, effective agenda.

Smart Economy: A smart economy that prioritizes technological advancement, waste recycling, environmental sustainability, and excellent social assistance. By utilizing modern technologies and data analysis, a smart city's main purpose is to enhance the quality of citizen life, optimize city operations, and promote economic expansions.

Smart Environment: A smart environment is a collection of sensors, actuators, and several computing components that offer services to enhance human life. Here, the Internet of Things and Big data meet.

Smart People: In a smart city, engagement from the bottom-up is just as important as the idea of smart government. The people of the city are given a prominent position, and they are taken into account. Nobody is left behind when the society becomes really inclusive and responsive to everyone's needs.

Smart Living: Smart living refers to innovative ideas that are specifically crafted to make life better in terms of contractility, economy, productivity, integration, and sustainability. Daily tasks become faster, safer, and more intelligent.

Smart Mobility: All types of transportation and infrastructure are integrated by smart mobility, which uses sensors, software, and data platforms to simplify everything from automobiles, including ride-sharing and autonomous vehicles, to bikes, parking spaces, buildings, and emergency vehicles.

IoT, sensing technology, and cloud computing not only make the cities smart but also its applications make the cities more innovative, efficient, sustainable, and interconnected cities.

16.2 CLOUD INFRASTRUCTURE, MANAGEMENT, AND OPERATIONS

16.2.1 Cloud Infrastructure and Management

The hardware and software elements that support the computational needs of a cloud computing environment are referred to as cloud infrastructure. These elements consist of servers, storage, networking, virtualization services, software, and management tools. Cloud infrastructure frequently offers an interface (UI) for managing these resources (UI) [3]. A notable and widely used example of this technique is infrastructure as a service or IaaS. With IaaS, a group or company may purchase the computer infrastructure they require over the internet, including storage, processing power (whether on real or, more commonly, virtual machines), load balances, and firewalls, among many other requirements [4]. The procedures and equipment required for efficiently allocating and delivering essential resources when and where they are needed are included in cloud infrastructure management. It is capable of providing cloud service to both internal users and external users. To fulfill the goal of cloud computing as a whole, cloud infrastructure management is essential. In the smart city, many services have been expanded using cloud computing to make it simpler to access city services. The cloud, when properly managed and optimized, gives organizations more flexibility and scalability for their applications and infrastructure while lowering costs.

16.2.2 Cloud Infrastructure Management Tools

Enterprises may manage resources and services across hybrid and multiple clouds with the use of cloud management tools (on-premises, public cloud, and edge). These tools provide some variations of the following characteristics:

Provisioning and Configuration: These tools enable (1) installing a new server, (2) installing software or an operating system, and (3) distributing storage resources and meeting other criteria for cloud infrastructure.

Resource Allocation: Resource allocation capabilities, which are connected to cost optimization, give consumers fine-grained control over how they use cloud infrastructure, including self-service provisioning.

Cost Optimization: One essential feature of cloud infrastructure management solutions is cost control.

Automation: Automation features for a variety of operational activities, including configuration management, auto-provisioning, and auto-scaling, are occasionally available in cloud infrastructure management solutions.

Security: Another element of a comprehensive cloud security plan is cloud infrastructure management tools. They are techniques for correctly setting a cloud provider's native security measures depending on specific setup and requirements.

16.2.3 Cloud Operations

Controlling the delivery, tuning, optimization, and performance of IT workloads and services utilized in a cloud environment, including multi-cloud, hybrid, in the data center, and at the edge. In the smart city, many services have been expanded using cloud technology to make it simpler to access city services. Systems for cloud computing combine data from several sources through a network. Infrastructure, software, and platforms may all be used with a single set of service logic thanks to cloud computing systems [5].

16.3 CLOUD-BASED IoT SOLUTIONS

Information technologies are the core stones of smart cities. Without communication networks, we would still be "chained up" to our desktop computers, and the internet or mobile networks would not even exist. Technology is highly used in healthcare, education, research, the economy, and even social connections. Even individuals who are not particularly interested in technology have come to rely on services that would not be feasible without the advancement of ICT. The entire cloud environment can be disrupted; that is why cloud-based IoT solutions are built to use strictly enforced rule validations and hierarchical rights management to prevent unauthorized incursion. The IoT cloud is a gigantic network that connects IoT devices and applications. Several renewed cloud-based IoT solutions are used all over the world [6].

16.3.1 Thingworx 8

One of the top IoT solutions for industrial businesses, Thingworx 8, offers simple device connectivity. It makes it possible to enjoy today's linked world. The power to create, distribute, and develop industrial projects and applications is available on the better, faster, and simpler Thingworx 8 platform.

16.3.2 Microsoft Azure IoT Suite

Multiple services are available through Microsoft Azure to build IoT applications. With ready-made linked solutions, it improves your efficiency and profitability. In order to revolutionize the company, it evaluates untapped data. This offers the answers for a quick PoC to roll out your concepts. With Azure Suite, new data can be easily examined and used [6] (Figure 16.1).

16.3.3 Google Cloud's IoT Platform

One of the best platforms now accessible is Google. A complete platform for IoT solutions is offered by Google. It makes managing, connecting, and storing IoT data straightforward [6]. This platform aids in business expansion. Making things quick and simple is their major priority. In comparison to competing platforms, Google Cloud charges on a per-minute basis, which is less expensive.

16.3.4 IBM Watson

IBM Watson, a hybrid cloud PaaS (platform as a service) development platform, and IBM Watson together constitute a dynamic platform. They make IoT services accessible to beginners by offering simple example apps and interfaces. It distinguishes itself from other platforms by making it simple for you to test out its sample to see how it functions [7].

We can get some exciting features including data service and data exchange, and real-time APIs in IBM Watson which are reflected in Figure 16.2 [6].

16.3.5 AWS IoT Platform

For developers, Amazon greatly simplified the process of gathering data from sensors and other internet-connected devices [6]. These assist you in gathering information,

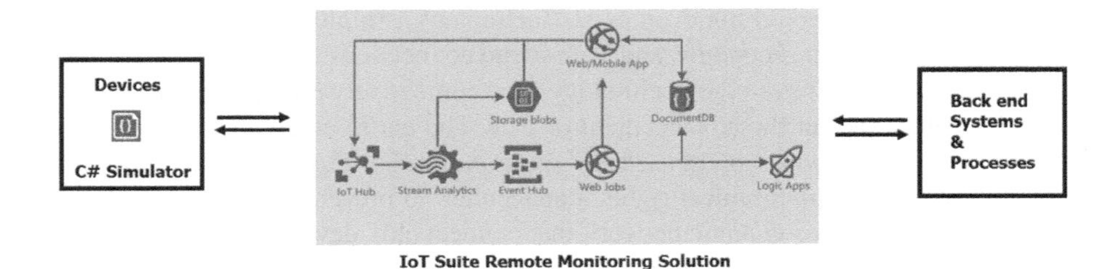

FIGURE 16.1 Microsoft Azure IoT suite.

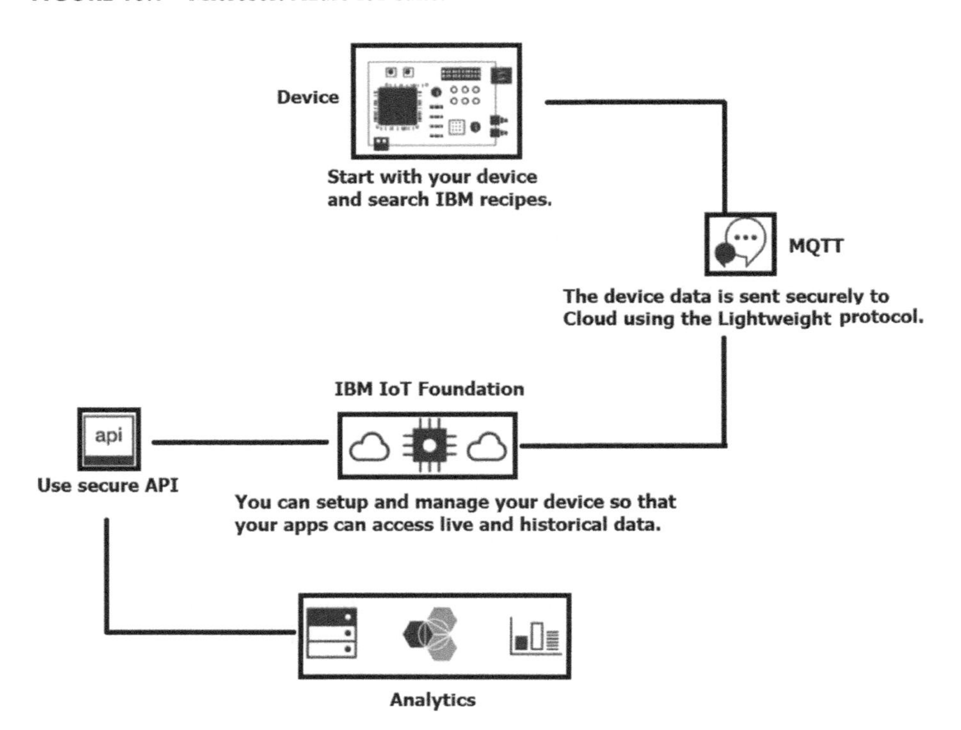

FIGURE 16.2 IBM Watson.

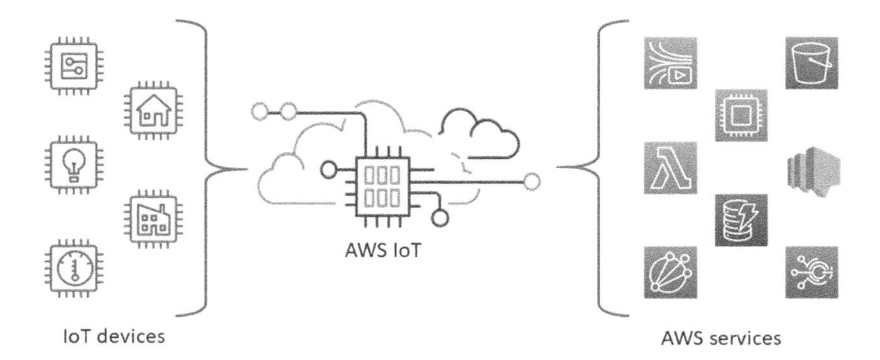

FIGURE 16.3 AWS IoT platform.

sending it to the cloud, and then analyzing it to make it possible to control devices. The AWS IoT platform's primary attributes include device administration, secure device gateway, encryption and identification, and gadget shadow. Figure 16.3 [6] shows a relationship between IoT devices and AWS services.

16.3.6 Cisco IoT Cloud Connect

The digital transformation and actions based on your data are accelerated with the Cisco IoT. A mobile, cloud-based suite is Cisco IoT Cloud Connect. It provides mobile carriers with ways to deliver amazing IoT experiences. It gives your devices broad deployment choices. Its primary features are voice and data connectivity, reporting on devices and IP sessions, billing is customized, and solutions for flexible deployment.

16.3.7 Sales Force IoT Cloud

Salesforce Thunder is the engine that drives Salesforce IoT Cloud. It takes information from gadgets, websites, apps, and partners to set off reactions in real time. IoT and Salesforce work together to give better customer service.

16.3.8 Kaa IoT

Kaa is a free, open-source middleware platform that may be used to construct smart devices and the IoT from beginning to finish. Cost, risk, and market time are all decreased. Besides, Kaa provides a selection of IoT solutions that may be quickly plugged in and used in IoT use cases [6].

16.3.9 Thingspeak

You may gather and store sensor data in the cloud using the open-source software Thingspeak. It offers you the app to use MATLAB to analyze and visualize your data. To transmit sensor data, you may use an Arduino, Raspberry Pi, or Beaglebone. To save data, you can build a different channel here [6].

Smart devices are connected with MATLAB in Thingspeak, and it analyzes data through algorithms. Figure 16.4 is an interface overview of Thingspeak IoT solution.

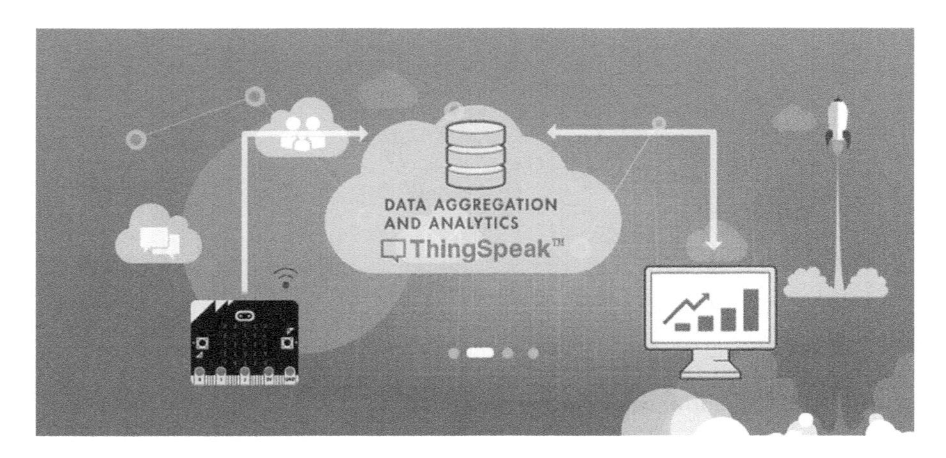

FIGURE 16.4 ThingSpeak.

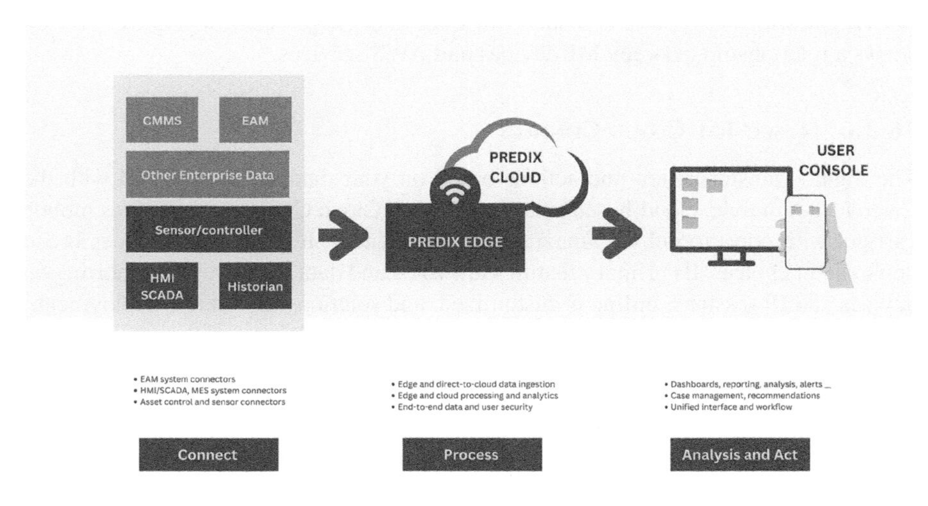

FIGURE 16.5 GE Predix IoT.

16.3.10 GE Predix IoT

The first industrial platform in history is called Predix. Predix offers a straightforward environment and was created with factories in mind. It can immediately assess and save data coming from the machine. For its cloud platform, GE hopes to provide the expanding industrial IoT. This platform may grow and is safe [6]. Figure 16.5 shows a working overview of GE Predix IoT.

16.4 APPLICATIONS OF IoT, CLOUD COMPUTING, AND SENSING TECHNOLOGY FOR SMART CITIES

Smart cities are becoming an excellent application area for cloud technologies in the context of all the aforementioned developments, and the cloud is a key factor in their growth. The core component of the cloud is the IoT. IoT makes it possible to create and use smart devices to handle problems and difficulties in the real world. Have we ever thought, 'How do they become smart?' The secret is using a variety of sensors to gather and interpret data in real time. Sensors gather information and data from the outside world and replace it with a signal that both people and machines can recognize. To collect real-time data, IoT cloud servers and devices rely on sensors. This is how sensor technology plays a key role when we use it in IoT [8]. As previously said, IoT technology is one of the most significant technological developments that have emerged in the last few years [8]. IoT has shaped several corporate practices, technological capabilities, and consumer trends throughout the years. There are several industries where this new class of cloud-based IoT applications is present. IoT applications currently encompass a broad range of challenging industries, including manufacturing and the industrial sector, agriculture, the healthcare sector, smart cities, security, and emergency services, among others which is shown in Figure 16.6 [8].

16.4.1 Work from Home with IoT

Because of the use of apps like Siri, Alexa, Google Home, and other IoT-enabled home products, IoT has emerged as one of the most important aspects of our daily lives. These tools are also becoming an essential part of our modern work ethic, which involves working remotely. It is now simpler to manage all of the responsibilities related to working without losing personal time thanks to the features of these devices, which include automatic scheduling and highly advanced calendar functions. Additionally, virtual conference rooms and interactive video conferencing have increased your total productivity when working from home [9].

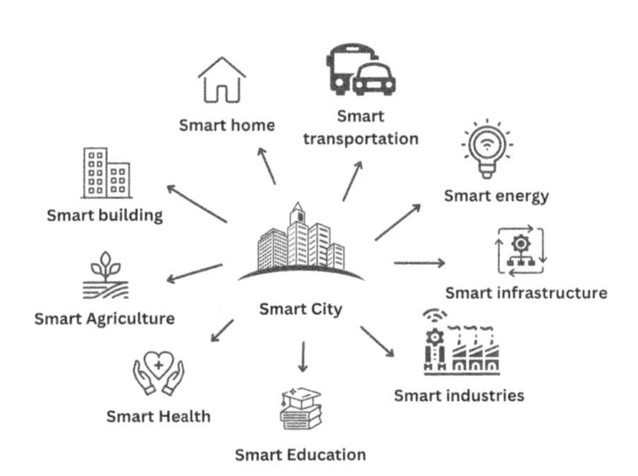

FIGURE 16.6 Applications in smart city.

16.4.2 Smart Healthcare

IoT is undeniably changing the healthcare sector by changing how people and devices interact while delivering healthcare solutions. Applications of IoT provided benefits of improvements in healthcare that are felt by patients, physicians, hospitals, and health insurance companies [10,11].

IoT for Patients: Wearable technology, such as activity trackers, and other Bluetooth connection medical tools, such as blood glucose monitoring and pulse and heart rate monitor cuffs, can help patients receive tailored treatment. These tools may be set up to remind users to monitor their blood pressure fluctuations, consultations, and a variety of other tasks [12].

IoT for Doctors: Doctors use wearable devices and other home monitoring devices to monitor a person's health. This information can help the doctor know if the person is following the treatment plan and if they have any urgent medical needs. With IoT, healthcare personnel can now be more engaged with the patient and help to get the best results [13].

IoT for Hospitals: There are many different ways that IoT devices can help hospitals. For example, the location of medical equipment, such as wheelchairs, defibrillators, nebulizers, oxygen pumps, and other monitoring devices, can be tracked in real time. Additionally, real-time analysis of medical personnel placement at various sites is possible using IoT devices [2].

IoT for Health Insurance Providers: Health insurers have a variety of options for using IoT-connected intelligent devices to improve their operations. Underwriting and claims departments can use data from health monitoring devices to identify candidates for underwriting, uncover fraud claims, and make better judgments in all operational processes. This increased transparency between insurers and their customers helps ensure that customers have a complete understanding of the decisions being made and the outcomes of the process [14].

16.4.3 IoT in Retail

Stores and supermarkets will likely become safer and more effective with the aid of smart devices and linked equipment. Retailers may also save operating expenses while still developing novel use cases to improve consumer experiences. They may improve their competitiveness in their industry by (1) improved supply chain management using RFID tags and GPS sensors; (2) minimized overstocking, stockouts, and shrinkage, smart inventory management based on consumer purchasing behavior is necessary; (3) with the help of IoT, retailers can automate inventory visibility and create purchase plans; and (4) automate and even customize consumer checkouts.

16.4.4 Smart Education

IoT in education is the application of intelligent IoT technologies in learning settings like schools and universities. The transition from conventional to digital teaching methods has been made easier by IoT developers. Notably, individuals may study any discipline using graphics and animation, from languages to medical sciences. In other words, IoT is enabling education to move beyond the classroom and become more applicable. Some examples are smart classrooms, safety in establishments, automated attendance tracking, accommodating disabled students, distance learning, monitoring students, and improving productivity.

For students, IoT features make studying easier, more inexpensive, and more accessible. Even if it may take some time before IoT fully penetrates the education industry, it will considerably raise teaching standards. For both students and instructors, IoT developers have created technologies that are worthwhile to invest in.

16.4.5 Smart Agriculture

Agriculture is already experiencing disruption as a result of the Industrial IoT, which has already had a negative impact on a variety of industries. At the end of 2018, the linked agriculture sector was worth USD 1.8 billion and was still expanding. At a CAGR of 19.3%, by 2023, it is projected to reach USD 4.3 billion. Applications for the IoT in agriculture concentrate on traditional farming tasks to meet increasing demand and minimize output losses. Robots, drones, remote sensors, computer images, and ever-improving machine learning and analytical tools are utilized in IoT in agriculture to monitor crops, survey and map fields, and give farmers the information they may use to make time- and money-saving farm management decisions [12].

16.4.5.1 Climate Conditions

The climate has a significant impact on farming. Furthermore, inadequate climate knowledge substantially degrades the quantity and quality of agricultural output. IoT technology, on the other hand, provides you with weather information right now. Both inside and outside of the agricultural fields are equipped with sensors. The farmers use sensors to collect environmental data about climatic conditions. This information is used to choose the best crops for a particular area. The IoT ecosystem includes many sensors that can monitor real-time climatic factors very accurately. For each of these elements, a variety of sensors are available that can be configured to match your demands for smart farming. The health of the crops and the local weather are monitored by these sensors. If any suspicious weather is found, an alarm is generated. Reducing the need for human intervention during bad weather conditions eventually increases output and enables farmers to achieve higher agricultural advantages [15].

16.4.5.2 Precision Agriculture

Precision agriculture, commonly referred to as agricultural technology, is perhaps the most well-known application of IoT in farming. It enhances the precision and control of agricultural operations by employing smart farming technologies, such as animal tracking, GPS tracking, survey, and stock control. Precision farming aims to analyze sensor data and take relevant actions. In precision farming, sensors are used to help farmers quickly acquire data, analyze it, and make choices. Regulating animals, keeping an eye on cars, and managing irrigation are a few strategies to improve precision agriculture. All of this increases productivity and efficiency. Precision farming can be used to assess the condition of the soil and other relevant parameters to improve operational effectiveness. Additionally, you can verify the connected devices' operating conditions to look at water and nutrient levels [16].

16.4.5.3 Smart Greenhouse

Thanks to the IoT, which enables weather stations to automatically change the environment's parameters in line with a detailed set of instructions, our greenhouses are now wiser. Because IoT has been implemented in greenhouses, there is no longer any need for human interaction, which lowers costs and improves accuracy throughout the process, creating modern, affordable greenhouses, for instance, using IoT sensors driven by solar energy. Real-time data are gathered and transmitted by these sensors, and it is then used to precisely track the greenhouse's status at all times. The sensors allow for the tracking of greenhouse conditions and water use via emails or SMS warnings. Intelligent and autonomous irrigation uses the IoT. The information provided by these sensors includes pressure, humidity, temperature, and light levels [16].

16.4.5.4 Data Analysis

The data gathered by IoT devices are too large for the traditional database system to handle. The end-to-end IoT platform and cloud-based data storage for the smart agriculture system are crucial components. It is thought that these mechanisms are crucial for enabling improved task performance. In the IoT era, sensors are the main way to collect significant volumes of data. Analytics tools are used to examine the data and transform it into useful information. The examination of livestock, agricultural conditions, and the weather is made easier by the application of data analytics. Utilizing the gathered data and technological advancements could lead to better decisions. You may get real-time information about the health of the crops by using IoT devices to gather data from sensors. Utilizing predictive analytics can help you learn information that can help you make better harvesting decisions [16].

16.4.5.5 Agriculture Drone

A major change in agricultural operations has been brought about by technological breakthroughs in the form of agricultural drones. Agribusiness uses drones for spraying, crop monitoring, field analysis, and crop health evaluations. With correct strategy and planning based on real-time data, the agriculture business has experienced a tremendous increase and transformation owing to drone technology. The locations where irrigation needs to be modified can be found using drones equipped with thermal or multispectral sensors. The vegetation index is calculated by sensors once the crops have begun to grow and reflect their state of health. Eventually, intelligent drones contributed to lessening the environmental impact [16]. The amount of chemical that penetrates groundwater has significantly decreased as a result.

16.4.6 Smart Transportation

In the transportation sector, the IoT makes use of a complex web of sensing devices, actuators, gadgets, and other technologically advanced devices. This network collects data on the current situation and transmits it using specialized software to provide useful information. The operations of the transportation sector have changed as a result of IoT-enabled technologies and smart solutions. In addition, the complexity of the urban transportation system keeps growing as the number of cars on the road increases. This demonstrates how communities must incorporate IoT into transportation to get more extensive and secure transportation advantages [17].

16.4.6.1 Traffic Management

Traffic management is where the transport is located industry where IoT technology adoption is thought to be most prevalent. Millions of billions of terabytes of data related to traffic and automobiles are generated by CCTV cameras. The purpose of providing this information to traffic management facilities is to enable them to keep a closer check on the vehicles and punish violators of traffic regulations. Smart parking, autonomous traffic signal systems, and smart accident assistance are some of the few IoT applications that help the police successfully regulate traffic and reduce the chance of accidents.

16.4.6.2 Automated Toll and Ticketing

The conventional tolling and ticketing methods used in transportation systems are outdated and ineffective. The congestion caused by the increased number of vehicles on the road has made it difficult for drivers to enter toll booths, and the toll booths don't have the equipment or staff to handle a large number of cars quickly. However, IoT has many advantages over traditional tolling and ticketing systems, including automated tolls. Traffic police personnel now have an easier time-handling tolls and tickets thanks to RFID tags and other smart sensors [18].

Nowadays, the majority of modern automobiles have IoT connectivity. Any car, even if it is a kilometer distant from the tolling station, may be easily spotted using IoT technology. This allows the traffic barriers to be lifted, allowing vehicles to pass through. However, older automobiles lack IoT connectivity, but car owners' cell phones can serve the same purpose, namely, accepting automated payments via phones linked to a digital wallet. This shows that IoT in transportation is considerably more adaptable and compatible with new cars, as well as demonstrating easy connectivity with older vehicles for automated toll and ticketing systems.

16.4.6.3 Self-Driven Cars

The majority of modern automobiles now have IoT connections. Any car, even if it is a kilometer distant from the tolling station, may be easily spotted using IoT technology. This allows the traffic barriers to be lifted, allowing vehicles to pass through. However, older automobiles lack IoT connectivity, but car owners' cell phones can serve the same purpose, namely, accepting automated payments via phones linked to a digital wallet. This shows that IoT in transportation is considerably more adaptable and compatible with new cars, as well as demonstrating easy connectivity with older vehicles for automated toll and ticketing systems. Sensors in self-driving cars use IoT to continually capture data about their surroundings in real time and communicate it to a central unit or the cloud. The technology analyzes the data in a matter of seconds, allowing self-driving cars to function in accordance with the information presented. This means that IoT links the sensor network for self-driving cars and allows them to function properly [18].

16.4.6.4 Transportation Monitoring

Many businesses now employ supply chain management and vehicle-tracking systems to effectively manage their fleets [18]. Transportation businesses may easily acquire real-time vehicle positions, information, and numbers thanks to GPS trackers. This allows transportation businesses to monitor their critical assets in real time [18]. Aside from tracking location, IoT devices may also monitor driver behavior and provide information on driving style and idle time. IoT has reduced operational and fuel costs, as well as maintenance costs, in fleet management systems [17].

16.4.6.5 Security of Public Transportation

The security of public transportation is one of the major areas in which it is discovered that the IoT in transportation is most beneficial. Municipalities may track traffic infractions and take the necessary action by monitoring every mode of transportation with the aid of IoT devices. Beyond security, IoT in transportation enhances public transportation management by offering a variety of intelligent solutions. Integrated ticketing, automated fare collecting, improved vehicle logistic solutions, and passenger information systems are some of these. These methods aid in the control of traffic congestion and public transportation. IoT has made it feasible to handle public transportation in real time. Because of this, transportation providers can now communicate with passengers more effectively and give them the information they need via mobile devices and passenger information displays.

16.4.7 Smart Infrastructure

Analysts estimate that between 20 and 50 billion linked devices will exist globally by 2020, as IoT continues to expand globally. It is a crucial enabler for a hyper-connected company and a major force behind technological advancement and commercial innovation.

i. Cities should establish the circumstances for ongoing development, with urban infrastructure and buildings needing to be developed more sustainably and effectively as digital technologies become more vital.
ii. The amount of CO_2 emitted should be maintained as minimal as possible, for instance, by purchasing autos and self-propelled vehicles.
iii. Intelligent technologies are used in smart cities to create an infrastructure that is both environmentally friendly and energy efficient.
iv. In order to save energy, smart lights should only turn on when someone really walks past them. This may be done by controlling brightness levels and monitoring daily usage.

Keeping this in mind, we can divide the IoT infrastructure necessary to build an IoT system into the following components.

All components of smart infrastructure and their usage are shown in Table 16.1. Companies will need to obtain the necessary capabilities inside their organization to use and manage IoT technology as it develops and becomes a more commonplace business tool. Numerous businesses provide IoT solutions for various sectors of the economy, such as manufacturing, maintenance, retail, logistics, and mobility. These solutions may help with IoT operations and provide the necessary business results.

TABLE 16.1
Components for Smart Infrastructure

Components for Smart Infrastructure	Description
Sensors	Measures physical amounts that the IoT device transmits via the network
Controller	The device's processing and storing are done onboard by the device's brain, which also serves as a link between the sensor and the network
Network	The method utilized to transfer data to or from the cloud or other system components
Cloud	Computing, storage, and gateway resources available through the internet
User facing application	Applications for mobile and the web that let users connect with IoT systems and see data
Data analytics	The instruments and resources that allow users to get an understanding of the data received by an IoT system (typically on the cloud)

16.4.7.1 IoT Devices – Sensors and Actuators

In order to keep track of IoT devices, sensors offer information. Information in this context can include both the device's identity and measures of its physical status. Any equipment that is being watched can have sensors attached to or incorporated in it, or sensors can be installed in the surrounding area to monitor the item covertly. Actuators have the ability to change a device's physical state, move (translate, rotate, etc.) basic devices, or activate/deactivate features of more complicated ones.

16.4.7.2 Edge Gateways and IoT Connectivity

Before transmission, an IoT gateway gathers, converts between sensor protocols, and analyzes sensor data. The networking of devices, protocol translation, data filtering and processing, security, and administration are just a few of the crucial tasks that IoT gateways carry out. The application code that analyzes data and develops into an intelligent component of a system with connected devices may run on IoT gateways. IoT gateways are located where edge systems—sensors, devices, and controllers—meet the cloud (Figure 16.7).

16.4.8 Smart Energy

The generation of power, the transportation of products, and numerous other essential human activities are all supported by the main sector of energy. IoT has advanced the way this industry functions by stepping into the world of energy generation and consumption.

16.4.8.1 Optimization of Energy Resources

The ability to optimize renewable energy sources is one of the main uses of IoT in the energy industry. Energy users may optimize and perhaps lower their energy use by installing smart monitors that automatically regulate the room temperature in a factory, hall, or even a room. Additionally, companies and organizations may install smart sensors in structures to control energy use based on the occupancy of a room

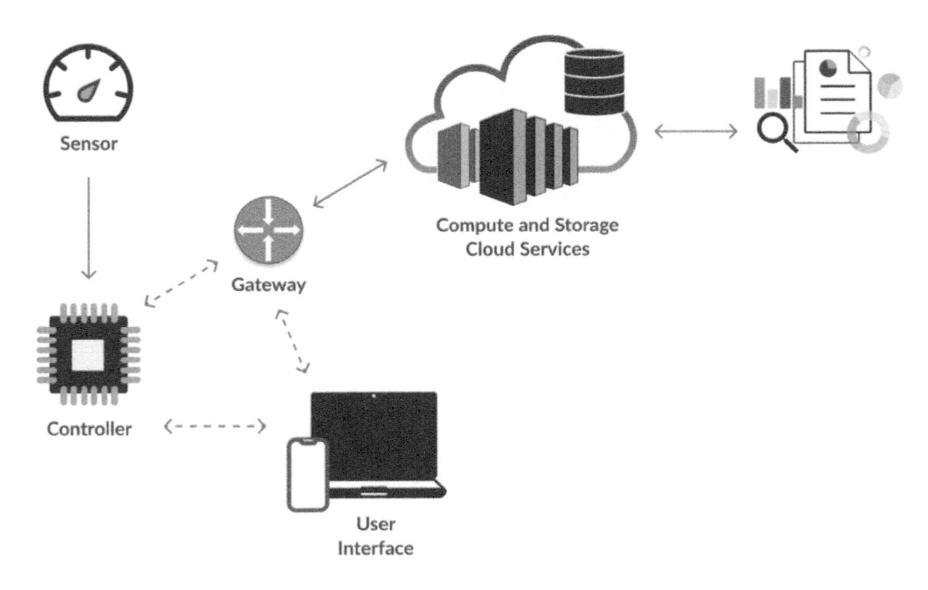

FIGURE 16.7 Components of infrastructure for a smart system.

or building. This specific application is highly helpful for the energy industry since it not only prevents enormous quantities of energy from being wildly lost but also connects the entire industry to the internet for better operations over time. Despite the fact that mass drives for energy consumption are increasing, they are insufficient to regulate how humans use energy. Perhaps, the solution to such issues is an internet-controlled network of energy devices in industrial and commercial facilities. Furthermore, it is unquestionably a powerful approach to conserve energy and direct it only where it will be put to use.

16.4.8.2 Empowering Microgrids

To introduce IoT in this industry, however, merely lowering the energy footprint is insufficient. Perhaps, more advanced energy grids are needed to support the power of IoT. Smart grids are a system of electrical circuits that can self-heal when there is a maintenance problem and provide a two-way energy flow in the energy industry. Nevertheless, since the unused energy is returned to its source, smart grids also make it possible for customers to participate actively. In this manner, the end-user circuit is finished and made livelier for intelligent consumption. Let us use an example to better grasp this. Let us say a microgrid is built in a neighborhood of ten homes, and each home is linked to the grid. These homes may now self-generate electricity for their own usage thanks to solar panels on their individual roofs. The amount of energy a single home generates could, however, exceed the quantity it really uses. This implies that any excess energy is squandered. In contrast, microgrids operate differently.

16.4.8.3 Smart Meter Technology

Next, by integrating smart meter technology, IoT applications in smart energy may also be applied in the energy industry. The smart meter installation technology may assist in analyzing the regions where energy is spent the most and the locations where energy can be significantly conserved. This is in reference to the monitoring and analysis of energy consumption by a specific household, building, or organization. Having said that, installing IoT-enabled smart meters in homes and other structures would only benefit consumers by enabling them to obtain an in-depth report on their energy usage trends. Additionally, it will aid in the detection of gaps, and ultimately, the smart meter technology will increase the efficiency and transparency of the operation of the energy sector. Smart meters let utilities and customers communicate with each other more easily so that the two parties have a better relationship from the start. It is crucial for people to comprehend the miracles that IoT can do in order to reduce energy consumption and encourage it toward better consumption and conservation habits in order to integrate it with this industry.

16.4.8.4 Proactive Repair Mechanism

There are several networks and cables in the energy industry that support energy flow and enable consumption from one point to the other. Yet, for months and days, energy authorities and representatives are unaware of what goes wrong in between these enormous networks. However, since IoT technology and the energy industry have partnered, it is now feasible to spot flaws in an energy circuit, let the authorities know, and take care of repair work as soon as possible. The use of IoT in the energy industry may allow for the installation of a preventative repair mechanism that issues alerts and warning signals in the event that a circuit is tripped or that equipment needs to be maintained. This specific application is a quick way to solve issues and has been shown to significantly reduce maintenance costs in the industry. Modern smart circuits are completely capable of alerting the energy maintenance booths about the repair work, unlike previous models where repairs to energy circuits might take up to a week or more. Furthermore, self-healing smart circuits that facilitate energy flow eliminate the need for any sort of human intervention.

16.4.9 Smart Parking

Smart parking systems can detect when a car has departed from its parking slot. The vehicle receives notifications of open, free parking spots from sensors in the ground via a smartphone. Others use vehicle input to locate the openings and point awaiting cars in the direction that will encounter the least amount of opposition. Smart parking is ideal for a mid-sized smart city because it already exists, does not require complicated infrastructure, and does not require a lot of money [19].

16.4.10 Smart Waste Management

Solutions for waste management help reduce operational costs, increase waste collection efficiency, and better manage the environmental issues resulting from inefficient garbage collection [20]. The management platform of a truck driver tells them via their smartphone when a specified level is reached by a garbage container equipped with a level sensor. The notification appears to empty a full container, preventing drains from being only partially full.

16.4.11 Water Quality Management

In order to promptly identify any contamination and take appropriate action, water monitoring sensors analyze the quality of the water. Additionally, communities can monitor critical water sensors in real time thanks to IoT, decreasing water loss and spotting concerns before they become serious ones. IoT-based sensors are frequently used to monitor variables like turbidity and total dissolved solvents (TDS) [21].

16.4.12 Crime Reduction

By lowering crime rates, IoT-based smart city technologies make cities safer for their inhabitants. Police officers may work more effectively when using IoT technology for smart cities, such as internet-linked body cams and connected crime scenes. In reality, several communities are already implementing clever methods to build safer neighborhoods.

16.5 THE IMPORTANCE OF CLOUD-BASED IoT FOR SMART CITIES

By 2030, it is anticipated that cities will be home to more than 60% of the world's population. It is a risky prognosis that, if the appropriate steps are not taken, may spell tragedy. Resources are necessary for the support of huge populations. Along with having access to clean air and water, residents will also need to manage their garbage in a practical manner. Future cities will be able to effectively and efficiently meet resident expectations thanks to inventive smart city principles implementation and widespread IoT technology adoption [22]. One of the key advantages of IoT for smart cities is the availability of a massive quantity of priceless information. By carefully evaluating this information and data, city authorities can improve the city and the quality of life for its citizens [8]. IoT and cloud-based big data analytics for smart future cities can help cities more effectively detect high-risk locations and dispatch police. Similar to how residents' wants and interests may be better identified and met by local leaders.

Companies utilize IoT to manage and monitor widely dispersed operations in order to be more inventive. Since information is constantly being sent into apps and data storage, they can even manage the latter even from a distance. The benefit of understanding things beforehand is provided by IoT. Activities that were previously out of reach may now be monitored and managed because of the cheap cost of IoT. The biggest benefit is economics since this new technology might take the place of people who now oversee and maintain supply. Costs can be greatly decreased and optimized as a result. IoT also enables the acquisition of entirely new information, such as the correlation between weather impacts and industrial outputs. A smart city's busier sections can use cloud-based IoT solutions since they are adaptable and can handle the needs of a large amount of information. A key component of the smart city solution is the automatic expansion that cloud computing provides to handle a large number of devices and individual data. Additionally, cloud technology has the flexibility to handle all data and sensor streams in real time at any moment.

A smart city's busier sections can use cloud-based IoT solutions since they are adaptable and can handle the needs of a large amount of information. A key component of the smart city solution is the automatic expansion that cloud computing provides to handle a large number of devices and individual data. Additionally, cloud technology has the flexibility to handle all data and sensor streams in real time at any moment. Using cheap computers, the cloud, big data, analytics, and mobile technologies, real things can exchange and collect data with the least amount of human involvement. In the hyper-connected world of today, digital technologies have the power to record, watch, and modify every interaction between connected objects. The physical and digital worlds work together even when they overlap [23]. Smart solutions can be produced by connected gadgets and massive data. These remedies have the power to resolve issues, improve urban dwellers' quality of life, and use fewer resources. The IoT is required for a city to be called really smart [8].

16.6 THE FUTURE OF CLOUD-BASED IoT TECHNOLOGY

IoT applications have a better chance of being ingrained in future businesses and our daily lives because of digital transformation, quicker connections, and advancements in machine learning and artificial intelligence. IoT technology is currently in its prime. By 2021, more than 94% of enterprises are expected to be utilizing IoT, according to a Microsoft analysis from 2019. Statistics claims that the global revenue for technology crossed $100 billion for the first time in 2017 and is projected to reach $1.6 trillion by 2025. Cloud computing means that you can access a pool of re-configurable computing resources from anywhere, at any time, without having to wait for someone to set up a server for you. Businesses are seeking solutions to increase agility, business continuity, profitability, and scalability as they travel a much faster digital transformation route. Cloud computing technology will be the cornerstone of every strategy used in the new normal to accomplish these objectives [8]. Due to cloud-based IoT and sensing technology, which also fosters corporate agility, streamlines procedures, and reduces costs, businesses can grow and respond quickly. In addition to helping firms get through the current crisis, this might also foster long-term prosperity. The future impact of cloud-based IoT technology is anticipated to be as follows [24].

16.6.1 Increased Storage Capacity

Right now, data generation is smashing records, and it will continue to do so. Many businesses store important data, like client information and company information, in real data centers. But cloud server vendors are working on creating cheaper, cloud-based data centers that will be more popular. There are a lot of cloud service providers available, so the prices will be competitive. This will be good for businesses because it will make it easier for them to find the best deal. Data storage would become streamlined and space-effective with the aid of this invention [24].

16.6.2 IoT in the Automobile Industry

The rise of automobiles with integrated wireless connectivity, which gives users access to cloud services for voice recognition, picture detection, artificial intelligence (AI), and other applications, is evidence of this new trend. BMW recently revealed at the Mobile World Congress (MWC) in Barcelona that they will continue working with Microsoft to develop next-generation capabilities and support their current features, which rely on Azure's Cognitive Service. The computational need of the car's technology is significantly decreased since physical sensors on the vehicle transmit data remotely to Azure, where all the processing takes place.

16.6.3 Better Security

Data currently kept on cloud servers is largely secure. Smaller cloud service providers might not be able to offer or comprehend all the security precautions required for adequate data security. Improved cyber security safeguards will be used by future cloud services, and stronger safety rules will be implemented. Businesses will not have to worry about data security or alternate data storage techniques, allowing them to focus on more important duties.

16.6.4 IoT Advanced Forecast

It is also possible to link the rise of IoT devices to the potential of wireless technology. Forbes estimates that by 2021 the combined IoT markets will reach $520 billion, while Cisco projects that by 2022 IoT traffic would make up almost 5 zettabytes (1 ZB is equal to 1 trillion GB) of annual traffic. This traffic will come from a range of IoT devices, such as smart locks and gas station pump controls, and it will only grow as cloud providers develop easier ways to integrate low-cost IoT devices. With their operating systems created especially for micro-controllers and other low-power devices, Amazon and Microsoft are now dominating the market in IoT adoption.

16.6.5 Smart Eye

The smart eye technology and Google Glass, their most ambitious effort, are quite comparable. In order to present alternatives and accessible features directly in front of your eyes without being distracting, the smart eye is outfitted with sensors, Wi-Fi, and Bluetooth. Reading texts, using the internet, and other things are all made possible by technology.

16.6.6 Short-Term Growth and Explosive Long-Term Growth

As new advancements in wireless speed and power supply are made, the application of IoT and cloud-based connectivity will keep expanding tremendously in the near future. IoT will see a boom in the near future in more locations than was previously considered conceivable. We can only speculate about the possible consequences and uses of this sort of technology while we wait for these advancements. Who knows, perhaps, the day will come when we will be able to manage our Marty McFly hoverboards from a distance. In the interim, get in touch with us to see how we can support you in utilizing this technology to enhance your business processes.

16.6.7 The Impact of the Cloud and IoT on the Economy

If cloud computing advances at its current rate or faster, the requirement for hardware will decrease. The vast majority of operations and business activities will use virtualization, cloud computing, the IoT, and virtual machines (VMs). Due to the dramatic reduction in the price of installing physical infrastructure and software, this trend will lead to decreased hardware utility. Furthermore, with the advancement of cloud computing, fully automated and virtualized data processing and interpretation will no longer require human input. The concept of "economy" is vast and subtle; it refers to all economic activity that takes place within and between nations, including production, trade in goods and services, management, financial operations, and more. Utilizing statistics is the easiest way to illustrate how the growth of the IoT software has affected the world economy.

- In 2020, the market for the IoT was valued at $389 billion, and by 2030, analysts predict that it will reach $1 trillion.
- Businesses invested a total of $128.9 billion in IoT development in 2020, and they project an annual growth of 26.7%.
- There will probably be 30 billion connected devices by the end of 2025, up from the $10 billion number in 2019.
- According to a McKinsey analysis, IoT applications would together have an annual economic impact of $3.9 trillion to $11.1 trillion by 2025.

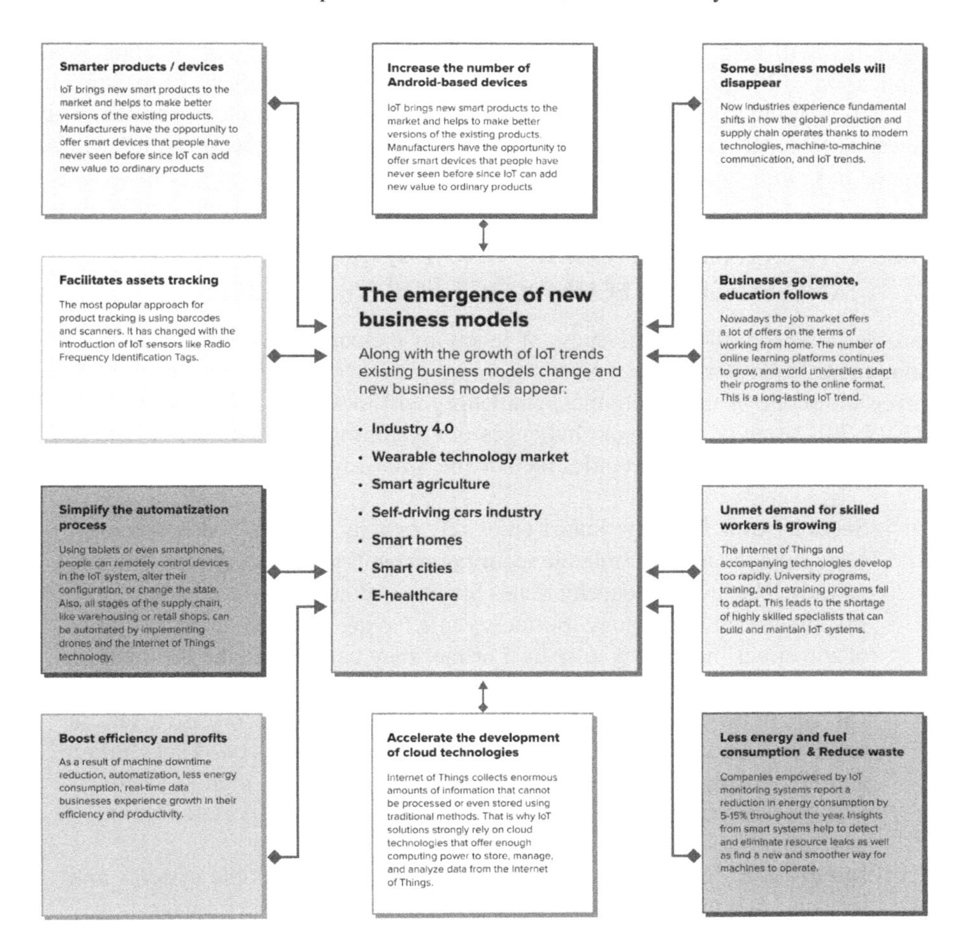

FIGURE 16.8 Economic impact.

- According to a McKinsey analysis, IoT applications would together have an annual economic impact of $3.9 trillion to $11.1 trillion by 2025.
- The market for smart products in the healthcare industry is anticipated to grow at an average yearly rate of 8.84% by 2023. The market capitalization will rise as a result, rising from $31.71 billion to $57.85 billion by 2023.
- The global automotive IoT market is projected to grow at a CAGR of 16.4% from 2019 to 2025, reaching $541.73 billion [25].
- IoT has fostered Industry 4.0, the wearable technology market, smart agriculture, the self-driving car industry, smart homes, smart cities, and E-healthcare as well. Figure 16.8 [25] shows how these industries are connected and affected by cloud computing and IoT.

16.6.8 Robotics

From being only a work of fiction, robots are now a reality. They now play a significant role in IoT operations going forward. Robots have the potential to improve productivity and efficiency in many facets of life. Increasing its machine-learning capability will help in gathering and analyzing enormous amounts of data. This application will improve future operations to their best use, whether it be assisting in production at industrial locations or something as simple as monitoring traffic. This powerful force has the potential to alter the planet and boost productivity. IoT will be crucial in keeping track of and managing these robots.

16.7 CHALLENGES OF COMBINING IoT, CLOUD COMPUTING, AND SENSING TECHNOLOGY FOR SMART CITIES

In order to ensure the effectiveness of functions including government, transportation, energy, and healthcare, smart cities use smart technology, sensors, and data. Given the scope of these activities, challenges will inevitably arise for smart cities [26,27–30]. Some of the major challenges of combining IoT, cloud computing, and sensing technology for smart cities include the following:

i. **Security and Privacy**: Smart cities rely heavily on data and technology, increasing the need for reliable security protocols to protect citizen data and privacy. The level of security issues increases along with the deployment of IoT and sensor technology. Besides, while having cameras on every corner of the street may assist to reduce crime, they can also make law-abiding residents fearful and paranoid [31].
ii. **Infrastructure**: Smart cities need to be built on a robust infrastructure in order to support the applications and services they rely on. Major urban regions already struggle in addition to putting in place high-speed internet and replacing outdated infrastructure including transportation tunnels, water pipelines, and underground wires.
iii. **Integration**: Smart cities require the integration of multiple systems and services, making it difficult to ensure compatibility and performance.

iv. **Cost**: Smart cities require significant investment in order to develop the infrastructure, applications, and services needed to power them. Sometimes, funding for new projects is limited and requires a long time to become successful.

v. **Sustainability**: Smart cities must be designed to be sustainable in order to reduce their environmental impact and conserve resources.

16.8 CONCLUSION

A smart city should promote economic growth while also providing its residents with a high quality of life. In order to build prosperous, livable cities in the 21st century, the smart city will be crucial. Technology advancements in the fields of sensing, IoT, and cloud computing are advancing us further than we could have ever anticipated. Therefore, it is critical to increase funding for initiatives that make use of cutting-edge technology. To improve our corporate operations and function in the connected world of today, we need a full IoT framework, cloud computing, sensor applications, machine learning, artificial intelligence, and embedded systems.

Through numerous forms of public engagement, residents may express their opinions and save time thanks to the capabilities provided by smart cities. Crowdsourcing is another way to take use of the crowd's strengths and skills. We can take advantage of the smart functionality, features, and productivity of the connected IoT ecosystems by making the most of this powerful technology. In the upcoming years, it will significantly contribute to the advancement of every industry, whether it be in manufacturing or healthcare. It will also play a key role in aiding efficient disaster management operations. IoT will also build a shield over the internet to safeguard cyberspace. The main applications of IoT, cloud computing, and sensing technology are all briefly covered in this chapter's conclusion. With widespread adoption, deliberate deployment, and careful management, cloud-based computing, the IoT, and sensor technology could transform our metropolitan areas into intelligent, sustainable, and effective settings.

The success of every sector, from healthcare to business, from transportation to education, depends on the effective sharing of information. The potential of smart cities is essentially limitless, and in the next few days, their growth will be steadily accelerated.

REFERENCES

1. Gupta, Deepak, Victor Hugo C. de Albuquerque, Ashish Khanna, and Purnima Lala Mehta. 2021. *Smart Sensors for Industrial Internet of Things: Challenges, Solutions and Applications*. Cham: Springer.
2. Alam, Tanweer. 2021. "Cloud-based IoT applications and their roles in smart cities." *Smart Cities* 4(3): 1196–1219. https://doi.org/10.3390/smartcities4030064.
3. Target, Tech. n.d. "Cloud Computing Definitions." Cloud Computing (blog). https://www.techtarget.com/searchcloudcomputing/definitions.
4. Innocent, Anasuya Threse A. 2012. "Cloud infrastructure service management: A review." *IJCSI International Journal of Computer Science Issues* 9: 287–292.

5. Iyoob, Ilyas, Emrah Zarifoglu, and A. B. Dieker. 2013. "Cloud computing operations research." *Service Science* 5: 88–101. https://doi.org/10.1287/serv.1120.0038.
6. Rana, Diksha Rana. 2022. "Top 11 cloud platforms for Internet of Things (IoT): DZone." Top 11 Cloud Platforms for Internet of Things (IoT) (blog). May 19, 2022. https://dzone.com/articles/10-cloud-platforms-for-internet-of-things-iot.
7. Hagarty, Rich, and Einar Karlsen. 2019. "Introduction to IBM Watson Studio." Learn the basics of IBM Watson Studio and machine learning services (blog), September 2, 2019. https://developer.ibm.com/articles/introduction-watson-studio/.
8. Deshmukh, Surbhi. 2021. "Internet of Things (IoT) and Its Applications in Today's World." October 4, 2021. https://www.evosysglobal.com/blog/iot-and-its-applications-in-todays-world.
9. Guan, Allen Lim Chong, Mani Manavalan, Alim Al Ayub Ahmed, Mir Mohammad Azad and Md. Shelim Miah. 2022. "Role of Internet of Things (IOT) In enabling productive work from home (WFH) for environmental volatiles." *Academy of Marketing Studies Journal* 26: 1–11.
10. Babu, Sobhan B., K. Srikanth, T. Ramanjaneyulu, and Inampudi Lakshmi Narayana. 2013. "IoT for Healthcare." *International Journal of Science and Research* 5(2). https://www.ijsr.net/archive/v5i2/NOV161096.pdf.
11. Matthews, Kayla. 2020. "6 Exciting IoT Applications in Healthcare." June 26, 2020. https://www.iotforall.com/exciting-iot-use-cases-in-healthcare.
12. Ponnusamy, Vijayakumar, and Sowmya Natarajan. 2021. "Precision agriculture using advanced technology of IoT, unmanned aerial vehicle, augmented reality, and machine learning." In *Smart Sensors for Industrial Internet of Things: Challenges, Solutions and Applications*, edited by Deepak Gupta, Victor Hugo C. de Albuquerque, Ashish Khanna, and Purnima Lala Mehta, pp. 207–229. Cham: Springer International Publishing. https://doi.org/10.1007/978-3-030-52624-5_14.
13. Kodali, Ravi, Govinda Swamy, and Lakshmi Boppana. 2015. An implementation of IoT for healthcare. *2015 IEEE Recent Advances in Intelligent Computational Systems (RAICS)*, Trivandrum, 10–12 December 2015. https://doi.org/10.1109/RAICS.2015.7488451.
14. Singh, Vishakha, Sandeep S. Udmale, Anil Kumar Pandey, and Sanjay Kumar Singh. 2021. "IoT for Health Insurance Companies." In Sanjay Kumar Singh, Ravi Shankar Singh, Anil Kumar Pandey, Sandeep S. Udmale, and Ankit Chaudhary (Eds.), *IoT-Based Data Analytics for the Healthcare Industry*, pp. 139–147. Elsevier. https://doi.org/10.1016/B978-0-12-821472-5.00008-9.
15. Muslim, Muhammad Aziz, Raden Arief Setyawan, Achmad Basuki, Angger Abdul Razak, Fakhriy P. Hario, and Edward Fernando. 2021. "IOT based climate monitoring system." *IOP Conference Series: Earth and Environmental Science* 746(1): 012044. https://doi.org/10.1088/1755-1315/746/1/012044.
16. Venu, Nookala, Aarun Arun Kumar, and Allanki Sanyasi Rao. n.d. "Smart agriculture with internet of things and unmanned aerial vehicles." *NeuroQuantology* 20(6): 9904–9914. doi: 10.14704/NQ.2022.20.6.NQ22966.
17. Tyagi, Amit Kumar, and Niladhuri Sreenath. 2023c. "Introduction to intelligent transportation system." In *Intelligent Transportation Systems: Theory and Practice*, edited by Amit Kumar Tyagi and Niladhuri Sreenath, pp. 1–22. Disruptive Technologies and Digital Transformations for Society 5.0. Singapore: Springer Nature Singapore. https://doi.org/10.1007/978-981-19-7622-3_1.
18. Kuncara, Tommy, Arman Syah Putra, Nurul Aisyah, and Vh. Valentino. 2021. "Effectiveness of the E-ticket system using QR codes for smart transportation systems." *International Journal of Science, Technology & Management* 2(3): 900–907. https://doi.org/10.46729/ijstm.v2i3.236.

19. Tyagi, Amit Kumar, and Niladhuri Sreenath. 2023b. "Intelligent transportation system: Need, working, and tools." In *Intelligent Transportation Systems: Theory and Practice,* edited by Amit Kumar Tyagi and Niladhuri Sreenath, pp. 201–228. Singapore: Springer Nature Singapore. https://doi.org/10.1007/978-981-19-7622-3_9.
20. Tyagi, Amit Kumar, and Niladhuri Sreenath. 2023a. "Environmental sustainability for intelligent transportation system." In *Intelligent Transportation Systems: Theory and Practice,* edited by Amit Kumar Tyagi and Niladhuri Sreenath, pp. 123–148. Disruptive Technologies and Digital Transformations for Society 5.0. Singapore: Springer Nature Singapore. https://doi.org/10.1007/978-981-19-7622-3_6.
21. Pratchett, Morgan S., Tom C. L. Bridge, Jon Brodie, Darren S. Cameron, Jon C. Day, Michael J. Emslie, Alana Grech, et al. 2019. "Australia's great barrier reef." In *World Seas: An Environmental Evaluation*, pp. 333–362. Elsevier. https://doi.org/10.1016/B978-0-08-100853-9.00014-2.
22. Lokhande, Nishant. 2022. "Understanding the importance of IoT (Internet of Things): Pros, cons, and opinions." *Understanding the Importance of IoT (Internet of Things): Pros, Cons, and Opinions (blog),* September 7, 2022. https://techresearchonline.com/blog/importance-of-internet-of-things-iot-pros-cons/.
23. Gillis, Alexander S. 2022. "What Is the Internet of Things (IoT)?" March 2022. https://www.techtarget.com/iotagenda/definition/Internet-of-Things-IoT.
24. Fingent, Roshna R. 2022. "How cloud computing has changed the future of internet technology." Https://venturebeat.com/, April 15, 2022. https://venturebeat.com/datadecisionmakers/how-cloud-computing-has-changed-the-future-of-internet-technology/.
25. Williams,Brain.n.d."UpcomingGlobalInternetofThingsMarket."GlobalInternetofThings Market (blog). https://vocal.media/journal/upcoming-global-internet-of-things-market.
26. Pierce, Paul, and Bo Andersson. 2017. "Challenges with smart cities initiatives: A municipal decision makers perspective." *Conference: HICSS 2017.* https://doi.org/10.24251/HICSS.2017.339.
27. Jawed, M. S., and M. Sajid. 2022. "A comprehensive survey on cloud computing: Architecture, tools, technologies, and open issues." *Journal: International Journal of Cloud Applications and Computing (IJCAC)* 12(1): 77.
28. Haidri, R. A., M. Alam, M. Shahid, S. Prakash, and M. Sajid. 2022. "A deadline aware load balancing strategy for cloud computing." *Concurrency and Computation: Practice and Experience (Wiley)* 34(1): e6496.
29. Qasim, M., M. Sajid, and M. Shahid. 2022. "Hunger games search: A scheduler for cloud computing." *International Conference on Data Analytics for Business and Industry*, Sakhir, Bahrain, pp. 170–175.
30. Bilal, K., M. Sajid, and J. Singh. 2022. "Blockchain technology: Opportunities & challenges." *International Conference on Data Analytics for Business and Industry*, Sakhir, Bahrain, pp. 519–524.
31. Hamid, Bushra, Nz Jhanjhi, Mamoona Humayun, Azeem Khan, and Ahmed Alsayat. 2019. "Cyber security issues and challenges for smart cities: A survey." In *2019 13th International Conference on Mathematics, Actuarial Science, Computer Science and Statistics (MACS)*, pp. 1–7. Karachi, Pakistan: IEEE. https://doi.org/10.1109/MACS48846.2019.9024768.

17 Utilization of Artificial Intelligence in Electrical Engineering

Shailesh Kumar Gupta

CONTENTS

17.1 INTRODUCTION

In 1956, John McCarthy invented the term "artificial intelligence," when he organized the first scholarly meeting on the topic. The search to see if robots are capable of thinking for themselves began far earlier. In his intellectual work, as we may think [1], he initiated a framework for improving people's individual understanding and expertise. Five years later [2], he published a study on the idea of robots being able to emulate humans and perform complex tasks such as playing Chess. Nobody can disagree that

DOI: 10.1201/9781003438588-17

a computer is capable of logic processing. However, many individuals doubt are here that a machine can think. Because there has been a lot of debate over whether this idea is even viable, the definition of think is crucial. This branch of research arose from the belief that human intellect can be defined that it is exactly enough for a computer to mimic it. Innovative area of study in artificial intelligence (AI) combines psychology, information science, system science, thinking, perceptional knowledge science, and biology to research and create theory, techniques, and application systems for imitating and increasing human intellect. Changing demand and supply patterns, rising demand, efficiency, and a lack of analytics required for successful management are just a few of the challenges faced by the global energy sector. Emerging market countries face more challenges. Consumers do not pay for energy responsibly because they have no incentive to utilize it; hence, a significant amount of electricity is not metered or paid, resulting in losses and increased CO2 emissions. In industrialized countries, AI and comparable technologies are already being utilized in electrical engineering to improve communication between smart meters, smart grids, and IoT devices. These approaches can assist with power management, transparency, and efficiency, as well as promoting the use of renewable energy.

In the subject of electrical engineering, AI has created a lot of promise and room for improvement. This can result in considerable improvements not just in terms of cost, but also in terms of safety and operational management. AI has been widely utilized in various fields, including robotics and computer programming, because it has progressed to the point where it can be used in a variety of businesses to make labor easier. AI has recently been shown to be a viable alternative in the area of electrical engineering, notably as a solution to several long-standing power system difficulties where traditional approaches have failed. In the disciplines of power system planning, operation, diagnostics, and design, powerful computer tools are becoming more and more important tools for addressing difficult problems. AI is one of these computer technologies that has gained popularity in current years and has been used in various electrical engineering applications.

This article discusses AI at first, then goes on to explore various AI technologies and how they are used in electrical engineering. An ANN control in a dual 2-l three-phase inverter system to produce multi-level output is shown for better comprehension. Results and a conclusion are given at the end.

17.2 ADVANTAGES AND LIMITATIONS OF ARTIFICIAL INTELLIGENCE

AI, as a relatively new technology, has both benefits and drawbacks. We may determine how AI and electrical engineering can be combined and related areas to develop a better system by learning more about its benefits and limits. AI applications have huge benefits and have the potential to change any industry [3].

1. AI may fully eliminate human mistake as the quantity of data available increases, resulting in a significantly more accurate result when compared to human findings. This is a significant benefit since precision is critical in

the area of electrical engineering, where a single error may not only interrupt the power supply but also result in disastrous consequences.

2. Another key benefit of AI is that people are not compelled to contact directly with circumstances that might be harmful or dangerous to their careers. In the case of a fire in a flat, AI-powered robots can stroll around the building and reach out the fire, reducing the danger of people's death while still achieving the goal.
3. Tedious and boring jobs include sending the identical email to a bunch of individuals or set up a certain application on each and every employee's PC. AI may be utilized to undertake such jobs with ease, guaranteeing that not only the task is completed, but also that human effort is not squandered. Manufacturing and automation of various electrical components might be a potential use.
4. AI is typically faster due to the fact that it is eventually carried out by machines. Human intelligence differs from AI in that humans consider all variables pragmatically and emotionally before making a judgment based on what they believe is right. As a result, the conclusion might differ greatly from one individual to the next. AI, on the other hand, produces a result that takes into account all circumstances and is dependent on how it was designed. This method dramatically reduces the time it takes to solve problems and make decisions. This can be critical in the event of power outages or interruptions in power distribution to customers.
5. Problems might strike at any time. A system must be capable of dealing with any issue that may arise at any moment. Because humans cannot be available 24 hours a day, to satisfy the demand for supervision during the day, employees must labor in shifts. AI is accessible at all hours of the day because it is a computerized program that analyses the multiple available inputs to the system. AI-enabled machines and robots do not wear out like people, allowing them to operate for longer periods of time.
6. The conventional classical controller is often required to design in accordance with the model with controlled scheme, but the model building will typically include several unpredictable aspects, such as parameter changes and mathematical kind, making the design more challenging. Controlling AI is simple, furthermore, the AI function approximator doesn't require control over the object's model. Outcomes can be quickly improved by properly modifying related parameters. The controller with fuzzy logic technique, for example, responds quicker than the ideal PID controller and has a reduced overreach.
7. The AI controller is more flexible to noble data or information than the conventional controller and is easier to alter. As a result, it is quite convenient to use.
8. The typical control technique is tailored to the unique item; therefore, only for object control, effect is excellent, but inconsistent for other control objects. Whether for defined or unknown input data, the AI control technique can get acceptable consistency estimation.

The numerous benefits that AI provide may be sufficient to persuade someone that it can be a viable alternative for usage in the sector of electrical engineering. However, it is equally critical to comprehend AI's flaws.

1. Once it comes to if AI should be applied in different field, this is one of the most contentious issues. Some individuals feel that the adoption of AI in many sectors would make work simpler, while others predict that AI will be able to substitute human workers in many sectors, therefore removing the requirement for humans. Many people who rely on this labor to pay for their food and expenses may face unemployment and income loss as a result of this.
2. AI is not only available 24 hours a day, but it also remains alert and productive throughout the day. As previously said, this minimizes the need for human administration, making us feel at ease and lazy because we assume AI is capable of handling the problems at hand.
3. AI is capable of solving problems and making decisions faster than humans, nevertheless at an amount. Humans also think sensitively, and this affects their decision-making in some ways, but AI does not take this into consideration. AI can only do what it has been taught, thus programming emotions will necessitate a significant deal of research, study, and effort from specialists all around the world.
4. Developing and testing a competent AI to deal with real-life events would take a long time due to a scarcity of highly experienced programmers and the difficulties of establishing a good model. Integration of AI takes huge effort and time, as well as handsome amount of money. This implies that a large amount of money must be invested in AI in order for it to be a feasible option.

Now that we have a firm handle on the various benefits and drawbacks of AI, we can consider how to integrate AI into engineering sector especially electrical branch, as well as the myriad utilities and challenges that may be resolved as an outcome.

17.3 AI TECHNOLOGIES IN ELECTRICAL ENGINEERING

17.3.1 Expert Systems

Expert systems use an inference engine that pulls information about a specific domain from a knowledge base, mostly in the form of if-then rules, to solve issues. In simple terms, an expert system is a software system that has the decision-taking capabilities of a human expert, showing that these systems are experts in a certain area of study. This can be a concern because these systems are incapable of dealing with new problems or situations. Because expert systems are essentially computer programs, it may be utilized to calculate and decide features and parameter in the generation, distribution, and transmission of electricity. These systems, which have been around since the 1970s, are less flexible than modern AI systems, but they are typically easier to develop and maintain [4,5].

17.3.2 Artificial Neural Network

Artificial Neural Network (ANN) is a sort of machine learning system that is made up of artificial synapses that are meant to mimic the structure and function of the brain. As synapses communicate data to one another, the network watches and learns, digesting input as it goes through many levels. An ANN is made up of artificial neurons, which are interconnected nodes or units that are modeled after biological brain networks. The aim of ANN is to mimic the behavior of a human brain and, as a result, manage problems in the same way that a human brain does by employing different approaches such as problem-solving and decision-making. Variations on this theme include speech recognition and computer vision. By sending the input across a network of neurons, ANNs turn a collection of inputs into a consequent number of outputs. Based on the input, each neuron in the neural network generates its own output. ANNs are utilized in power system stabilizers, load modeling, load forecasting, and state approximation [6,7].

17.3.3 Machine Learning

Machine learning (ML) is a set of procedures and statistical approaches that allow computers to recognize patterns, draw inferences, and learn to do tasks without being explicitly instructed. Using the information and skills learned, ML reorganizes the current knowledge system and improves its performance over time. In the implementation of expert systems, ML techniques are frequently utilized to alleviate the knowledge acquisition bottleneck [8]. Experts are better at gathering and storing examples than they are at formally stating their expertise and encountered scenarios into manufacturing rules, resulting in the KA bottleneck [9]. When ML approaches are applied to deal with this bottleneck, performance is automatically gathered from the data [10]. Symbolic data can be included in an ANN learning algorithm as well as the learning system can support extraction and knowledge modeling.

17.3.4 Fuzzy Logic Systems

In all fields, fuzzy logic (FL) is becoming a fundamental way of problem-solving. It has a huge impact on the design of self-driving intelligent systems. Unlike Boolean logic, which only has two output values, 0 and 1, FL can be written as a multi-parameter logic. FL generates truth parameters that can be any real number between zero and one in order to encompass the concept of partial truth. There are a few instances where the output is entirely true or absolutely false, but the most of the time, there is some truth or deceit. The word fuzzy comes from the fact that humans frequently make decisions based on sloppy and non-numerical facts.

Stability control, voltage management, stability analysis, power flow control, load forecasting, and transmission line performance improvement are all applications where fuzzy logic is applied [11].

17.3.5 Deep Learning

ANNs are used in deep learning (DL) to tackle problems with various hidden layers. DL is a sort of machine learning that uses data representations rather than task-specific methods to learn. Data science, which also confines predictive modeling and statistics, accommodates deep learning as a key component. Data scientists who collect, analyze, and interpret vast volumes of data find deep learning to be of great value since it speeds up and makes the process easier. Supervised, unsupervised, and partially supervised learning are the three forms of learning. ANN research, a relatively new discipline of machine learning, gave rise to the phrase "deep learning." [12]. Some of the models, such as neural coding, which aims to find a link between various inputs and linked neuronal responses in the brain [13], are built on perceptions of information dispensation and communication arrangements in the biological nervous system. The self-driving car is the greatest example of deep learning in the world of electrical engineering.

17.3.6 Pattern Recognition

Pattern recognition methods are well suited to proceed monitoring due to the expected correlation between defect classes and data patterns while disregarding underlying process states or structures [13]. ANNs are one of the most extensively utilized pattern recognition approaches. The focus of pattern recognition research is split into two areas: the first is the perception of an object, which is covered by the category of scientific knowledge, and the second is pattern recognition by computer when the case's objective is known. Electrical pattern recognition is useful in a variety of applications; it is typically used to detect specific events or anomalies in the signal under examination, as well as to identify precursors, particularly in electrophysiology. Each application necessitates its own set of algorithms and signal processing skills. A pattern recognition-based short-term load forecasting approach that acquires input sets from a multi-layered fed-forward neural network, recent applications in the electrical area include automatic pattern recognition on electrical signals used to neutron gamma discrimination.

17.4 UTILIZATIONS OF AI IN ELECTRICAL ENGINEERING

With each new day, we progress technologically. Progress in the field of AI has improved our lives. These beautiful technologies are now employed to optimize systems and satisfy the objectives of the desired organization. AI and machine learning not only improve the system's performance, but they also solve business challenges like never before. Furthermore, problems are dealt with more effectively and quickly than previously. Overall, incorporating the most recent AI and machine learning technologies may prove to be a way to greater heights. Electrical engineering systems generate huge volumes of data. As a result, data mining can be used to find novel relationships in these systems. With the introduction of deep neural networks, we can now learn new mappings between the inputs and outputs of these systems, thanks to technological advancements. On that point, have a look at some of the most useful AI and ML applications in the area of electrical engineering that have made our lives easier. As previously said, traditional approaches cannot resolve numerous issues in power systems since the criteria for these ways to work are not viable at the moment. AI might be utilized in a variety of fields as depicted in Table 17.1.

TABLE 17.1
Application-wise AI Techniques [14–25]

S. No.	Example	Applications	Techniques
1	Hydrothermal coordination, unit commitment, maintenance scheduling, economic load dispatch, and power flow	Power system operation	Artificial intelligence
2	Planning—Power system consistency, Planning—Generation expansion, Planning—Reactive power and Planning—transmission expansion	Power system planning	
3	Stability control, power flow control, load frequency control, and voltage control	power system control	
4	Power plant	Power plant management	
5	Recovery, management, network security, and fault diagnostics	Power system automation	
6	Demand side management and response, distribution system operation and planning, network setup, and operation and control of smart grid	Distribution systems	
7	Renewable energy resources Planning-Distributed generation, control of solar PV power plant, and wind turbine plant	Distributed generation applications	
8	Long-term and short-term load forecasting, solar power and wind power forecasting, and electrical market forecasting	Forecasting	
9	Allocation of FACTS devices, economic load dispatch, automatic generation control, hydrothermal scheduling power system protection, Relays for stability analysis	Power system	Heuristic search algorithms PSO, GSA, GA, EP, BFO
10	Motor conditioning and speed control, online unsupervised diagnosis, real-time, dynamic updating of the network without retraining the circuit. Filtering out transients, noise and disturbances during faults, fault prediction in incipient stages on account of working operation, Detection of stator and rotor inter-tum faults, steady state and transient analysis of electrical machines, short circuit detection, speed torque, flux, position estimation, vector and DTC control of induction motor	Electrical machines an drives	Artificial neural network

(Continued)

TABLE 17.1 (*Continued*)
Application-wise AI Techniques [14–25]

S. No.	Example	Applications	Techniques
11	Evaluation of performance indices utilizing linguistic flickering data, predict abnormal operation and locating defective part, system modeling nonlinear mapping, using human expertise reflected to FL, optimizing diagnostic structure parameters, Fault classification DTC on induction motor	Electrical machines and drives	FUZZY logic
12	Smart grid, battery management, fault detection, solar power plant, autonomous vehicle, control system	Power system, control system and electrical machine	Machine learning
13	Islanding detection	Power system	Pattern recognition

Because electrical engineering is such a specialized area, it needs the utilization of expert systems to resolve issues by making decisions, solving problems, and archiving knowledge using logic, judgment, as well as heuristics. The principle application of fuzzy logic systems is fault diagnostics. Assume there has been a transmission line failure. The flaw detector's information could be sent into the FL system. After that, the fuzzy system analyses the data to provide us a clear picture of the situation.

ANNs can be utilized to improve the working efficiency of transmission lines. The transmission lines can be connected to a variety of sensors to monitor the environment and other ambient parameters in real time. These events may be sent into an ANN, which will analyze them and adjust the line's parameters to improve performance. The efficiency of the ANN is directly proportional to the improvement in performance. To improve the ANN's efficiency and operation speed, the count of hidden neuron layers can be raised. It is self-evident that AI can have an impact on power systems. Our current systems will be more efficient and reliable, thanks to AI.

17.4.1 Utilization of AI in Electrical Component and Machine

The electrical architecture for electrical automation control is where AI first appears. As we all know electrical machinery is a complicated construction. Electrical equipment design is a complication because it requires not only knowledge of electromagnetic fields, electronics, motors, circuits, automation, as well as other sectors, but also an understanding of sensors, generators, and other devices of the role and principle, as well as a high professional level and work experience for the designer [26]. The functioning of the electrification setup in electrical automation gadget is a very sophisticated topic that encompasses several fields and disciplines. Its needs for control and operation necessitate a great level of knowledge conserves and improved quality. AI technology is a much better approach to realize the regular operating of electrical automation equipment. It is possible to achieve the automatic functioning of electrical device and to substitute human work through computer programming and operation, resulting in a significant reduction in labor costs. Simultaneously, it dramatically enhances the speed by utilizing AI technologies.

17.4.2 Utilization of AI in Control of Electrical Machinery

The control of electrical automation is essential in the electrical organization; if it is achieved, production efficiency may be raised while generation costs and human resource expenditures are lowered. Electrical automation control uses FL, expert systems, ANNs, and other AI technologies. AI may aid not only general operation progress in the area of automation control in electrical, but also the creation of automated control progress, in the development of automation. As a result, AI technology is being utilized to increase human awareness of mechanical capacity and strengthen electrical automatic control in electrical engineering field. [27]. When it comes to computer engineering, one of the finest uses of AI has been in power systems. AI has covered everything from detecting defects to predicting. AI has done a fantastic job of decreasing the strain of human operators by taking over duties like data processing, regular maintenance, and training, to name a few.

17.4.3 Utilization of AI in Fault Analysis

AI can be utilized in the fuzzy "neural network" expert systems logic to notice errors quickly and accurately, to detect the reason, nature, and place of the failure along with to control the fault repair process in a timely manner, which is an upright sign of electrical apparatus's long-term viability [28–30].

Case-based reasoning (CBR), rule-based reasoning (RBR), and fault-based tree fault diagnosis are examples of AI-based fault diagnostic methodologies. A mechanical defect detection system depends on CBR and RBR reasoning is built based on the fundamental composition and basic concept of the traditional expert system.

Once a problem with electrical equipment arises, the symptom and practical difficulties of its significance are quite complicated, which makes it difficult to access and determine whether the employment of AI systems would solve this tough problem. Expert systems, FL, and neural networks are all leveraging AI technology to assess fault systems. The transformer is a highly frequent and popular component in power systems, and there is a lot of study on it. Currently, transformer fault analysis and accessing the degree of the defect is done mostly in the transformer oil, because there is a gas breakdown. Generator failure detection employing AI technologies is also quite widespread in the generator and motor industries. In the fuzzy "neural network" expert systems logic, AI may be employed in a timely manner. This allows not only for precise fault detection, but also for determining the failure reasons, the kind and place of the failure, and prompt fault repair control.

17.5 ANN CONTROL IN DUAL 2-L THREE-PHASE INVERTER SYSTEM TO ACHIEVE MULTI-LEVEL OUTPUT

Variable-speed ac drives are now the industry's workhorse. Inverters are the most common way to get a variable voltage and frequency supply. When it comes to multi-level inverter networks, a lot of work has done on multi-level converters. Power electronics technology has progressed to the point that it is now possible to achieve high safety voltage with lower harmonics. Switching within the inverters may be one of the reasons for harmonics in output voltage. For 2-L three-phase voltage source inverters, a number of pulse width modulation (PWM) methods have been proposed but due to better dc bus usage and simpler digital execution and, space vector PWM is the most common alternative and implemented earlier. Here, the author proposed ANN-based PWM technique of AI to achieve same results with low harmonic distortion. Validation of the findings is provided by the entire Simulink model of a three-phase VSI for 3-L operation and simulation results.

17.5.1 Dual 2-L Three-Phase VSI

There are six switching devices in each three-phase inverter, but only three of them are independent since the action of two power switches on the same leg is complimentary. When these three switching states are coupled, there are eight possible space voltage vectors. Six of the eight vectors (states 1–6) create non-zero output voltage, while the other two (states 0 and 7) produce zero output voltage and are referred to as zero voltage vectors. As illustrated in Figure 17.1, the space vectors create a hexagon with six unique sectors, each of which spans 60. A set of three vectors (one zero and two active) can be selected in space vector PWM to synthesis the required voltage in each switching period and provide a reference vector at any point in time.

Space vector equation for three phase is presented as

$$v_{\text{ref}} = \frac{2}{3}\left(v_a + \bar{a}v_b + \bar{a}^2 v_c\right) \tag{17.1}$$

Notation $\underline{a} = \exp(j2\pi/3)$ and $\underline{a}^2 = \exp(j4\pi/3)$. The space vector is an illustration of all three-phase standards at the same time. In contrast to the phasors, it is a complex data that is a function of time. To generate the PWM signals that produce the rotating vector, the PWM time intervals for each sector is

$$T_P = \frac{\sqrt{3}}{V_{\text{DC2}}} V_s T_s \sin(60 - \theta) \tag{17.2}$$

$$T_N = \frac{\sqrt{3}}{V_{\text{DC2}}} V_s T_s \sin(\theta) \tag{17.3}$$

$$T_0 = T_s - T_P - T_N \tag{17.4}$$

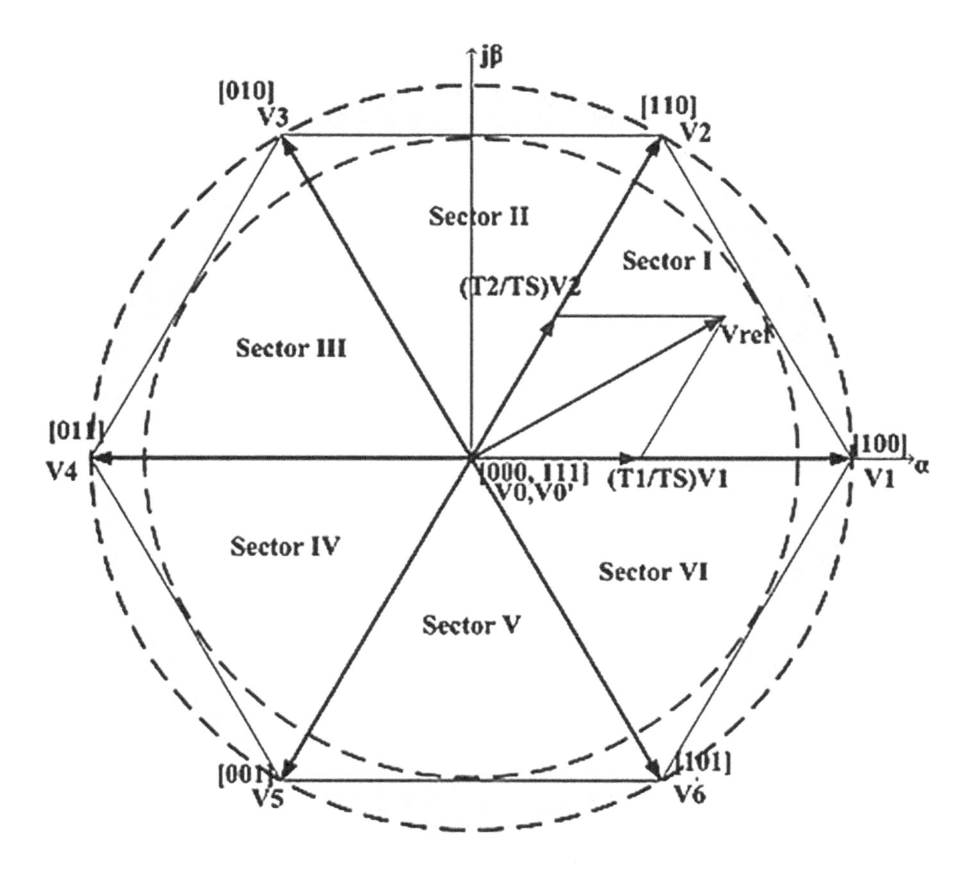

FIGURE 17.1 Space vector illustration of the switching positions.

where all symbols signify their usual meaning. The reliance of time spans T_1, T_2, and T_0 are just on the peak value and angle α of the reference vector.

17.5.2 Three-Level Operation Using Dual 2-L VSI

There are two, three-phase voltage source inverter which will generate two reference voltage vectors each i.e. V_{ref} & V'_{ref}. These two reference voltages are used to add three voltage vectors nearest to the reference vector during sampling time to generate an average voltage vector. Both reference voltages rotate in a hexagon with six sectors (I–VI) [31].

There are a total of 12 active voltage vectors for inverters in both systems during three-level operation, (V_1–V_6) for inverter-1 and ($V_1' - V_6'$) for inverter-2, and two zero voltage vectors (V_0 and V_0'), which are located at the center of origin as shown in Figure 17.2. It is necessary to reduce switching states to reduce switching losses along with total harmonic distortion for inverter operation. The SVPWM approach is already implemented on three-phase and five-phase dual 2-L inverter system [32–39]. Table 17.2 displays the sequence pattern for both V_{ref} & V'_{ref} reference voltage vectors in all six sectors to recognize a three-level operation using a dual 2-L VSI.

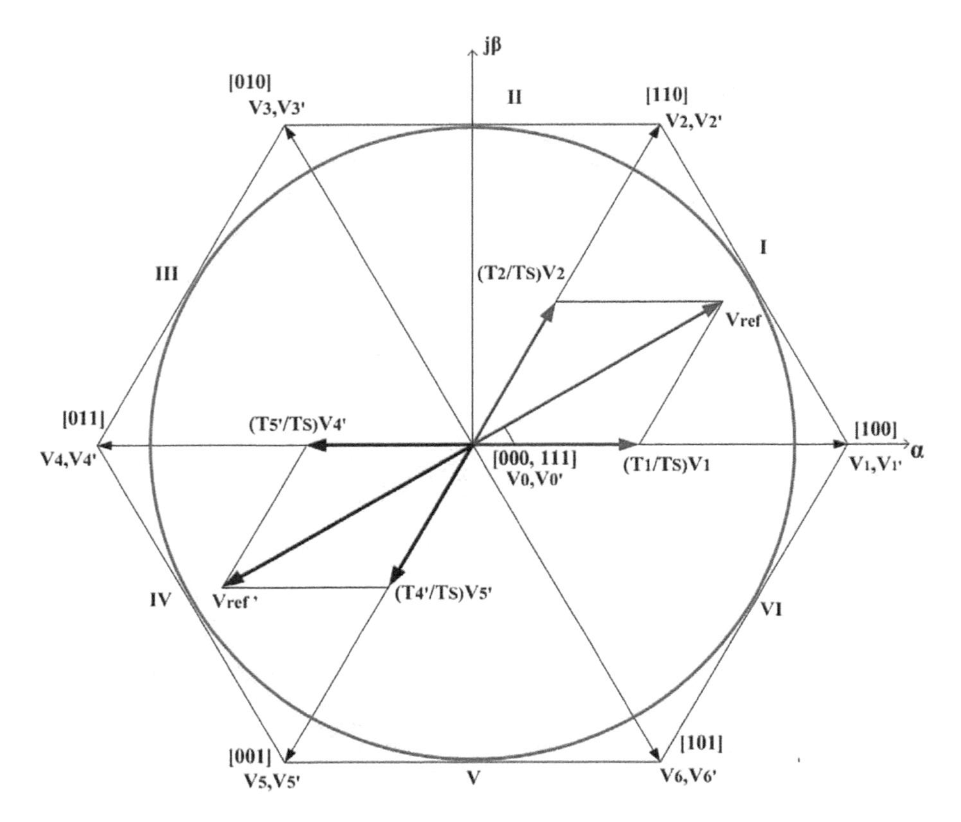

FIGURE 17.2 Space vector diagram for 3-L operation.

TABLE 17.2
Switching Order for the Dual 2-L VSI for a 3-L Output

Sector	Switching Order						
I	V_0	V_1	V_2	V_0'	V_2	V_1	V_0
	000	100	110	111	110	100	000
II	V_0	V_3	V_2	V_0'	V_2	V_3	V_0
	000	010	110	111	110	010	000
III	V_0	V_3	V_4	V_0'	V_4	V_3	V_0
	000	010	011	111	011	010	000
IV	V_0	V_5	V_4	V_0'	V_4	V_5	V_0
	000	001	011	111	011	001	000
V	V_0	V_5	V_6	V_0'	V_6	V_5	V_0
	000	001	101	111	101	001	000
VI	V_0	V_1	V_6	V_0'	V_6	V_1	V_0
	000	100	101	111	101	100	000

The general switching sequence and output voltages of the two inverters for both reference voltages V_{ref} & V'_{ref} indicated in sectors I and IV are shown in Figure 17.3 for switching pattern. Blue color represents inverter 1 and green color represents inverter-2 have been used for simplification. The voltage vectors V_0, V_1, V_2, and V'_0, V'_4, and V'_5 are used to analyze the voltage vectors for inverter 1 and inverter 2, respectively. Figure 17.3 shows the V_{ab} and $V_{a'}V_{b'}$ line voltages for both inverters in sampling time T_s.

17.5.3 ANN-Based Pulse Width Modulation Technique

An ANN is a data processing model that simulates the way organic nerve systems operate, and its structure is similar to how human nervous systems process information in the brain. Several tiny particles known as neurons are interconnected to tackle a specific challenge in this system. The capacity to drive meaning from difficult or imprecise data is built and learned by or for a specific example, and then habituated to extract pattern and spot trend for a complicated scenario that is not easily attainable in a normal method. These well-trained neurons are put to work in new situations to produce solutions. These are the most significant benefits of utilizing ANN [33].

a. An ANN learns by seeing data that is accessible. Random function approximation tools are used to estimate cost effectiveness. Instead of using full data set, a sample data are used by ANN to compute the solution. This saves money, space, and time.
b. During the learning phase, the ANN constructs a framework and a simple mathematical model so that it may quickly generate a solution for a complicated problem.
c. The capacity to do parallel and real-time operations is simple to implement.
d. Fault Tolerance aids in the retrieval of network output, although performance suffers when the ANN is fed redundant information.
e. ANNs is used to work with nonlinear statistical data, and a nonlinear data modeling tools are available to work.
f. ANN only takes part of data from the full sample to accomplish solutions of complete dataset. The system then creates structure and patterns for larger data sets or challenges.

The use of AI in the field of power electronics has lately exploded. Many research papers have utilized ANN to implement space vector PWM. While the ANN-SVM controller is incredibly quick, it does have one drawback: it is difficult to train the nonlinear modulation approach. The modeling and simulation of ANN-based SVPWM are configured using the Simulink/MATLAB package application. Detail discussion is provided in Ref. [33].

17.5.4 Simulation Results and Discussion

Figure 17.4 shows a Simulink model of a three-phase VSI built up of numerous blocks. This section makes you walk through a Simulink/MATLAB model for a

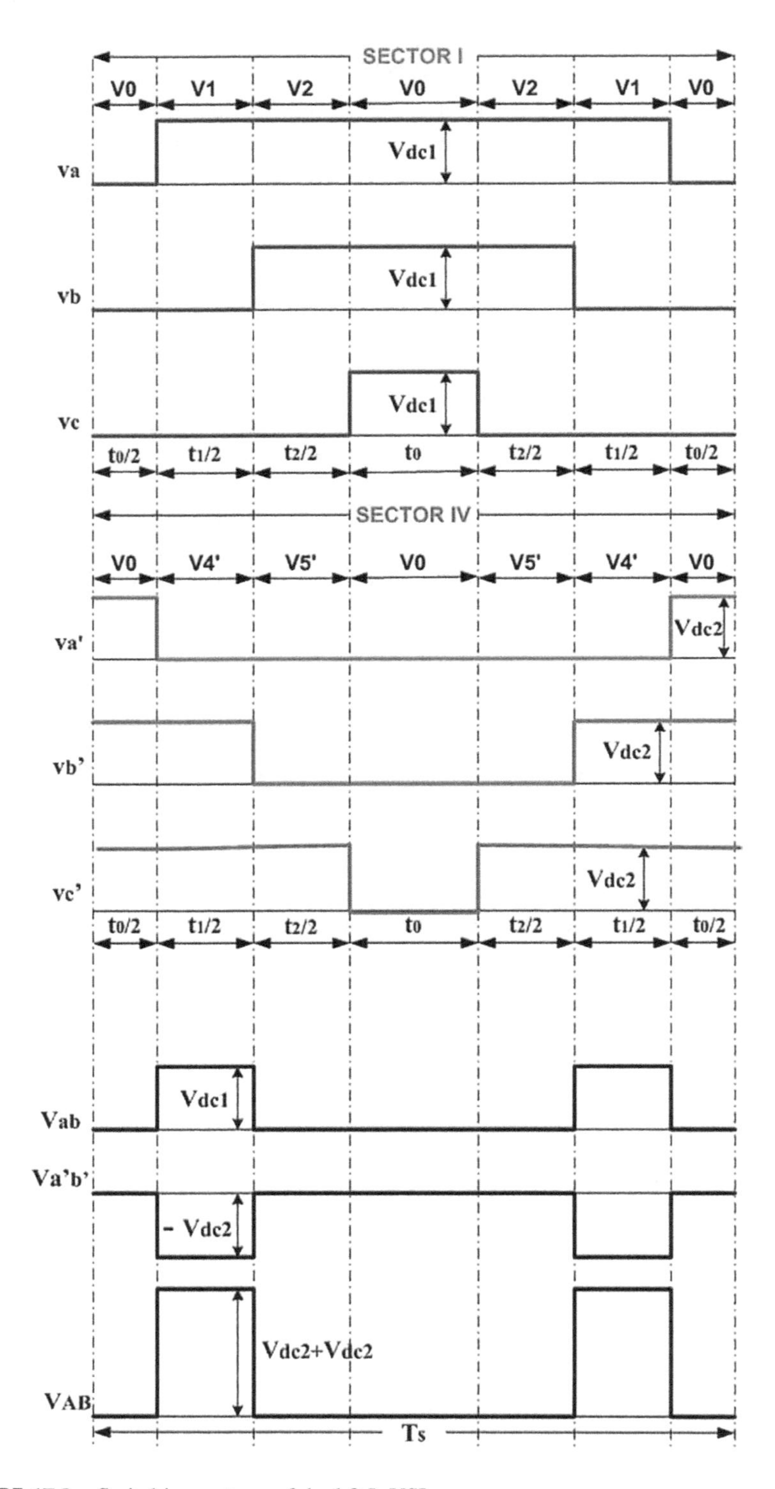

FIGURE 17.3 Switching pattern of dual 2-L VSI.

three-phase VSI that uses ANN-based PWM. The input neuron values in SVPWM are the reference voltage vector magnitude and angle, while the target values are created during switching SVPWM. The MATLAB neural network tool is then used for training and simulation, after which the ANN block is generated for the foresaid design. There are ten neurons in each of the input and output layers, as well as ten neurons in the hidden layer. The training function "trainlm" is used to simulate the network, which is of the feed-forward back propagation kind. The adaptive learning function is LEARNGDM, and the performance function is mean squared error (MSE). TANSIG (tansigmoidal) is a form of transfer function. Figure 17.5 depicts the neural network plan for the simulation. Figure 17.5a shows the input layer, which has three neurons, and Figure 17.5b shows the hidden layer, which has ten neurons. Figure 17.5c shows an input layer with six neurons and an output layer with six neurons. A three-phase VSI block generates the phase voltages. The data is filtered using a first-order filter with a time constant of 0.8 ms. The output of the voltage acquisition block is saved in the workspace. Figures 17.6–17.10 illustrate the outcomes of the simulation.

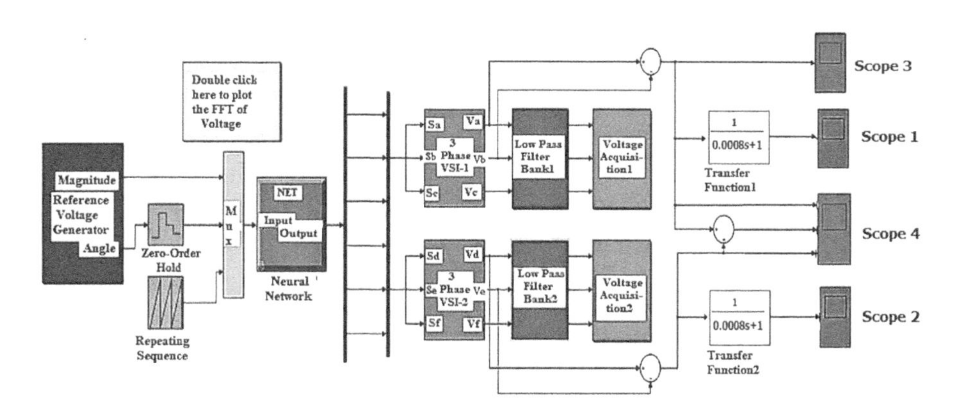

FIGURE 17.4 MATLAB/SIMULINK model of three-phase VSI.

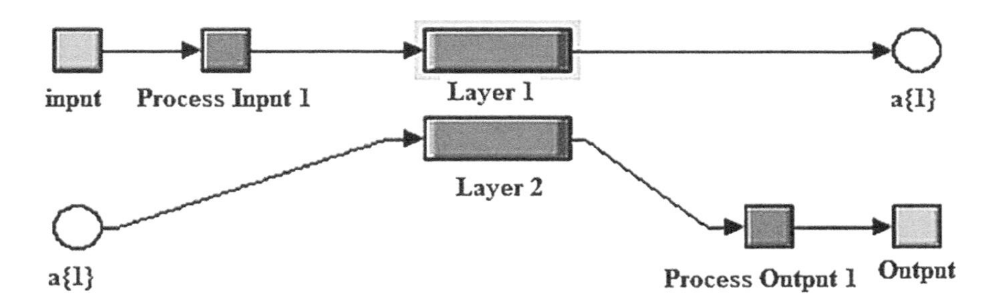

FIGURE 17.5 Artificial neural network. (a) Input layer having three neurons. (b) Hidden layer having ten neurons. (c) Output layers having six neurons.

(Continued)

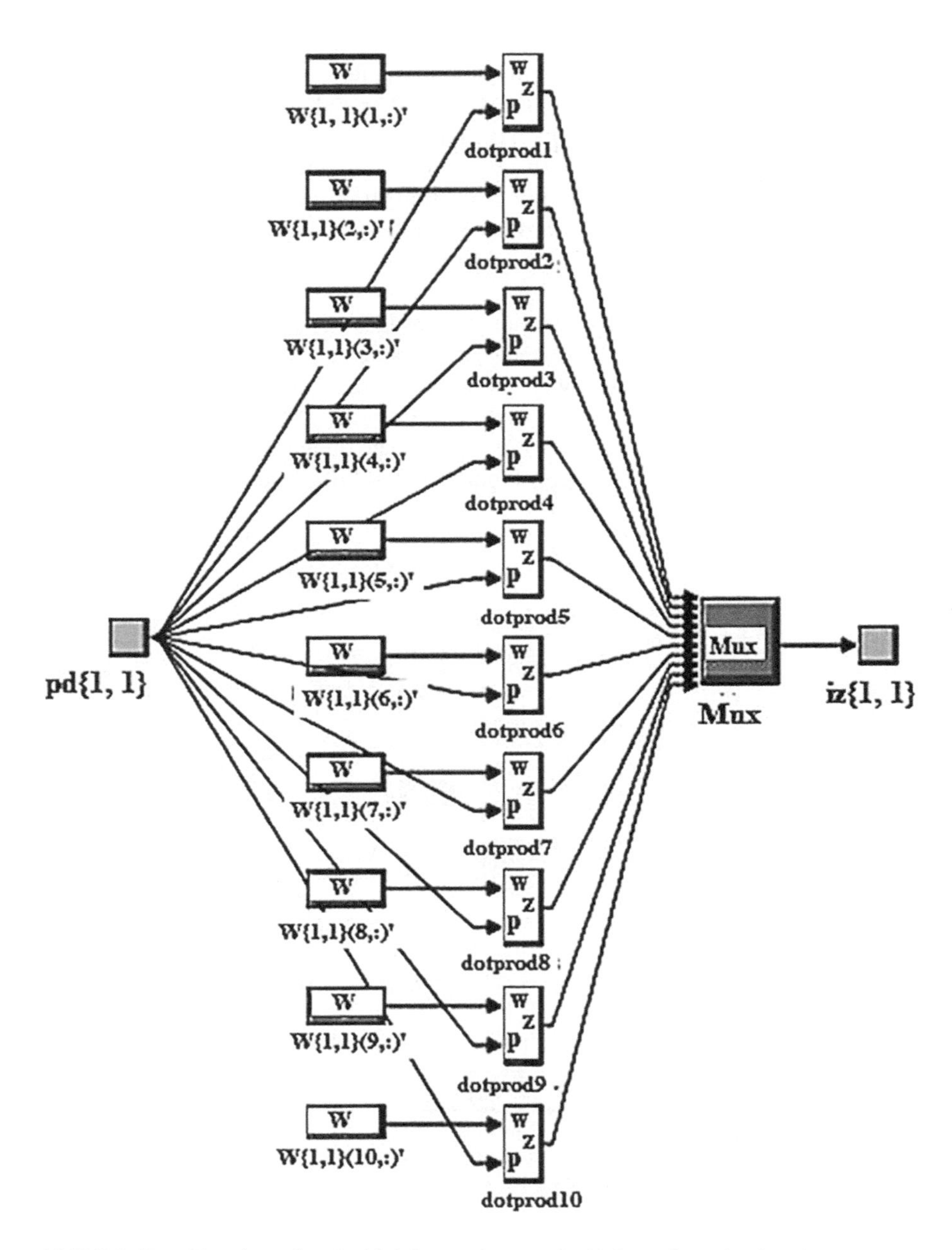

FIGURE 17.5 ***(Continued)*** Artificial neural network. (a) Input layer having three neurons. (b) Hidden layer having ten neurons. (c) Output layers having six neurons.

(*Continued*)

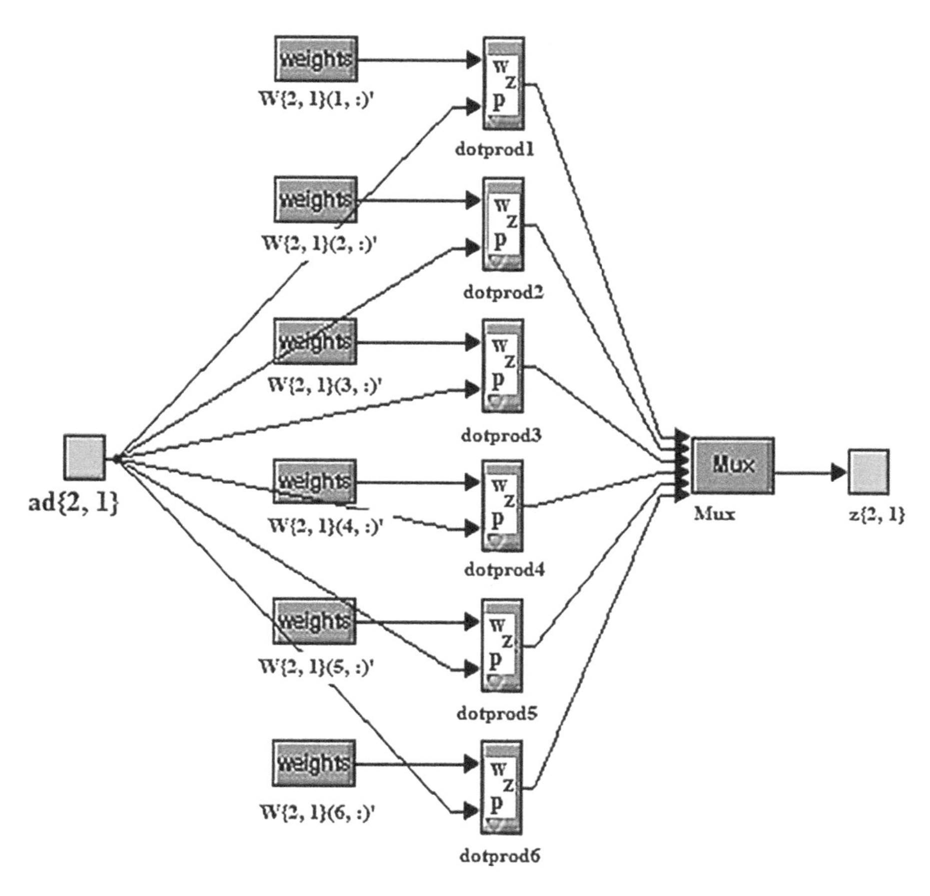

FIGURE 17.5 ***(Continued)*** Artificial neural network. (a) Input layer having three neurons. (b) Hidden layer having ten neurons. (c) Output layers having six neurons.

Figure 17.6 represents the three-phase sinusoidal input signal, whereas Figure 17.7 depicts the unfiltered phase voltages of INV1 and INV2, while Figure 17.8 depicts the filtered phase voltages. Figure 17.9 displays the line voltages of INV1 and INV2. All the results are simulated in per unit value having dc link voltage, supply frequency, and switching frequency 1, 50 Hz, and 5 kHz, respectively. Figure 17.10 shows the output voltage of the inverter along with its harmonic spectrum for the output phase 'A' voltage, showing that the output has a single fundamental voltage with a magnitude of 0.0.33549 p.u. rms at a frequency of 50 Hz. The output voltage has a total harmonic distortion of 4.81% while weighted total harmonic distortion of 5.53% of the fundamental voltage, which is near to the sinusoidal. $1 \cdot e^{-06}$ s is the sampling time during simulation.

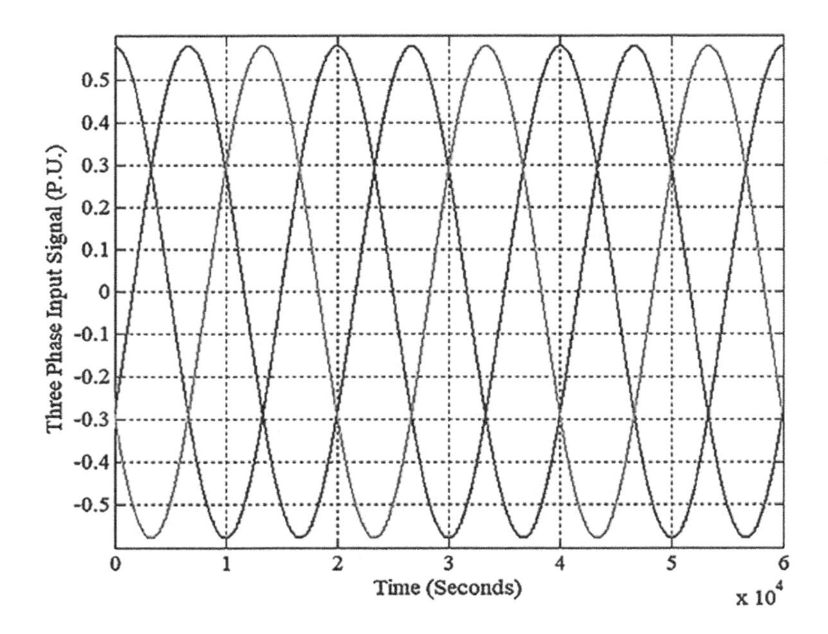

FIGURE 17.6 Three-phase sinusoidal input signal.

17.6 CONCLUSION

As in this modern era, there are variety of problems arising, which require modern solutions. Similarly, electrical engineering is also facing this problem. However, it has been proven challenging to adapt traditional manual control to the modern problems. Hence, the development of AI technologies has aided electrical engineering innovation to a large extent. The application of AI techniques have widen due to less human interference, faster response, accurate, and precise. In this book chapter, the author first discussed the benefits and limitations of AI in various fields of electrical engineering. Following that, the author discussed the utilization of various AI technologies in electrical engineering. To achieve the three-level output in two-level three-phase inverter system, the author has proposed an ANN-based PWM control of AI technique, and it is found that ANN-based control is simple, reliable, and fast to get the sinusoidal results. The THD is very less at the output of inverter. After going through the analysis, it has been also found that sinusoidal results with least THD is obtained at time constant 0.8 ms. It can be stated that AI is a highly viable technology that can be used to a variety of electrical engineering fields with the aim of not only design any system easier, but also to boost its efficiency and reliability. The researchers working in this sector will find this paper quite useful.

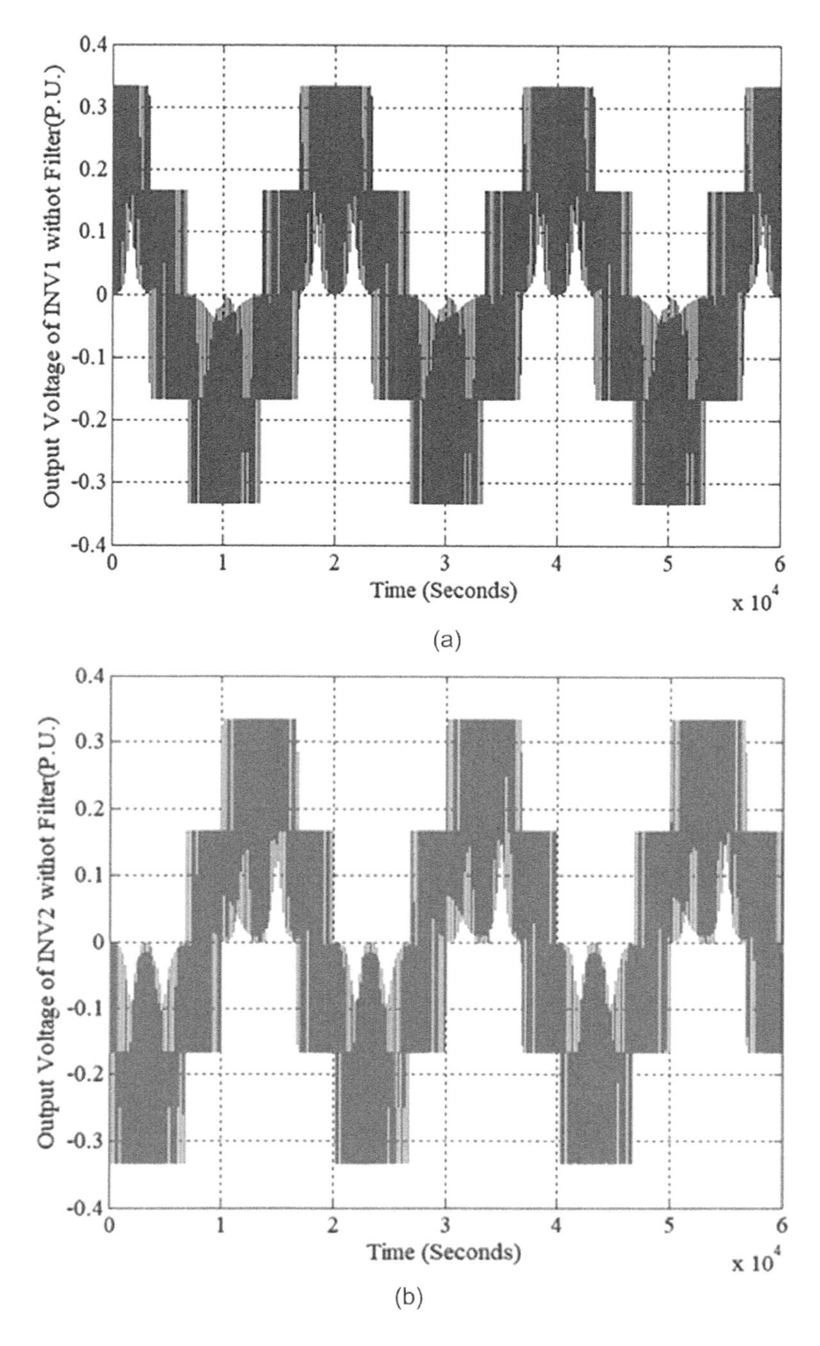

FIGURE 17.7 Output voltage without filter at INV_1 (a) and INV_2 (b).

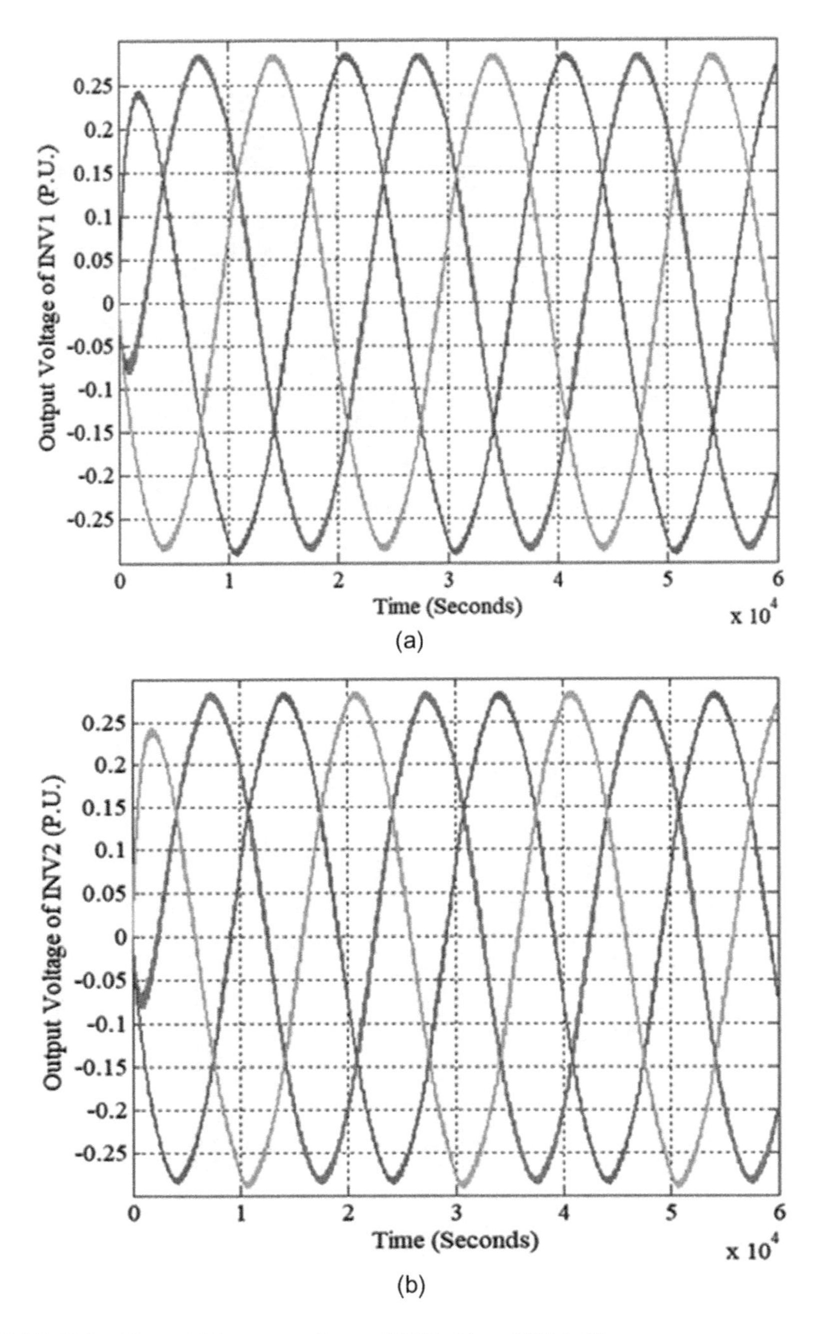

FIGURE 17.8 Sinusoidal output voltage of INV_1 (a) and INV_2 (b).

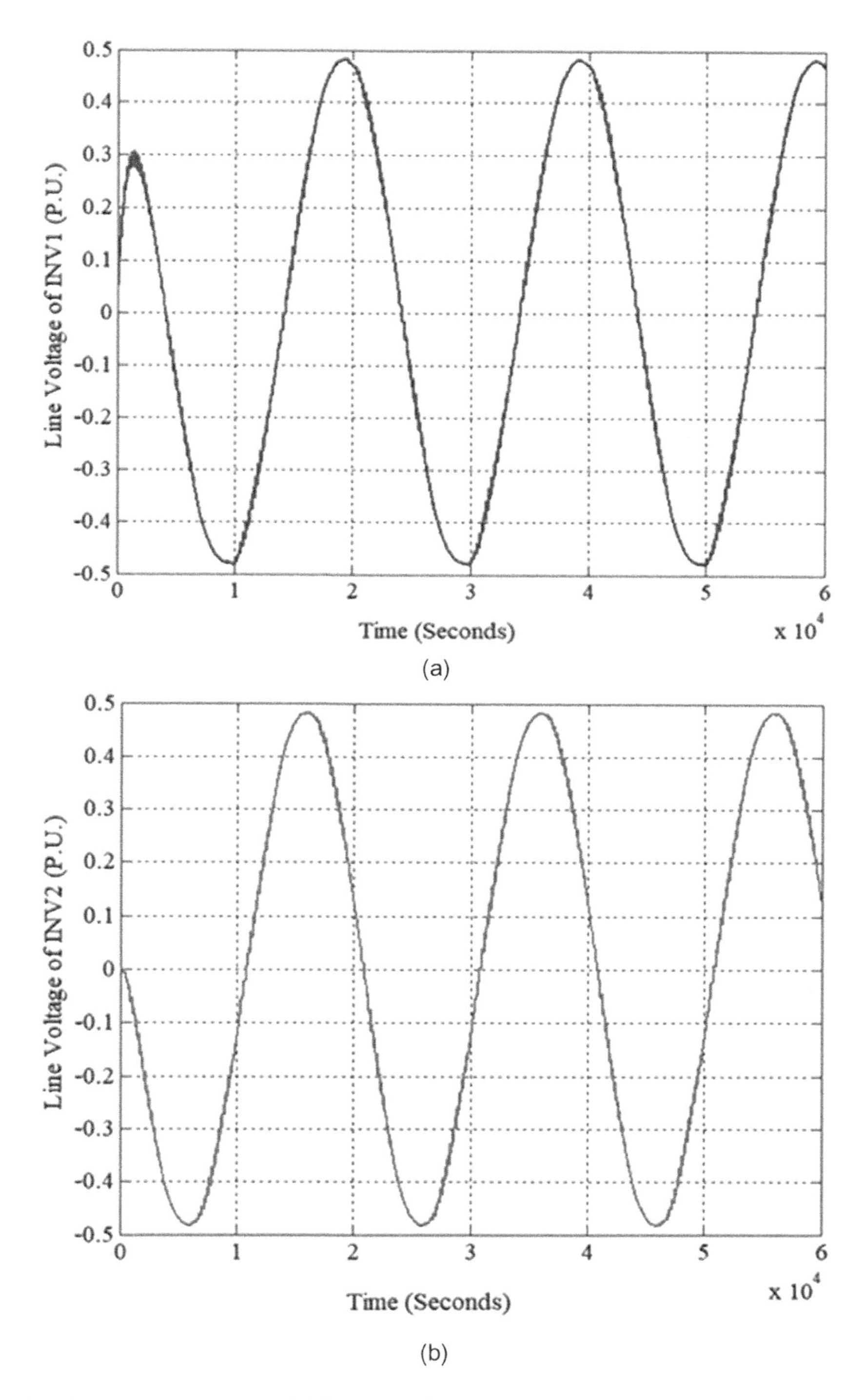

FIGURE 17.9 Line voltage of INV_1 (a) and INV_2 (b).

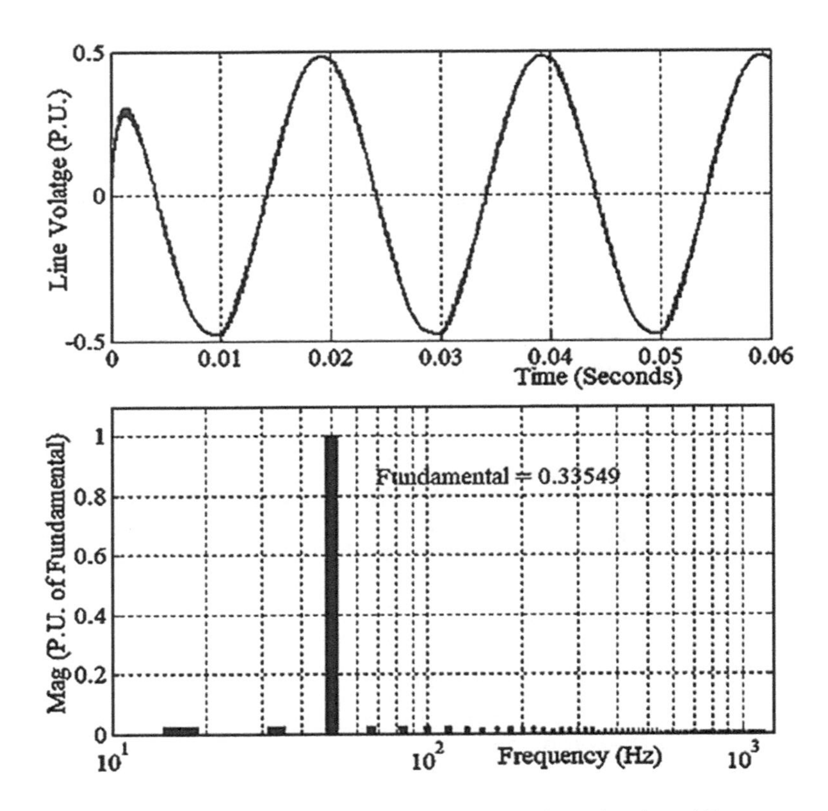

FIGURE 17.10 Harmonic spectrum of inverter output voltage for phase 'A'.

REFERENCES

1. V. Bush, "As we may think", *The Atlantic Monthly*, July 1945.
2. A. M. Turing, "Computing machinery and intelligence", *Mind*, vol. 49, issue 236, pp. 433–460, 1950.
3. S. S. Gopal, "Artificial intelligence in the field of electrical engineering", *International Journal of Engineering Research & Technology*, vol. 9, issue 7, pp. 155–158, 2020.
4. V. Uraikul, C. W. Chan, and P. Tontiwachwuthikul, "Artificial intelligence for monitoring and supervisory control of process systems", *Engineering Applications of Artificial Intelligence*, vol. 20, issue 2, pp. 115–131, 2007.
5. S. H. Liao, "Expert system methodologies and applications: A decade reviews from 1995 to 2004", *Expert Systems with Applications*, vol. 28, issue 1, pp. 93–103, 2005.
6. M. H. Hassoun, "Fundamentals of artificial neural networks", *Proceedings of the IEEE*, vol. 84, issue 6, p. 906, 1996.
7. S. Walczak and N. Cerpa, "Artificial Neural Networks". In R. A. Meyers (Ed.), *Encyclopedia of Physical Science and Technology* (Third Edition). Academic Press: Cambridge, MA, pp. 631–645, 2003.
8. D. E. Goldberg and J. H. Holland, "Genetic algorithms and machine learning", Machine Learning, vol. 3, issue 2, pp. 95–99, 1998.
9. A. K. C. Wong and Y. Wang, "Pattern discovery: A data driven approach to decision support", *IEEE Press*, vol. 33, issue 1, pp. 114–124, 2003.

10. B. R. Bakshi and G. Stephanopoulos, "Representation of process trends-III. Multiscale extraction of trends from process data", *Computers & Chemical Engineering*, vol. 18, issue 4, pp. 267–302, 1994.
11. E. Dadios, *Fuzzy Logic Algorithms, Techniques and Implementations.* IntechOpen: London, 2012.
12. Y. Lecun, Y. Bengio, and G Hinton, "Deep learning. Nature, vol. 521, pp. 436–444, 2015.
13. V. Uraikul, C. W. Chan, and P. Tontiwachwuthikul, "Artificial intelligence for monitoring and supervisory control of process systems", *Engineering Applications of Artificial Intelligence*, vol. 20, issue 2, pp. 115–131, 2007.
14. M. A. Awadallah and M. M. Medhat, "Application of AI tools in fault diagnosis of electrical machines and drives-an overview", *IEEE Transactions on Energy Conversion*, vol. 18, issue 2, pp. 245–251, 2003.
15. Analytics Insight, "Applications of AI and Machine Learning in Computer Science and Electrical Engineering," Report, September 11, 2021.
16. B. K. Bose, "Neural network applications in power electronics and motor drives: An introduction and perspective", *IEEE Transactions on Industrial Electronics*, vol. 54, issue 1, pp. 14–33, 2007.
17. Ibrahim, W. R. A. and M. M. Morcos. "Artificial intelligence and advanced mathematical tools for power quality applications: A survey." *IEEE Transactions on Power Delivery*, vol. 17, issue 2, pp. 668–673, 2002.
18. O. Singh, S. K. Gupta, S. Urooj, and J. Sagar, "Pattern recognition technique based islanding detection scheme in grid-connected PV System", *17th IEEE India Council International Conference (INDICON-2020), NSUT*, New Delhi, 11–13 December, 2020.
19. B. K. Bose, "Expert system, fuzzy logic, and neural network applications in power electronics and motion control." *Proceedings of the IEEE*, vol. 82, issue 8, pp. 1303–1323, 1994.
20. P. Vas, *Artificial-Intelligence-Based Electrical Machines and Drives: Application of Fuzzy, Neural, Fuzzy Neural, and Genetic-Algorithm-Based Techniques*, vol. 45. Oxford University Press: Oxford, 1999.
21. O. Abdelkhalek, C. Benachaiba, M. Haidas, and T. Benslimane, "A new technique applied to a fuzzy regulator to control the shunt active filter DC bus voltage." *Information Technology and Control*, vol. 37, issue 3, 2008.
22. B.F. Wollenberg and T. Sakaguch, "Artificial intelligence in power system operations" *Proceedings of the IEEE*, vol. 75, issue 12, pp. 1678–1685, 1987.
23. S. Madan, and K. E. Bollinger, "Applications of artificial intelligence in power systems", *Electric Power Systems Research*, vol. 41 issue 2, pp. 117–131, 1997.
24. S. K. Perera, Z. Aung, and W. L. Woon, "Machine learning techniques for supporting renewable energy generation and integration: A survey", *International Workshop on Data Analytics for Renewable Energy Integration*, Springer, Cham, 2014.
25. A. S. Kalogirou, "Artificial intelligence in renewable energy systems modelling and prediction," *World Renewable Energy Congress VII*, Cologne, Germany, 2002.
26. W. G. Ji., "Application of artificial intelligence technology in the analysis of electrical automatic control", *Electronic Test*, vol. 3, pp. 137–138, 2014.
27. S. Q. Xiao and J. C. Peng, "The application of artificial intelligence technology in electrical automation control", *Automation & Instrumentation*, vol. 530, pp. 1049–1052, 2013.
28. H. Yang, J. Mathew, and L. Ma, "Intelligent diagnosis of rotating machinery faults: A review", In J. Mathew (Ed.), *Proceedings of the 3rd Asia-Pacific Conference on Systems Integrity and Maintenance*, pp. 385–392. Queensland University of Technology Press: Brisbane, Qld, 2002.

29. L. B. Jack and A. K. Nandi, "Fault detection using support vector machines and artificial neural networks, augmented by genetic algorithms", *Mechanical Systems & Signal Processing*, vol. 16, issue 2–3, pp. 373–390, 2002.
30. A. Siddique, G. S. Yadava, and B. Singh, "Applications of artificial intelligence techniques for induction machine stator fault diagnostics: Review", *IEEE International Symposium on Diagnostics for Electric Machines*, vol. 49, issue 3, pp. 29–34. 2003.
31. K. Yuttana and S. Watcharin, "A space vector modulation strategy for three level operation based on dual two-level voltage source inverters", *2014 International Power Electronics Conference (IPEC-Hiroshima 2014- ECCE ASIA)*, Hiroshima, Japan, 18–21 May 2014.
32. S. K. Gupta and M. A. Khan, "Space vector modulation strategy for three level operation of five-phase two-level dual voltage source inverter system", *17th IEEE India Council International Conference (INDICON-2020), NSUT*, New Delhi, 11–13 December, 2020.
33. M. A. Khan, S. K. Gupta, and O. Singh, "Pulse width modulation switching analysis for three phase dual inverter system using artificial neural network", *28th Australasian Universities Power Engineering Conference*, Auckland, 27–30th November 2018.
34. S. K. Gupta, A. Khan, and D. K. Chauhan, "Performance analysis of time equivalent space vector pulse width modulation scheme for three phase VSI at inductive load", *International Conference on Innovation in Cyber Physical Systems (ICICPS-2020)*, New Delhi, 22–23 October 2020.
35. S. K. Gupta, O. Singh M. A. Khan, and D. K. Chauhan, "Pulse width modulation technique for multilevel operation of five-phase dual voltage source inverters", *Journal Européen des SystèmesAutomatisés*, vol. 54, no. 2, pp. 371–379, 2021.
36. S. K. Gupta, O. Singh, and A. Khan, "Pulse width modulation switching schemes for two level five-phase voltage source inverter", *European Journal of Electrical Engineering*, vol. 23, issue 2, pp. 137–142, 2021.
37. A. Khan and S. K. Gupta, "Simulation analysis of pulse width modulation schemes for three-phase impedance source inverter", *International Journal of Power Electronics and Drive Systems (IJPEDS)*, vol. 13, issue 3, pp. 1478–1485, 2022.
38. S. K. Gupta, "THD analysis of twin two-level 5-phase inverter system for multilevel performance using two switching schemes", *IEEE Delhi Section International Conference on Electrical, Electronics and Computer Engineering (DELCON-2022)*, NSUT, New Delhi, 11–13 February, 2022.
39. S. K. Gupta, O. Singh, A. Khan, and D. K. Chauhan, "Simulation study of five-phase dual two-level inverter system using unified space vector PWM", *10th IEEE International Conference on Communication Systems and Network Technologies (CSNT 2021)*, Oriental Institute of Science and Technology, Bhopal, India, June 18–19, 2021.

18 Major Security Issues and Data Protection in Cloud Computing and IoT

S. Thavamani and C. Nandhini

CONTENTS

DOI: 10.1201/9781003438588-18

18.1 INTRODUCTION

The Internet of Things (IoT) and cloud computing are working together to build a system with lots of data storage demand where data protection is most critical in terms of security. The term "cloud computing" refers to services that provide extremely scalable, flexible, and compensated computer resource storage. As cloud computing services for compute and storage become more and more popular, growing numbers of enterprises are preferring to shift their data from internal communications infrastructure to cloud storage providers. As more IoT devices are purchased, more apps will be developed, which will increase the security risks associated with connecting to cloud services. In a world where there are a lot of smart things, more fine-grained data collecting will enable a fresh era of IoT software systems to provide more complex services. The use of IoT devices that facilitate user job completion has increased recently [1]. Examples include improved building management systems, public monitoring, smart city services, and even participatory sensing. It will be challenging to handle time-varying workloads and massive amounts of data in IoT applications using both traditional processing and cloud computing. The third-party auditors responsible with processing the data and guaranteeing its integrity are still in the early stages of development, and the encryption techniques are not as effective as they ought to be. We frequently employ two very distinct technologies in our daily lives: cloud computing and IoT. When they are more widely adopted and used, they will play an important role in the Future Internet. It is expected that a novel paradigm fusing cloud and IoT would be disruptive and open up a wide range of application scenarios. Every year, the quantity of data that numerous government agencies collect on individuals increases more quickly, making it harder for people to comprehend the ramifications. The potential of privacy problems increases with this kind of data. Ignorance of these problems may have long-term detrimental effects, including a lack of acceptance for technology, the failure of innovative technology as a result of issues with credibility and costly legal lawsuits [2]. The chapter examines the development of IoT in cloud computing, newly discovered security problems, and potential fixes.

The major objective is to recognize the security concerns and approaches in the integrated cloud and IoT, which can be used to provide secure data storage and movement. Numerous literary works have independently investigated the cloud and IoT, as well as their key traits, features, enabling innovations, and unresolved problems. To our greatest expertise, these studies do not fill the gap by providing a thorough study of the cloud-based paradigm, which encompasses completely new applications, challenges, and research topics. The enormous variety of resources available on the cloud may be of tremendous use to the IoT, and as the cloud becomes more widely used, it will be able to expand beyond its current physical item boundaries in a more dynamic and dispersed fashion. The literature on the integrated cloud and the IoT is examined and reported to analyze the advantages and challenges of integration. The complementarity of cloud computing, IoT, and other related issues, as well as the reasons that are now supporting their integration, have been thoroughly studied in the literature.

18.2 CLOUD-BASED IoT

The combination of IoT with cloud computing is referred to as "cloud-based IoT," and it significantly changes how high performance is attained in the global computing environment [3]. The IoT connects devices and equipment for administration and monitoring and is cloud-based. It is required due to the significant volume of data that IoT devices collect, which calls on cloud technology for processing and storage [4]. IoT powered by the cloud collects data from machine-mounted sensors [5]. Security risks are expanding as a result of the frequent sharing on computing and networking resources across the Internet by IoT devices and users.

Security and privacy are an important problem with cloud-based IoT [6]. It is currently vitally necessary to learn more about privacy and security-related concerns. While focusing on privacy and security concerns, we also called attention to the myriad dangers and difficulties that must be overcome. To achieve this, we look at the several layers, apps, and cloud-based IoT architecture. In our work, we also talk a lot about concerns, problems, and unmet challenges related to security and privacy.

18.3 CLOUD-IoT APPLICATIONS

IoT can benefit from the pay-per-use, portability, security, and performance features of cloud computing infrastructure. The use and analysis of many applications are enhanced by the incorporation of cloud and IoT technologies, creating a continuously improving IoT environment with prospects for affordable on-demand scaling.

A network of gadgets known as the "Internet of Things" transmits, trades, and uses data from the real world to offer services to people, businesses, and the general public. The things have individual IDs and can act alone or in cooperation with other things or people (identifiers). The IoT has a great amount of uses in the areas of energy, environment, transportation, and health (wearable).

The IoT and cloud computing are now two very connected Internet technologies that collaborate in sophisticated IoT implementations. Since the bulk of current IoT ecosystems are cloud-based, cloud computing technology stimulates the development and adoption of scalable IoT models and applications. The combination of the two technologies has brought about a number of notable benefits, including improved work processes, greater resource use, and cheaper prices. IoT and cloud computing are being utilized for a variety of uses, including those that are medical and health-related, related to home and building automation, connected to intelligent transportation systems, linked to network-controlled manufacturing, and related to agriculture.

18.3.1 Health Care

Cloud-based numerous advantages and opportunities have been introduced to the healthcare industry by the IcT. It may effectively create and enhance healthcare services while also preserving innovation in the industry (for instance, intelligent drug/medicine control and hospital administration). Communication lines, incorporated internal and exterior sensors, IoT servers, and cloud storage are all components of smart healthcare monitoring. The actions involving the health parameters take place at different stages of refinement, including the application layer, management layer, network layer, and device layer. Several data sensors have been acquired using wireless media via nodes. It is stored as an unstructured dataset on the cloud. A username and password are used to create and secure a patient database. Those who have been granted access can monitor cloud sensor data in data logs, analog inputs, digital inputs, and digital outputs.

18.3.2 Smart Cities

Distribution and construction of cloud-based systems are advantageous for green infrastructure. Urban mobility, smart transportation controls, smart flood control, and other IoT modules are anticipated sustainable urban characteristics. They could also offer more info in general. Through intelligent cloud integration, the global grid now can manage these documents and applications.

The deployment of new products and the expansion of existing ones, both of which have historically raised serious issues about the accessibility of the necessary computational resources, may be simultaneously accelerated by the cloud. By granting other variety of appropriate to its architecture and enabling them to integrate IoT information as well as computer resources running on IoT systems, a public cloud computing provider may broaden the IoT ecosystem. The organization could provide access to IoT information as well as associated solutions.

This demonstrates the need for and applicability of cloud computing and IoT infrastructure upgrades. The competing IoT and cloud architecture, however, have always made integration difficult. Before data and their services are integrated into the cloud, sensors and devices are built. This allows them to be distributed over all cloud resources and minimizes inconsistencies.

18.3.3 Smart Homes

Future living will center on smart dwellings, as depicted in Figure 18.1. Smart houses are currently being implemented in numerous communities throughout the world as part of modernization projects. These always-on homes produce enormous volumes of valuable information from connected IoT systems and smart appliances [1].

It is possible to find out a range of information that has a significant influence on the economics, health, and safety of our society by having the capacity to study these data both online and offline. For instance, the healthcare system in a smart city may evaluate a patient's status by observing how they use equipment and recognizing any routine or peculiar actions that may be signs of a health problem [2].

This study describes a number of critical processes in the creation of the Smart Home system. The first step in developing the project's problem statement is to outline the limits of the system, any interfaces that must be created between the system and outside forces, the customer's needs and constraints, and the requirements and specifications generated by the project team members. Then, a comprehensive design is produced. The functions of the system are outlined, broken down into easier-to-manage activities, and solutions are found to complete these tasks. To create a network for the IoT, the server software that was selected is then investigated, paying attention to the server implementation that was selected to permit device connectivity [7,8].

In order to understand how homeowners behave, a utility company may examine a significant quantity of information about energy use from equipment within the home [9]. Based on these analyses, the company may suggest to customers various solutions for reducing their electricity bills. A situation like this results in lower costs for utility companies as well as for consumers. Manufacturers may continuously evaluate data, develop or predict a timetable for appliance maintenance, and quickly replace failing equipment thanks to real-time IoT applications.

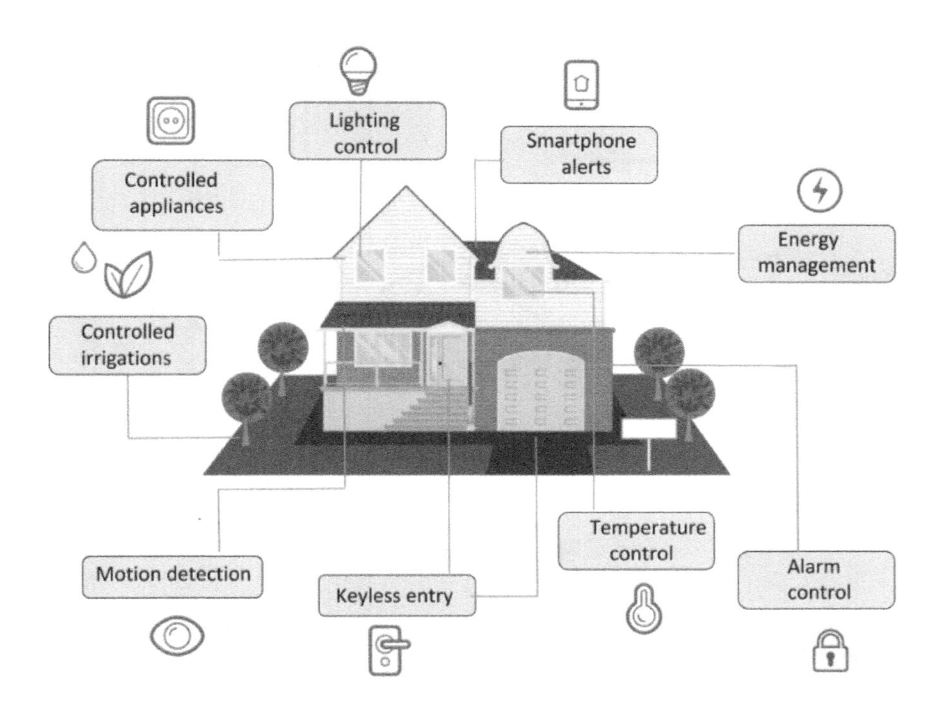

FIGURE 18.1 Smart home appliances.

These IoT application examples highlight the benefits of home automation data gathering. These data provide important insights into the dynamics and behavior of smart homes and the people who live in them, but they also provide significant challenges for data administration, preservation, and processing. Systems that can handle, analyze, and turn this volume of data into insights that can be deployed for sustainable urban applications that demand rapid reflexes and high standards are necessary to prevent users from drowning in floods of data. Both the gaining substantial and the spatial precision of choices, whether they are made offline or in close to real-time, must be handled by these platforms, and they must be additionally extensible.

18.3.4 Smart Energy and Smart Grid

IoT and cloud computing can successfully work together to give users more intelligent energy management. The smart grid is made up of the electrical network and communication infrastructure, smart meters, smart appliances, and renewable energy sources seen in Figure 18.2. A smart grid has the potential to distribute electricity more reliably and effectively than a traditional power system by employing bidirectional communication and power flows.

An energy network containing "intelligent" components that can function, communicate, and interact with one another on their own is referred to as a smart grid. This allows for the efficient delivery of electricity to the customers. Because a smart grid's architecture can vary, it stimulates the use of cutting-edge technology

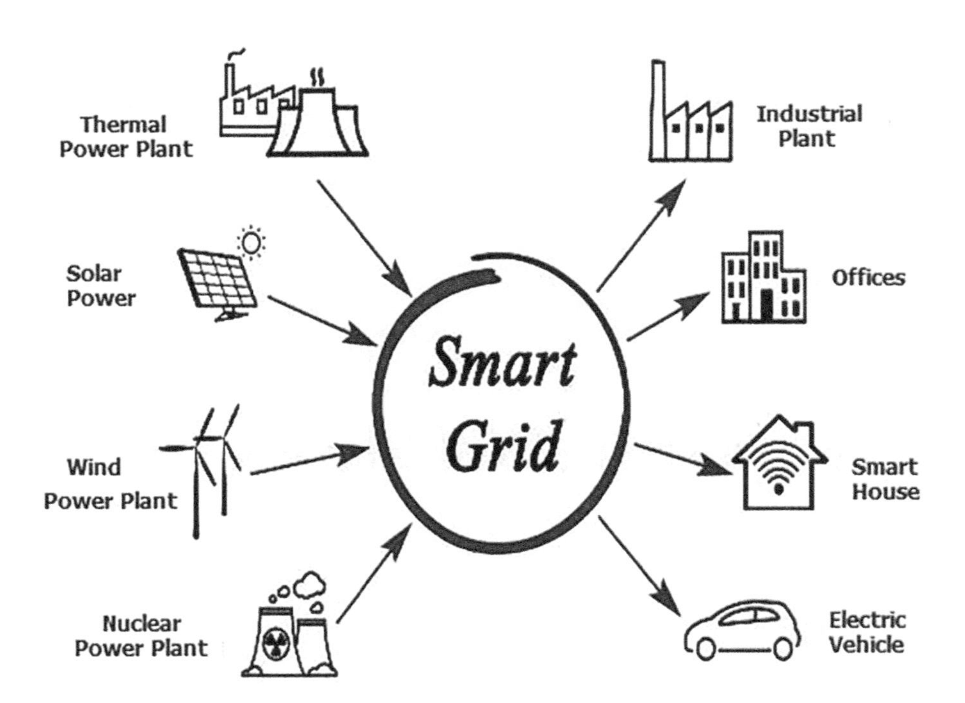

FIGURE 18.2 Smart energy and smart grid.

to address a variety of technological challenges at various levels. Any intelligent network design should be able to support real-time, two-way interaction between providers and their customers as well as the ability for software systems at both the manufacturer and consumer sides to control and monitor energy consumption.

Electric grids are the complex web of networks that move energy from its source, like power plants, to consumers like residential and business customers. The traditional American electric grid, which has been in use for more than a century, relies on a one-way flow of power from the source to the destination.

The production, storage, and use of energy on the grid are currently being changed by technology, which is also paving the way for the growth of a smart energy infrastructure. Microgrids, the IoT, digitization, distributed renewable energy sources, automation, and other technologies are being combined to build the new smart energy ecosystem, of which the smart grid is a component.

The phrase "IoT" refers to a broad category of World Wide Web devices that facilitate data transmission in real time, data pipelines, and data collection. Smart grids are a representation of IoT technology adoption in the energy sector. When properly implemented, smart grids can solve a variety of problems with conventional grids, such as outages, security concerns, high carbon emissions, and other factors.

References to specific IoT and smart grid application-integrated solutions are provided in the list below.

- Battery monitoring systems
- Smart meters
- Solar farm monitoring
- IoT-based electric vehicle charging

18.3.5 Automotive and Smart Mobility

Several of the current issues, such as congestion status forecast and reporting, autonomous cars, and other difficulties, might potentially be addressed by combining cloud computing with GPS and other transit technology. The way we live will likely change significantly as a result of the IoT. IoT is a revolutionary technology that is upending society and has the power to fundamentally alter the course of history. People are connected through Internet and are now linking "Things" in order to promote facilitate effective communication and the sharing of intelligence. It leverages affordable Internet-connected gadgets and sensors to give new opportunities. Recently, the automotive industry believed that linked cars, driverless automobiles, and IoT applications in the car ecosystem, which includes parking, the environment, supply chains, and transportation regulatory agencies, were futuristic possibilities.

18.3.6 Smart Logistics

Smart logistics enables and facilitates automated control of the flow of commodities among suppliers and users and also monitoring of items while they are in route (for example, tracking shipments in the logistics sector). One of the first ICTs was the IoT, which is now seen as a crucial area of future technology. When mobile communication develops quickly in the realm of new wireless communications, IoT is gaining traction. The idea of IoT is continually evolving, starting with an early focus on machine-to-machine (M2M) communication and applications and moving on to the "ubiquitous aggregation" of data. As a result, the IoT has produced enormous amounts of data, and a variety of quantitative analytic techniques may be used to continually explore the complex connections between the activities indicated by this data (Figure 18.3).

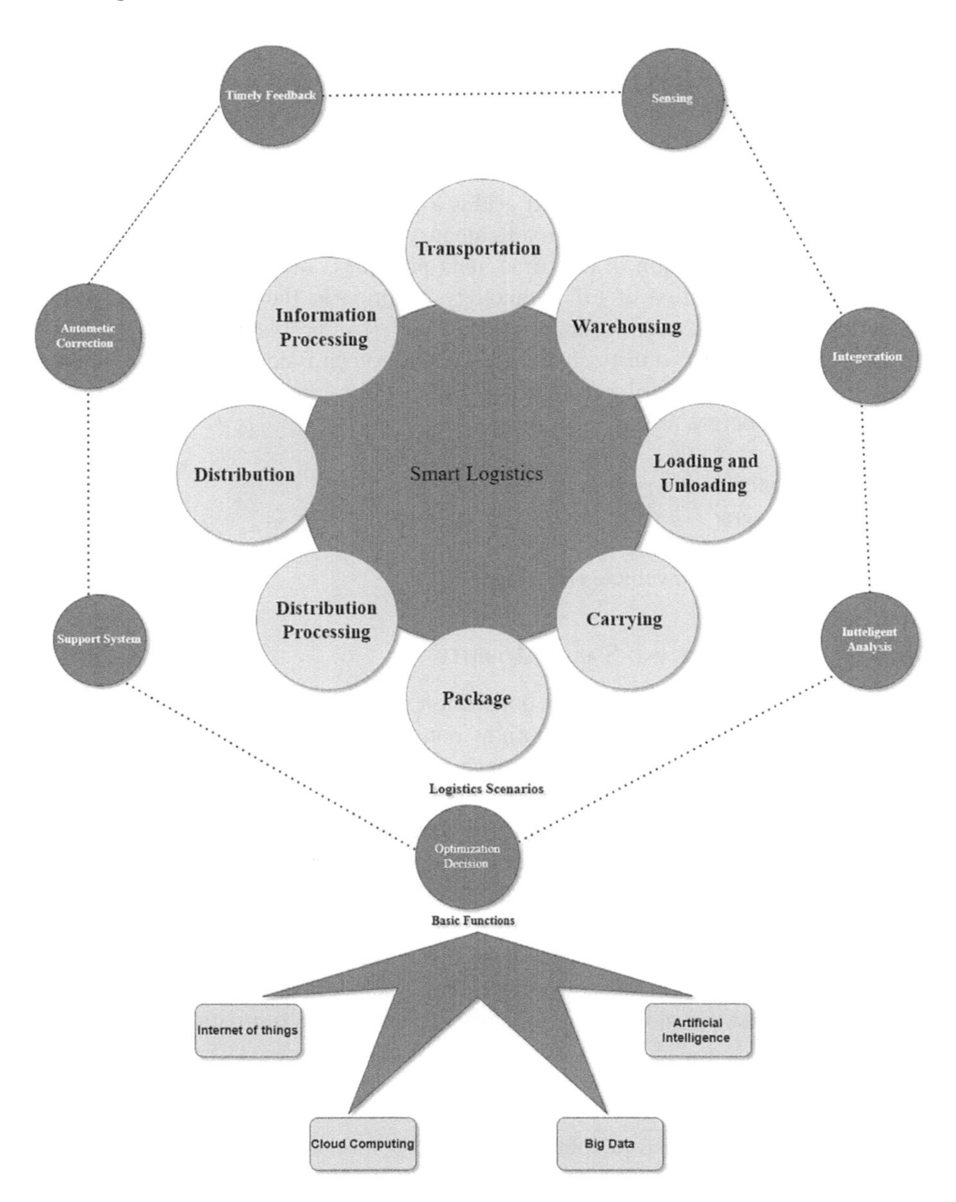

FIGURE 18.3 Smart logistics.

The deployment of smart logistics, which will radically change both the management of logistics and the structure of the logistical system, will surely require IoT. There are still a lot of considerations to make before IoT-based smart logistics become a reality, such as acceptable circumstances, existing challenges, and future directions. To help individuals who are interested in the growth and development of this industry, we engage in this activity. Although there have been several studies on IoT and smart logistics published in the literature, they typically only cover one of these two topics or a particular application of IoT in logistics.

18.3.7 Environmental Monitoring

Monitoring and management of the environment are becoming more crucial as more legislation and initiatives to minimize pollution are driven by environmental management and an awareness of its effects. IoT sensors are able to provide precise, real-time data on the environment around us, enabling us to better understand how our actions affect the environment and improve urban quality of life. The establishment of an increased data network connecting the organization that monitors wide-area settings with sensors that have been put effectively in the region is made possible by merging the cloud with IoT (for example, pollution source monitoring, water quality monitoring, and air quality monitoring).

The size and price of environmental sensors have decreased. The ability to put them at numerous locations around a city where it would have been impracticable only a few years ago is now widely available. Mobile IoT, a reliable connectivity option made possible by advancements in mobile technology, enables municipal governments and other entities to install these sensors quickly and effectively. Low-power wide-area networks from mobile operators, such as NB-IoT and LTE-M, are examples of mobile IoT technology.

The ubiquitous sensor communication networks that properly capture our general surroundings have raised awareness of the importance of the IoT in modern life.

A framework like this enables the plan to keep track of fundamental event-produced data that may be relocated and kept in the cloud, where it is possible to share this data by way of usage and decide what to do about an event that has already happened. The temperature and moisture of the area are measured using sensors in the Ecological Monitoring framework. This information could be used to animate cyclical activity, such as a device warming up or cooling down, as well as other long-term insights into the gadget.

The IoT is positioned to transform our environment through the usage of devices that can gather data, analyze it, and wirelessly send it to a storage server, like the cloud, which collects, assesses, and offers this data in a useful way.

This information can be retrieved from the cloud via a number of front-end user interface designs, including mobile and Internet applications, depending on the requirement and suitability. The Internet, which is at the hub of this transformation, makes data transmission between fog, apps, and end users secure, effective, and speedy. The idea of a typical Internet end user or server is modified by this new paradigm (Figure 18.4).

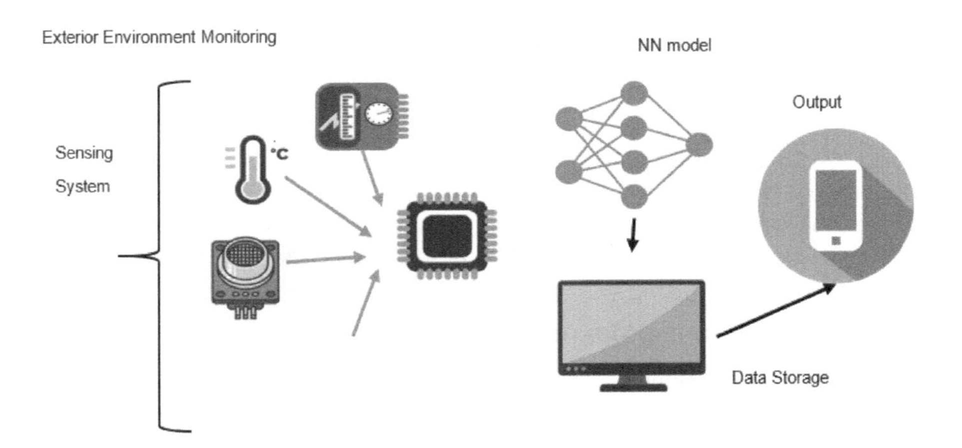

FIGURE 18.4 Smart environment.

18.4 ADVANTAGES OF IoT AND CLOUD INTEGRATION

Because of its limited storage and processing capability, the IoT faces problems with dependability, performance, security, and privacy. The bulk of these problems may most effectively be solved by integrating IoT with the cloud. The cloud can even gain from IoT by enhancing its capacities with physical objects in a more flexible and dispersed fashion and by supplying new services for billions of devices in a range of real-life scenarios [10,11]. Also, the cloud reduces the cost and complexity of using apps and services for end users. The cloud provides rapid, affordable installation and integration for difficult information processing and deployment, designed to simplify the IoT data flow [12].

18.4.1 Communication

The cloud-based IoT paradigm's two key components are application and data interchange.

The IoT may be used to supply global applications, and automation can be used to provide low bandwidth collecting and distribution. By using integrated apps and personalized interfaces, the cloud is a robust and affordable technology that can be used to connect, manage, and track everything [13]. Real-time data access, remote object control, and dynamic monitoring are made possible through quick network connectivity. It is important to remember that the cloud still has some limits in a number of areas, despite the fact that it may considerably enhance and support IoT connectivity. So, when a significant volume of data needs to be transmitted from the Internet to the cloud, there may be practical limitations [13,14].

18.4.2 Storage

The IoT has a huge number of information sources that generate a lot of semi-structured or unstructured data because it can be utilized on billions of devices [15].

Three features of big data are diversity (such as different sorts of data), velocity (such as regularity of data production), and volume (e.g., data size). The cloud is one of the acceptable and useful solutions for managing the enormous volume of data generated by the IoT. It also opens up new doors for the collection, distribution, and integration of data [16].

18.4.3 Processing Capabilities

Major on-site data processing is impractical with IoT devices due to their limited computing capabilities. Instead, the gathered information is sent to elevated nodes for analysis and consolidation. Yet, attaining scalability remains difficult in the absence of a strong infrastructure required.

The cloud, which offers limitless virtual computer capacity and an on-demand consumption paradigm, offers an option [16]. The IoT can leverage probabilistic algorithms and data-driven decision-making to boost income and minimize risks at a reduced cost [10].

18.4.4 Scope

The world is quickly moving toward the Internet of Everything (IoE) realm, a web of networks with gazillions of things that create novel possibilities and hazards [16]. As a result of billions of people communicating with one another at once and a wealth of data being gathered, huge amounts of things will be connected to the IoE realm. The core of the IoT strategy, which offers new applications and services, is the emergence of the cloud through IoT devices, which in turn allows the cloud to work with a variety of new real-world situations and fosters the development of novel services [17].

18.4.5 Additional Abilities

- It is characterized by the variety of its technology, code of conduct, and equipment. Efficiency, interoperability, security, availability, and reliability all be quite challenging to achieve. The bulk of these problems may be solved by integrating IoT with the cloud [10]. Moreover, it is simple to use and obtain access to, and its installation costs are modest [17,18]. New opportunities for intelligent goods, programs, and services are made possible by cloud-based IoT integration [6,19]. The following is a list of a handful of the new models:
- SaaS (Sensing as a Service) [19], a system that provides access to sensor data;
- EaaS (Ethernet as a Service) [20], whose main purpose is to provide universal connection for controlling wireless controllers;
- IPMaaS (Identity and Policy Management as a Service) [20] allows access to identification and regulation administration, while SAaaS (Sensing and Actuation as a Service) [19] instantly generates control logic;
- SEaaS (Sensor Event as a Service), whose messaging services are disseminated as a consequence of sensor events; Database as a Service (DBaaS) [20], which provides global database administration services; SenaaS (Sensor as a Service), where remote sensor management is available;
- Data as a Service (DaaS) [20] gives global access to all data types.

18.5 CLOUD-BASED IoT ARCHITECTURE

According to existing research, cloud-based IoT architecture is frequently divided into three layers: application, network layer, and sensor layer. As shown in Figure 18.5, it is assumed that the cloud layer fulfills the Functionality of cloud-based IoT architecture.

Recognizing things and gathering data from the environment are done using the sensing layer. On the other hand, the network layer's main objective is to communicate the obtained data to the Internet or cloud. Moreover, the application layer provides access to a number of services [6].

18.6 MAJOR BENEFITS OF CLOUD-BASED IoT

18.6.1 Accessibility

Data from various devices can move seamlessly between different servers thanks to this infrastructure, which simplifies data storage, analysis, and overview. The cloud's on-demand functionality enables accessibility on any device and at any time.

18.6.2 Scalability

The number of IoT devices used by a company increases along with its size, which increases the requirement for more money to be invested in new networks and data storage to accommodate the extra data. IoT cloud enables businesses to easily connect new devices to an existing cloud without incurring additional expenses or needing to add more servers or hard drives to their IT infrastructure. The content of each new device will be logically joined to the other stuff in the cloud when more are added.

18.6.3 Fewer Cables, Papers, and Minerals

Utilizing the cloud can help conserve resources like the steel servers and cables' mined minerals, paper, electricity, and metal. There is no longer a requirement for an on-premise physical storage medium that requires cables and steels to function because the generated data from all IoT devices is saved in the cloud.

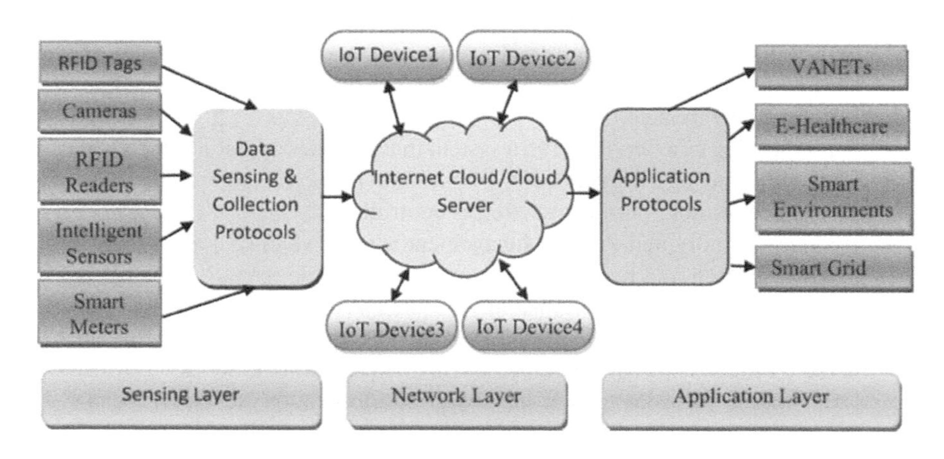

FIGURE 18.5 Cloud-based IoT architecture [1].

Employees do not need to print out the obtained data on paper in order to view it because it is available digitally.

18.6.4 Collaboration

The communication between various departments in an organization will be seamless with the storage of data from IoT devices in a cloud that is accessible to authorized employees at any time and from any location. The employees' interest, efficiency, productivity, and sense of teamwork will all rise as a result.

18.6.5 Disaster Recovery

Data deletion and recovery are not always possible due to IT disasters. IoT cloud platforms and cloud computing services in general provide quick data recovery in the event of any emergency, including power outages, human mistake, and natural disasters.

Research found that 20% of cloud users said their disaster recovery procedure was completed in 4 hours or less, compared with 9% of users who did not use the cloud.

18.6.6 Data Mobility

When data are processed and stored on a cloud server, it is no longer constrained by equipment and is accessible from any location in the globe. IoT is now used to provide data services by a wide range of devices, including automated factories, driverless automobiles, sensor networks, smart cards, smartphones, smart watches, etc. Most of these devices already have mobility or are working to create it. All of these IoT systems also need real-time device management of connected devices. You may manage, process, analyze, and update the data and the devices remotely and in real time using an IoT cloud-based service as opposed to on-premise servers, which can only be accessible inside the company's walls.

18.6.7 Data Security and Reliability

In today's data-centric world, businesses place a high priority on data security. Since the outset, the security and long-term viability of the data have been a top priority for IoT industries. Businesses have always struggled with the question of whether the finest IoT cloud providers can offer the same level of data security and dependability as on-premise technology. The accessibility of data for on-demand access is another significant challenge for businesses. Although on-premise IoT services can appear superior at first, they are actually superior in this area.

The majority of cloud service providers use high levels of security that are out of reach for most organizations. Additionally, the notion that objects that are physically closer to us are safer does not apply in the context of cyber security. When you have control over who has access to and uses your data, it is more secure. It is hence susceptible to on-premise infrastructure. Data saved in the cloud cannot be accessed by a bad actor since cloud solutions offer businesses trustworthy authentication and encryption procedures.

18.6.8 Cost-Effectiveness

The cloud services are very cost-effective because you only pay for what you really use. In contrast to on-premise storage, which falls under capital expenses, cloud storage costs are categorized as operating expenses. In terms of hardware installation, upscaling, and buying, on-premise network infrastructure demands a bigger financial commitment. Additionally, the cost of maintenance and IT assistance is incurred. Similar to cloud systems, there is no upfront cost, and pricing is adjustable based on what organizations need.

18.6.9 Data Storage

The cloud is where the majority of the data is located. Even edge computing solutions do not locally store a lot of data. It is safer and simpler to expand this.

18.7 IMPLICATIONS OF CLOUD-BASED IoT INTEGRATION

Many challenges may make it difficult for the cloud-based IoT model to be fully integrated. Some difficulties include:

18.7.1 Security and Privacy

IoT on the cloud enables data to be sent from the real world to the cloud. How to implement appropriate authorization rules and regulations while ensuring that only authorized personnel get access to confidential information is actually a crucial issue that has not been resolved yet. This is essential for preserving user privacy, and it becomes much more so when preserving data integrity is necessary [16]. Further problems that develop when significant IoT applications are moved to the cloud include a lack of confidence in the service provider, ignorance of service level agreements (SLAs), and the physical placement of data [21,22]. Another risk brought on by multi-tenancy is the disclosure of private information. Since IoT devices have processing power restrictions, public key cryptography is not employed at all tiers [16]. There are now additional problems that need special attention. A few examples of potential attacks on the distributed system include SQL injection, session hijacking, cross-site scripting, and side-channeling. Major problems like virtual machine escape and session hijacking are also seen as troublesome [16,23].

18.7.2 Heterogeneity

The vast array of devices, infrastructures, software platforms, and services that are available and might be utilized for novel or developing applications is a significant big problem with the cloud-based IoT method. Heterogeneity problems influence cloud systems; for instance, cloud services could have patent interfaces that offer resource integration based on certain vendors [16]. Moreover, a variety of providers would provide services that aim to increase application speed and resilience, and end users that use multi-cloud techniques may make the heterogeneity problem worse [24].

18.7.3 Big Data

With many predicting that big data will approach 50 billion IoT devices by 2020, it is imperative to invest more time in the transmission, access, storage, and interpretation of the enormous amount of data that will be produced. The IoT is likely to be one of the primary sources of big data, and the cloud can make it simpler to store this data for a long time and submit it for in-depth analysis [8]. This is in fact obvious given recent technological breakthroughs. Managing the vast amount of data collected is a major challenge since the features of this data management service are very important to the program's overall success.

18.7.4 Performance

To send the massive amounts of information that IoT devices produce to the cloud, high bandwidth is required. Because bandwidth expansion is not keeping up with improvements in storage and processing, securing an adequate network speed is the key problem when migrating data to cloud settings [16]. In a variety of circumstances, the supply of services and data should be done with high responsiveness [23]. This is valid given that real-time applications heavily depend on timeliness and operation smoothly, both of which might be harmed by unexpected events [16].

18.7.5 Legal Aspects

Recent research on particular applications has placed a lot of emphasis on the importance of legal issues. For instance, there are several international regulations that apply to service providers. On the other side, customers should make donations to help in data gathering [4].

18.7.6 Large Scale

Due to the cloud-based IoT paradigm, it is now feasible to create brand-new apps that integrate and analyze data from the real world into IoT goods. This necessitates communication with billions of devices positioned all over the world [25]. The difference between a normal versus the one you got, for instance, it is becoming more difficult to meet the needs for storage and processing capacity. Moreover, as IoT devices must cope with network issues and delay dynamics, the monitoring procedure has made distribution more difficult [16].

18.7.7 Dependability

Time-sensitive IoT devices' high reliance on cloud services will affect the performance of the application to be quickly reflected in the results, for instance, in automobiles, aviation, medical equipment, or the field of defense [26].

18.7.8 Data Storage

By 2025, there are projected to be 50 billion IoT devices in operation, which will make it extremely difficult for cloud service providers to give quick, easy, and safe access to the data [11].

18.7.9 Maintenance

As was mentioned in the previous section, extremely effective techniques and strategies are required to track and preserve safety and performance in the public cloud in order to fulfill the demand for up to 50 billion IoT devices [1].

18.8 THE STRATEGIES AND PROBLEMS OF CLOUD-IoT SECURITY

Privacy and the safety of data are the two main issues that every technology that processes or maintains information faces. The main concerns for cloud-based IoT are security and privacy, and they are intertwined with user confidence. Both well-known and lesser-known businesses in the market have in the past experienced significant losses as a consequence of hacking and forgery incidents. The security risk associated with IoT is much more serious than security concerns in the banking industry, which might result in significant financial loss since, for example, any attack on smart healthcare IoT apps or smart cars connected to the cloud is life-threatening [27,28]. The following list of serious security issues includes appropriate solutions.

18.8.1 Data Security

The administration of data and restrictions on what happens to it are referred to as data privacy. Everyone's life will be impacted by the widespread adoption of Cloud-IoT in the future. The privacy concerns of both individual users and businesses, however, are a barrier to this universal adoption [27]. To facilitate the anonymization of data usage and the place where the data are utilized, a small number of unique and novel solutions can provide anonymization with precise privacy restrictions [29]. Ref. [30] offers User-driven Privacy Enforcement for Cloud-based Services in the IoT (UPECSI) to address the problem of data privacy. Consumer technology is used to safeguard the privacy of IoT device-generated data that are stored or processed in the cloud. Model-driven privacy, user engagement, and privacy enforcement points make up the majority of it. Model-driven privacy aims to incorporate privacy into cloud services. Transparency is developed through providing customers with encounters to better understand their wants and expectations. Finally, the user-defined policies must be observed when the data are transferred to the cloud, and this is the responsibility of the IoT gateway's Security Enforcement Points [31]. One method being researched by developers is to satisfy the IoT object's privacy need even before it is released or during the development phase.

(1) Piracy prevention measures should be strictly enforced rather than implemented in response to an existing problem. (2) Piracy should be viewed as a requirement from the beginning rather than as a service with added value. (3) Piracy goals should not inhibit the object's anticipated performance. (4) Security methods must be implemented to safeguard data privacy at all times, up until the data are deleted.

18.8.2 Identity Verification and Privacy

The IoT is a network of networked objects, each of which may be given an address. These linked IoT devices frequently need to verify one another before sharing a communication request. The Cloud-IoT has the same identification problems. For the IoT, a multiple authentication security solution based on the Datagram Transport Layer Security (DTLS) protocol is introduced (IoT). Between the transport and application layers is where the DTLS is located. RSA created this mechanism to support WPANs. The authors' approach provides message authentication, secrecy, and validity while using the smallest amount of power possible to account for end-to-end delay and memory cost.

18.8.3 Access Control

With the large amount of IoT devices in our surroundings and the huge volume of data these devices transmit to the cloud for processing and storage, it is vital for data owners to have effective network access to their data on the cloud. It makes reference to the potential for giving various actors access to data use rights within the context of a sizable IoT network [32].

As the processing concentration is larger than that of a conventional database management system, performance and temporal restrictions with access control in the Cloud-IoT context are significant problems (DBMS) [3]. The data holders only provide the data collectors with the information that is required, and it is the duty of another equally skilled data collector to confirm the sources of the data and the identities of the data holders. Developing a cipher text-policy-based encryption approach for data storage and safe access in a cloud for IoT applications in order to reduce the storage overhead of public keys and provide authorized access for a particular user, the cloud server additionally builds a new user access control list (UACL).

18.8.4 Permissions

The definition of resource access rights is a phase in the authorization process. A three-step process is referred to as permission in the context of the cloud and the IoT. The first step would be to define the security policies, which are only a list of specified guidelines. The limitation on access despite the typically obvious aims of access control models, it could be difficult to comprehend these security guidelines throughout the authorization step. The authorizing procedures aid in bridging the gap among high-level security needs and low-level processes by making the interpretation of the rules easier. Using permission principles, the policies might be evaluated for accuracy and coherence. The attribute-based access control model (ABAC), also known as the BAC model and usage control, the discretionary model DAC, the mandatory model MAC, RBAC, and its several versions (UCON). When the model has been implemented, a set of regulations are then put into effect [27]. Approval models are frequently created using the concepts of people, things, and interactions between them. Nevertheless, other authorization models also take into consideration factors like context, knowledge, trust, and privacy. It is discovered that heterogeneity-supportive hybrid models exist to fulfill the demands of the company, particularly in the case of Cloud-IoT [33].

18.8.5 Secure IoT on Mobile

The Internet of Vehicles' (IoV) mobile nodes need identification, authentication, and personal privacy while they are in motion. They link to the cloud for computational and storage needs, and they form clusters or groups utilizing proximity as their main criterion. Such mobile IoT platform cloud services struggle to rapidly and precisely identify the nodes moving from one group to another [34,35].

The movable nodes created a hierarchy organization that could be utilized to facilitate effective network connectivity. The IoT has drawn businesses and researchers from all over the world due to its multitude of uses. The IoT enhances operations by enabling ubiquitous connection and giving all computer devices a connection to the Web. The emphasis has switched from basic IoT to smart, linked, and mobile IoT (M-IoT) devices and systems as a result of advancements in wireless infrastructure. Using sensors, machines, and even crowdsourcing, M-IoT devices and platforms can enable low complexity, low-cost, and efficient processing. These gadgets are all included in the "M-IoT" category as a whole. Even if applications have significantly improved, security, privacy, and trust remain the primary issues for such networks, and the threat posed by inadequate enforcement of these standards to M-IoT systems and devices is not insignificant. Understanding the variety of options out there is crucial for developing a reliable, secure, and privacy-compliant M-IoT mechanism. There has not been a direct assessment of the M-IoT handover protections, secure protocols, physical layer security, privacy, or security [36–38].

18.9 CONCLUSION

A large amount of data needs to be processed and saved in order for the IoT to become a more popular computing service. Because of the potential of the IoT, cloud integration with IoT is highly helpful in tackling these difficulties. Although there are benefits in terms of storage and processing speed, there are also issues with security, privacy, reliability, and performance. IoT and cloud computing, which have lately become research hotspots, are combined in the cloud-based IoT system. Yet, there are significant privacy and security concerns raised by cloud-based IoT application services that must be addressed. A revolutionary cloud-based IoT architecture with several layers is created with the aim of reducing different dangers, protecting privacy, and limiting illegal entry. With this design, a small number of trustworthy and secure processes provide practical solutions and protect the IoT's security. A significant quantity of test data demonstrates how our cloud-based IoT solution efficiently manages a variety of tasks while preserving data privacy. Future research will concentrate on how to optimize the design and make it able to deal with new problems.

REFERENCES

1. A. D. Jurcut, P. S. Ranaweer, L. Xu, *Introduction to IoT Security*, Wiley, 2019. DOI: 10.1002/9781119527978.ch2.
2. N. Alam, P. Vats, N. Kashyap, Internet of Things: A literature review, *2017 Recent Development in Control, Automation & Power Engineering*, IEEE, pp. 193–197, October 2017, https://doi.org/10.1109/RDCAPE.2017.8358265.
3. R. Roman, C. Alcaraz, J. Lopez, N. Sklavos, Key management systems for sensor networks in the context of the Internet of Things, *Computer & Electrical Engineering*, vol. 37, no. 2, pp. 147–159, 2011.
4. N. Zhang., Understanding IoT security through the data crystal ball: Where we are now and where we are going to be, 2017, arXiv preprint arXiv:1703.09809.

5. I. Bhardawaj, A. Kumar, M. Bansal, A review on lightweight cryptography algorithms for data security and authentication in IoTs, in *2017 4th International Conference on Signal Processing, Computing and Control (ISPCC)*, Solan, pp. 504–509, September 2017.
6. C. Formisano, D. Pavia, L. Gurgen, T. Yonezawa, J. A. Galache, K. Doguchi, I. Matranga, The advantages of IoT and cloud applied to smart cities, in *International Workshop on Learning Technology for Education in Cloud*, IEEE, pp. 325–332, 2015. DOI: 10.1109/FiCloud.2015.85.
7. H. Kim, E. A. Lee, Authentication and authorization for the Internet of Things, *IT Professional*, vol. 19, no. 5, pp. 27–33, 2017. DOI: 10.1109/MITP.2017.3680960.
8. M. Fenandez, J. Jaimunk, B. Thuaraisingham, Privacy-preserving architecture for cloud-IoT platforms, *2019 IEEE International Conference on Web Services (ICWS)*, pp. 11–19, 2019. DOI: 10.1109/ICWS.2019.00015.
9. L. Atzori, A. Iera, G. Morabito, Understanding the Internet of Things: Definition, potentials, and societal role of a fast evolving paradigm, *Ad Hoc Networks*, vol. 56, pp. 122–140, 2017.
10. A. M. Zarca, J. B. Bernabe, I. Farris, Y. Khettab, *Enhancing IoT Security through Network Softwarization and Virtual Security Appliances.* Wiley: Hoboken, NJ, pp. 1–18, 2018.
11. F. A. Alaba, M. Othman, I. A. T. Hashem, F. Alotaibi, Internet of Things security: A survey, *Journal of Network and Computer Applications,* vol. 88, pp. 10–28, 2017.
12. Q. Jing, A. V. Vasilakos, J. Wan, J. Lu, D. Qiu, Security of the Internet of Things: perspectives and challenges, *Wireless Networks*, vol. 20, no. 8, pp. 2481–2501, 2014.
13. Y. Zhang, Y. Shen, H. Wang, J. Yong, X. Jiang, On secure wireless communications for IoT under eavesdropper collusion, *IEEE Transactions on Automation Science and Engineering,* vol. 13, no. 3, pp. 1281–1293, 2015.
14. N. Agarwal, A. Rana, J. P. Pandey, A. Agarwal, Secured sharing of data in cloud via dual authentication, dynamic unidirectional PRE, and CPABE, *International Journal of Information Security and Privacy*, vol 14, no. 1, pp. 44–66, 2020.
15. K. Zhao, L. Ge, A survey on the Internet of Things security, in *2013 9th International Conference on Computational Intelligence and Security (CIS)*, Emeishan, *China,* pp. 663–667, IEEE, 2013.
16. SMARTBEAR BLOG. https://blog.smartbear.com/iot-2/how-to-protect-iot-gateways-from-securityvulnerabilities/ (Online; accessed on 04 May 2018).
17. A. Saroliya, U. Mishra, U. A. Rana, Improvement in routing techniques in P2P networks using a cloud service interface with secure multiparty computation, *Far East Journal of Electronics and Communications*, vol. 16, no. 3, pp. 673–683, 2016.
18. V. Kunwar, N. Agarwal, A. Rana, J. P. Pandey, Load balancing in cloud: A systematic review, *Advances in Intelligent Systems and Computing*, vol. 654, pp. 583–593, 2018.
19. C. Bekara, Security issues and challenges for the IoT-based smart grid, *Procedia Computer Science,* vol. 34, pp. 532–537, 2014.
20. S. K. Kesari, V. Kansal, S. Kumar, A systematic review of Qualityof Services (QoS) in Software Defined Networking (SDN), *Wireless Personal Communication*, Springer, 2020. DOI: 10.1007/s11277-020-07812-2.
21. I. Farris, J. B. Bernabe, N. Toumi, Towards provisioning of SDN/NFV-based security enablers for integrated protection of IoT systems, in *2017 IEEE Conference on Standards for Communications and Networking (CSCN)*, Helsinki, Finland, pp. 169–174, 2017.
22. F. A. Alaba, M. Othman, I. A. T. Hashem, F. Alotaibi, Internet of Things security: A survey, *Journal of Network and Computer Applications*, vol. 88, pp. 10–28, 2017.
23. A. Mosenia, N. K. Jha, A comprehensive study of security of Internet-of-Things, *IEEE Transactions on Emerging Topics in Computing*, vol. 5, no. 4, pp. 586–602, 2017.
24. J. Granjal, E. Monteiro, J. S. Silva, Security for the Internet of Things: A survey of existing protocols and open research issues, *IEEE Communications Surveys and Tutorials*, vol. 17, no. 3, pp. 1294–1312, 3rd Quart., 2015.

25. B. B. Zarpelão, R. S. Miani, C. T. Kawakani, S. C. de Alvarenga, A survey of intrusion detection in Internet of Things, *Journal of Network and Computer Applications,* vol. 84, pp. 25–37, 2017.
26. D. Miorandi, S. Sicari, F. De Pellegrini, I. Chlamtac, Internet of Things: Vision, applications and research challenges, *Ad HocNetworks*, vol. 10, no. 7, pp. 1497–1516, 2012.
27. J. Zhou, et al., Security and privacy for cloud-based IoT: Challenges. *IEEE Communications Magazine*, vol. 55, no. 1, pp. 26–33, 2017.
28. I. S. Alkhalifa, A. S. Almogren, NSSC: Novel segment based safety message broadcasting in cluster based vehicular sensor network. *IEEE Access*, vol. 8, pp. 34299–34312, 2020.
29. I. Mohiuddin, A. Almogren, Workload aware VM consolidation method in edge/cloud computing for IoT applications. *Journal of Parallel and Distributed Computing*, vol. 123, pp. 204–214, 2019.
30. M. Henze, et al., A comprehensive approach to privacy in the cloud-based Internet of Things. *Future Generation Computer Systems*, vol. 56, pp. 701–718, 2016.
31. K. A. Awan, et al., Robusttrust: A pro-privacy robust distributed trust management mechanism for Internet of Things. *IEEE Access,* vol. 7, pp. 62095–62106, 2019.
32. A. Ouaddah, H. Mousannif, A. A. Elkalam, A. A. Ouahman, Access control in the Internet of Things: Big challenges and new opportunities, *Computer Networks*, vol. 112, pp. 237–262, 2017.
33. C. Stergiou, et al., Secure integration of IoT and cloud computing. *Future Generation Computer Systems*, vol. 78, pp. 964–975, 2018.
34. K. Haseeb, et al., Intrusion prevention framework for secure routing in WSN-based mobile Internet of Things. *IEEE Access,* vol. 7, pp. 185496–185505, 2019.
35. M. S. Jawed, et al. XECryptoGA: A metaheuristic algorithm-based block cipher to enhance the security goals. *Evolving Systems*, 2022. DOI: org/10.1007/s12530-022-09462-0.
36. M. S. Jawed, et al. A comprehensive survey on cloud computing: Architecture, tools, technologies, and open issues. *Journal: International Journal of Cloud Applications and Computing (IJCAC)*, vol. 12(1), p. 77, 2022.
37. S. Sharma, et al. Integrated fog and cloud computing: Issues and challenges. *International Journal of Cloud Applications and Computing (IGI)*, vol. 11(4), p. 10, 2021.
38. M. S. Jawed, et al. Cryptanalysis of lightweight block ciphers using metaheuristic algorithms in cloud of things (CoT). *International Conference on Data Analytics for Business and Industry*, Sakhir, Bahrain, pp. 165–169, 2022.

Index

P

Q

R

S

T

U

V

W

X

Y